HEAT TRANSFER AND THERMAL CONTROL SYSTEMS

Edited by
Leroy S. Fletcher
University of Virginia
Charlottesville, Virginia

Volume 60
PROGRESS IN
ASTRONAUTICS AND AERONAUTICS

Martin Summerfield, Series Editor-in-Chief
Princeton University, Princeton, New Jersey

Technical papers selected from the AIAA 15th Aerospace
Sciences Meeting, January 1977, and the AIAA 12th
Thermophysics Conference, June 1977, subsequently
revised for this volume.

Published by the American Institute of Aeronautics and Astronautics

American Institute of Aeronautics and Astronautics
New York, New York

Library of Congress Cataloging in Publication Data
Main entry under title:

Heat transfer and thermal control systems.

(Progress in astronautics and aeronautics; v. 60)
"Technical papers from AIAA l5th Aerospace Sciences
Meeting, January 1977, and the AIAA 12th Thermophysics
Conference, June, 1977, subsequently revised for this volume."
Includes bibliographies and index.
1. Heat – Transmission – Congresses. 2. Heat pipes –
Congresses. 3. Temperature control – Congresses. I. Fletcher,
Leroy S., 1936- II. AIAA Aerospace Sciences Meeting, 15th, Los
Angeles, Calif., 1977. III. AIAA Thermophysics Conference,
12th, Albuquerque, N.M., 1977. IV. Series.
TL507.P75   vol. 60 [TJ260]   629.1'08s   [629.47'04'4]
ISBN 0-915928-24-8   78-5186

Table of Contents

The interdisciplinary field of thermophysics evolved during the rapid growth of space technology. Initially, this field dealt primarily with the heating and cooling problems of space vehicles and their components. More recently, thermophysics has expanded to include such diverse yet related areas as detection of air and water pollution, energy collection, conversion and storage, resource assessment by satellite, and thermal protection of space vehicles. Of particular importance is the application of thermophysics to energy conversion and the potential for applying space technology to the solution of the expanding energy dilemma.

As space exploration continues, more refined vehicles and instrumentation are required. The fundamentals of thermophysics are extremely important in the development of new and better thermal control components and systems. Space missions which require a uniform temperature environment for instruments or experiments must have precision thermal control.

Heat pipes have evolved as one of the major components in many thermal control systems, and the usefulness and effectiveness of these devices seem to increase each year. New analyses useful for predicting heat-pipe performance have been developed, and the performance of existing pipes has been improved. New heat pipes also have been developed for specific applications.

Application of fundamental heat-transfer characteristics has been advanced through new studies of conduction, convection, and radiation, and the effect of these phenomena on thermal control systems. Innovations include new space radiator systems, low-temperature phase change material packages, passive cooling systems, and temperature control of instrument containers. Although recent years have seen significant advances in these areas, many remaining problems invite continued investigation. The present volume contains a selection of recent studies dealing with heat pipes and thermal control systems. The papers were drawn from the AIAA 15th Aerospace Sciences Meeting in Los Angeles, Calif. in January 1977 and the AIAA 12th Thermophysics Conference in Albuquerque, N. Mex. in June 1977, and were revised and updated especially for this volume. They have been grouped into three chapters: heat pipes, heat transfer, and thermal control systems.

Chapter I consists of eight papers dealing with heat-pipe technology, from basic transport processes in heat pipes to the development of special-purpose heat pipes for actual applications. The basic concepts presented in this chapter will be useful in furthering heat-pipe design. The first paper, by Brennan, Kroliczek, and Jen, presents an overview of axially grooved heat pipes, summarizing the state-of-the-art and recent developments in analysis, design, and fabrication. A mathematical model is presented for the prediction of hydrodynamic behavior and losses along with supporting data for various fluids in the 100°-500° K range. In the paper by Kamotani, an analytical treatment of the performance of an axially grooved heat pipe with one-sided heat input and removal is presented, with results for various *g* conditions and fluid charge. In the next paper, Quadrini and McCreight report on the development of a cryogenic thermal diode heat pipe for spaceflight applications. They give details on pipe geometry, shutoff performance using the liquid blockage technique, and diode reversal performance. In the next paper, Saaski and Tower evaluate the characteristics of potentially stable two-phase, heat-transfer fluids for the temperature range 100°-350° C. They identify decomposition and corrosion characteristics through reflux heat-pipe tests with carbon, steel, and aluminum envelopes.

Excess liquid in heat pipes can degrade transport capacity, and Eninger and Edwards have developed a mathematical model to calculate the parameters governing the axial flow of liquid in fillets and puddles that form in vapor spaces. In the next paper, Kamotani presents another analytical treatment of heat-pipe performance, this time directed toward gravity-assisted heat pipes operated at small tilt angles, with the effect of vapor shear included. Results of the analysis indicate that fluid inventory and pipe tilt govern the performance. Roberts presents a new technique for varying heat-pipe thermal conductance in a zero *g* environment through use of a bubble pump to control return liquid flow. In the last paper in the chapter, Harwell, Kaufman, and Tower describe a theoretical and experimental investigation of an extruded axially grooved aluminum heat pipe with a re-entrant groove profile.

Recent developments in the fundamentals of heat transfer—conduction, radiation, and convection—are presented in the five papers in Chapter II. The analysis of radiative characteristics of materials is of increasing importance, and Armaly develops an approximate closed-form solution for the intensity, flux distribution, and emittance of an isothermal absorbing-emitting and isotropically scattering finite medium. In the next paper, Thornton and Wieting present a finite-element method for steady-state thermal analysis of convectively cooled structures, based on representing coolant passages by elements with fluid bulk temperature nodes. In the paper by Miller, the effect of a highly conducting wall of finite thickness on a stable, stratified fluid was investigated analytically and experimentally to ascertain the heat transfer across a thermocline in a fluid contained in an enclosure with convective currents. The last two papers deal with conduction across material interfaces, specifically thermal contact resistance. In the paper by Ogniewicz and Yovanovich, analytical solutions are developed for conduction in basic cells of regularly packed spheres, considering the influence of packing, mechanical lead, and gas pressure. The paper by Williams and Idrus develops a system for measuring contact surfaces to provide a three-dimensional surface description and presents results of experimental tests which trace the changing character of a surface during mechanical loading or thermal cycling.

Chapter III consists of seven papers dealing with thermal control systems, including space radiator, heat-pipe control systems, phase change materials, and the effects of radiation on these systems. The basic concepts and new developments presented in this chapter provide design information for the improvement of thermal design systems.

The paper by Leach and Cox describes design details and thermal vacuum test results for a lightweight flexible radiator system for on-orbit cooling of space payloads. In the next paper, Curran and Millard present flight temperature measurements on geosynchronous satellite thermal control surfaces, showing the long-term effects of contamination/degradation on radiator surfaces and calorimeter samples. The effects of space radiation on thin polymers and nonmetallic materials have been evaluated by Fogdall and Cannaday, and several candidate materials for advanced space system applications have been identified. Harwell, Haslett, and Ollendorf describe a thermal control canister which provides a uniform thermal environment for shuttle instrument payloads and utilizes longitudinal and circumferential heat pipes connected to feedback-controlled, variable conductance heat pipes. In the next paper, McKee and Steele report the results of an investigation of a cascaded, dry, reservoir variable conductance heat-pipe system with precise passive temperature control for a wide range of heat input and effective space environment temperatures. In the last heat-pipe paper, Lehtinen presents an analytical technique for steady-state and pseudo-transient control analysis of variable conductance heat pipes and feedback-controlled heat pipes, using a modified vapor temperature profile and a 5-mode network. In the final paper, Brennan, Suelau, and McIntosh report on the use of N-heptane, an *n*-parafin, as low-temperature phase change materials in an aluminum/aluminum honeycomb canister.

Throughout the planning and preparation of this volume, a number of people have provided guidance and assistance. Dr. Allie M. Smith, the chairman of the AIAA Thermophysics Technical Committee in 1977, provided advice and consultation. Dr. Robert K. MacGregor was responsible for organizing the thermophysics sessions at the AIAA 15th Aerospace Sciences Meeting and assisted in reviewing those papers for this volume. As General Chairman of the AIAA 12th Thermophysics Conference, from which a majority of the papers in this volume were selected, I was supported by Dr. Walter B. Olstad as Technical Program Chairman. He also was a most helpful Session Chairman and assisted me by handling the reviews of many of the papers. I am also grateful to Dr. Martin Summerfield, Editor-in-Chief of the AIAA *Progress in Astronautics and Aeronautics* series who provided thoughtful advice and consultation, and to Miss Ruth F. Bryans, Administrator of Scientific Publications for AIAA for invaluable assistance and support in the editing of the volume.

Leroy S. Fletcher
January 9, 1978

Chapter I—Heat Pipes

AXIALLY GROOVED HEAT PIPES: 1976

P. J. Brennan,[*] E. J. Kroliczek,[+] and H. Jen[≠]
B & K Engineering, Inc., Towson, Md.

R. McIntosh[§]
NASA/Goddard Space Flight Center, Greenbelt, Md.

Abstract

This paper summarizes the state-of-the-art of axially grooved heat pipes. Applications are identified, and the related heat-pipe design and performance are defined. Recent developments in the analysis, design, and fabrication of axially grooved hardware also are discussed. A mathematical model that predicts the hydrodynamic behavior and accounts for liquid recession, liquid/vapor shear interaction, and 1-g puddle flow also is presented. Performance data for various fluids in the 100-500 K range are compared to predictions from the Groove Analysis Program (GAP). Finally, a simplified closed-form solution that accounts for gravity effects, self-priming, and composite pumping by the grooves as well as all of the hydrodynamic losses also is discussed.

Nomenclature

A = cross-sectional area
A_ℓ' = area of a single groove with no meniscus recession
g = gravity
h = elevation
H = rise factor $H = \sigma/\rho_\ell g$
K = permeability
L = length
N = number of grooves

Presented as Paper 77-747 at the AIAA 12th Thermophysics Conference, Albuquerque, N. Mex., June 27-29, 1977. Copyright ©American Institute of Aeronautics and Astronautics, Inc. 1977. All rights reserved.
    *Manager.
    +Manager.
    ≠Project Engineer.
    §Member Thermal Systems Branch, Systems Division.

N_g = groove capillary flow factor
Q = total heat input or output
Q_x = axial heat flow rate
R_i = heat pipe internal radius
R_x = meniscus radius
Re = Reynolds number
W = groove opening
WP = wetted perimeter
X = axial location
α = groove aspect ratio
β = heat-pipe tilt angle
δ = groove depth
θ = angle
λ = heat of vaporization
μ = viscosity
ν = kinematic viscosity
ρ = density
σ = surface tension
ϕ = $\frac{1}{2}\alpha$
ψ = liquid/vapor shear parameter

<u>Subscripts</u>

a = adiabatic
c = condenser
e = evaporator
eff = effective
g = groove
ℓ = liquid
max = maximum
p = puddle
s = submerged
v = vapor
x = axial location

Introduction

The internal configuration of an axially grooved heat pipe consists of a series of flow channels fabricated as an integral part of a tube wall and parallel to its longitudinal axis. The large open-flow channel of such a design offers low resistance to liquid flow. As a result, performance of the axially grooved heat pipe in 0-g is exceeded only by the more complex and less reliable composite wick structure. Large open-flow channels, however, are sensitive to elevation and puddle flow contributions in gravity, as well as liquid/vapor shear interaction. Analytical models have been developed during the past year which incorporate these effects, and reliable performance predictions now can be obtained. Evaluations conducted point to new designs with reduced sensitivity to gravity and vapor flow effects. This paper presents an improved analytical model and new groove designs, together with the state-of-the-art of axially grooved heat-pipe technology in the cryogenic, ambient, and intermediate- to high-temperature ranges. Recent developments in the utilization of axially grooved heat pipes for thermal control applications also are identified.

Technology Review

Axially grooved heat-pipe technology has been developed extensively for fixed conductance applications in the cryogenic through ambient temperature range.[1,2] An aluminum axially grooved heat pipe flown aboard the Orbiting Astronomical Observatory-C (OAO-C)[3] still is functioning properly after more than four years in orbit. A total of 55 aluminum axially grooved heat pipes also were used to isothermalize the Applications Technology Satellite-6 (ATS-6).[2,4] Their extensive use and successful performance through almost three years of continuous flight operation have demonstrated the reliability of this design. Sounding rocket experiments[5,6] also have evaluated various performance parameters for these pipes.

Aluminum axially grooved heat pipes have been adapted for use as a gas-controlled variable-conductance (VCHP) system[7,8] within the past year. Both feedback and passive VCHP control have been domonstrated in the 180-300 K range.[9,10] The feasibility of axially grooved diodes and thermal switches is also currently under investigation.[10] Finally, fabrication with materials other than aluminim to accommodate higher temperatures and different working fluids, as well as low material conductance for thermal control applications, is currently under investigation. A copper/water axially grooved heat pipe for use in the isothermalization of radiator fins of a radioisotope thermoelectric generator (RTG) has been developed and operated at temperatures up to 500 K.[11] Axially grooved tubing also has been produced in stainless steel for potential thermal control applications.

Present and Future Requirements

Typical design and performance requirements for various areas of application are summarized in Table 1. In the cryogenic temperature range, a well-defined need for a variety of heat-pipe hardware has been identified. Potential applications include passive heat-pipe/radiant-cooler systems and hybrid coolers to augment or replace present coolers in order to achieve longer life and lighter weight. The use of heat

Table 1 Typical design and performance requirements

Temperature range	Potential applications	Designs	Heat load range	Heat transport requirements
Cryogenic to low temperatures (0-250 K)	Passive radiant coolers, hybird coolers, and remote satellite components coupling.	Fixed conductance, diode, and thermal switch.	Milliwatts to tens of watts	Up to 25 W-m
Ambient (250-350 K)	Satellite waste heat rejection, isothermalization, and temperature control. Various terrestrial cooling applications, including electronics; HVAC energy recovery systems and low-temperature solar collector systems.	Fixed-conductance and gas-controlled variable conductance (VCHP)	Ten to multihundred watts	50 W-m and up
Intermediate to high temperatures (> 350 K)	Isotope and solar space power supplies, high-temperature terrestrial energy recovery and solar collector systems.	Fixed conductance and gas-controlled	Hundreds to Kilowatt range	Up to kilowatt-meter range

pipes to couple remote components to centrally located active
coolers such as a Vuilleumier (VM) engine also has been considered. Current axially grooved heat-pipe designs are capable
of providing the anticipated performance requirements in 0-g
applications. However, higher static height performance
would be desirable to obtain more reliable 1-g performance
measurements. The need to develop variable-conductance designs
also has been identified, with particular emphasis on diodes
or thermal switches to protect against hot-sink conditions.
Axially grooved tubing fabricated from low-conductance
materials, such as stainless steel, will be required to
minimize heat leaks in the off-mode in thermal control designs.

The majority of spacecraft temperature-control applications have been near room temperature, and considerable
experience now exists with fixed-conductance axially grooved
heat pipes at ambient temperatures. In addition to a variety
of potential unmanned spacecraft applications, the advent of
the Space Shuttle and Spacelab has created an opportunity to
fly a wide variety of scientific instruments which will
require relatively low-cost temperature-control systems.
There is a well-defined need for both conventional fixed-
conductance and gas-controlled variable-conductance technology.
As with cryogenic applications, it is desirable to reduce
the groove's sensitivity to 1-g testing. Low-conductance
materials for thermal control applications are also desirable.
Finally, higher heat transport also will be required for large
space systems and ground applications.

Intermediate- to high-temperature requirements also
exist. Isothermalization of radioisotope thermoelectric
generator (RTG) radiator fins, for example, poses the need
for operation in the 350-500 K range. Development of axially

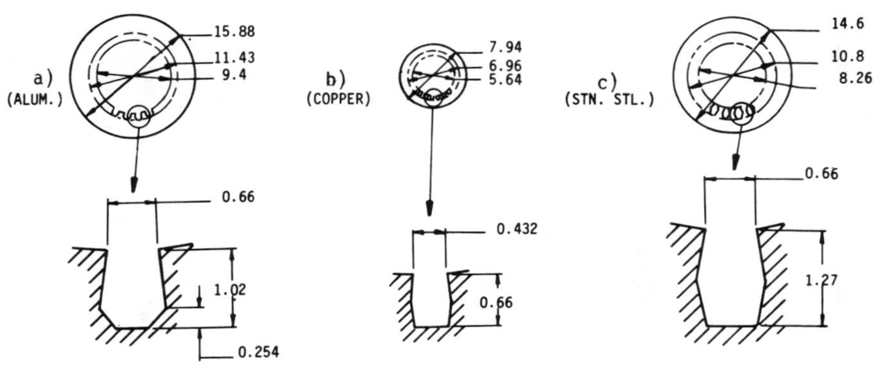

Fig. 1 Swaged groove forms.

Table 2 Measured performance summary for axially grooved heat pipes

Type fluid	Temp., K	0-g heat transport capability, W-m	Static height, cm	Film coefficient, W/m^2-C Evaporator	Condenser
Swaged aluminum					
OAO geometry					
Ammonia	295	130	1.09	7,265	9,480
Freon 21	295	28	0.51	1,135	1,700
Freon 23	295	12	0.46	653	1,135
ATS geometry					
Ammonia	310	145	0.89	5,676	8,515
Methane	150	18	0.52	1,362	...
Nitrogen	80	16	0.30	312	1,362
Swaged copper					
LCHPG geometry					
Water	363	67.6	2.8	...	...
Swaged stainless steel					
Approx. ATS geometry	——————— Performance forthcoming ———————				
Extruded aluminum					
ATS geometry					
Ammonia	273	143	1.6	7,000	13,600
Methane	125	33.4	1.1	1,730	6,100
Ethane	200	25	1.3	1,370	5,900
Lewis covert geometry					
Ammonia	293	143	2.51	7,300	20,500
Methane	120	28	2.13	...	...
Ethane	180	33	2.21	...	...

grooved tubing in materials such as copper and stainless
steel will be required to accommodate the higher temperatures.

Current State-of-the-Art

The axially grooved geometry has been utilized for more
than 10 years in a variety of applications, including high-
temperature liquid-metal heat pipes. Until recently, most
applications have been with fixed-conductance designs. A
number of different designs that were developed for aerospace
applications are shown in Fig. 1-2. Measured performance is
summarized in Table 2. Within the past few years, aluminum
axially grooved tubing also has been used in ambient temperature gas-controlled variable-conductance systems[7,8] and has
been tested in various thermal control modes at cryogenic
temperatures, including active and passive gas-controlled
variable-conductance (VCHP), gas-controlled diode, liquid
trap diode, and as a thermal switch.[10] Today, the state-of-
the-art can be summarized as follows.

Fixed-conductance heat pipes

Axially grooved tubing fabrication is the dominant
factor affecting the groove design. Initially, axailly
grooved tubing was produced by machining flat stock and
forming it into a tube or by broaching a thick-wall tube.
In aerospace systems, requirements for lightweight, cost-
effective designs led to the development of swaged aluminum
axially grooved tubing for the OAO-B and OAO-C spacecrafts.[3,12]
A modified design, as shown in Fig. 1a, which gave improved
performance and facilitated tubing fabrication was developed
for the ATS-6 spacecraft.[4] Both designs were produced from
6061 aluminum alloy with a patented swaging process of the

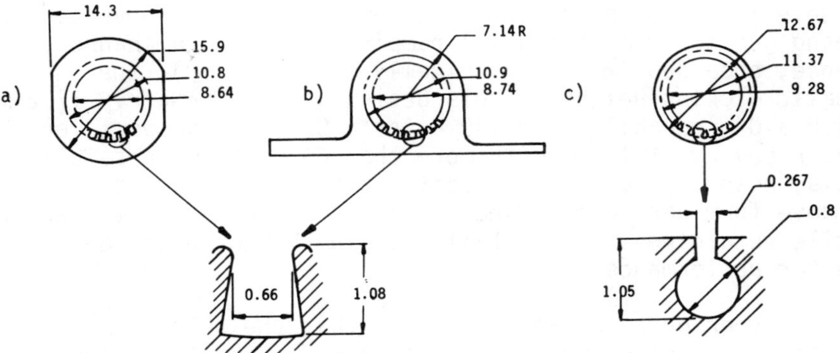

Fig. 2 Extruded aluminum groove forms.

French Tube Division of Noranda Metals, Inc. Tests results with various working fluids are summarized in Table 2.

Subsequently, extruded aluminum grooved tubing was developed which led to better control of the groove form as well as the ability to provide mounting flanges as an integral part of the tubing (Figs. 2a and 2b). Most of the extruded grooved tubing produced during the past few years duplicated the ATS groove form developed by NASA/Goddard Space Flight Center (GSFC). The tubing was produced in both 6063 and 6061 aluminum alloy[2] by Micro Extrusions, Division of Universal Alloy Corporation. Extensive testing with various working fluids has been performed as summarized in Table 2, and 0-g test data have been obtained from GSFC's Second[6] and the IHPE[13] Heat Pipe Sounding Rocket Experiments. Also, the GSFC extrusion design has been applied to the International Ultraviolet Experiment (I.U.E.) spacecraft platform, and an ethane heat pipe is being evaluated as part of the Heat Pipe Experiment Package (HEPP).[14] Finally, the GSFC extrusion has been employed as the vehicle to develop thermal control heat-pipe designs.

Recently the extrusion process was used to fabricate a closed-groove form (Lewis covert groove extrusion), which offers reduced sensitivity to gravity.[15] The NASA Lewis covert groove extrusion, shown in Fig. 2c, was produced from 6063 aluminim alloy by Mirco Extrusions. Testing was performed with ammonia as the working fluid. Measured performance is summarized in Table 2. As can be seen, a substantial improvement in static wicking height has been achieved with the heat pipe fully primed. However, as indicated by the test results, the heat pipe did not reprime at 48% of its static wicking height (12 mm) with heat loads greater than 36% (50 W) of its fully primed capacity at that elevation. This was to be expected, since the Lewis covert groove is a composite wick with pumping of the groove opening being greater than the axial pumping of the open channel. Hence, once deprimed (mechanically or thermally), the static wicking height of this design is approximately 17.2 mm, with a 0-g capability of 179 W for a 0.55-m transport length, and recovery at 12 mm with more than 60 W is not possible. This, however, does not present a serious problem, since the groove form can be modified to minimize the composite factor while retaining similar static wicking height with equal or better performance.

Experience to date indicates that the extrusion process is the best method for producing aluminum axially grooved tubing. Well-defined groove forms and good dimensional

control have been achieved. Mounting flanges can be
extruded as an integral part of the tubing which can simplify
interfacing in many applications. In addition, the ability to
produce complex groove forms has been demonstrated by the NASA
Lewis convert groove extrusion which should lead to higher
performance and greatly reduce sensitivity to 1-g testing.
For the intermediate- to high-temperature range, however,
axially grooved tubing of materials such as copper and its
alloys, stainless steel, carbon steels, and super alloys is
required, and the swaging process is the only known process
that can be used effectively today to produce axially grooved
tubing in these materials on a cost-effective basis.

During the past year, an extensive effort has been
conducted by B & K Engineering and Teledyne Energy Systems to
develop a small-diameter copper/water axially grooved heat
pipe for application to the isothermalization of radiator fins
of a radioisotope thermoelectric generator (RTG).[11] An internal
surface coating, designed to enhance surface wetting, developed
as part of the effort has demonstrated excellent stability in
continuous operation at a temperature of 500 K. The geometry
of the copper axially grooved heat pipe is shown in Fig. 1b.
Prototype test data are summarized in Table 2. Development of
stainless-steel axially grooved tubing (Fig. 1c) also has been
pursued for thermal control heat-pipe application. No
performance data are available at this time.

Thermal control heat pipes

Limited published data are available to ascertain
axially grooved heat-pipe performance in thermal control
designs. A number of ambient-temperature gas-controlled VCHP
designs that utilize the GSFC extrusion have been tested by
NASA/GSFC for potential application to Shuttle payload
thermal control canisters.[8] Unpublished data obtained from
Ref. 16 indicate performance degradation in the gas-controlled
VCHP's. However, the amount of degradation is not consistent,
with variations occuring mostly among the different vendor
designs. Degradation in performance also has been noticed
when noncondensible gases are introduced into a fixed-
conductance pipe using the Lewis covert groove extrusion.[15]
It is not clear at this point whether degradation is inherent
in gas-controlled axial designs or whether it is dependent on
the gas reservoir's design.

In other aspects, the GSFC extrusion has performed as
predicted.[10] Control to within $\pm$ 1°C was obtined with feedback control vs $\pm$ 10°C with the same system operating in a

passive mode. Tests conducted in a liquid trap diode mode indicate that the energy associated with shutdown is approximately equal to twice the latent heat associated with the inventory required to fill the heat pipe (1.6 W-hr shutdown energy for the configuration tested). Operation of a thermal switch also was demonstrated using a liquid trap as a second heat pipe. The simultaneous heat-piping action by the liquid trap had a negligible effect on the diode's shutdown.

Analysis

A number of analytical models have been developed over the years to predict axially grooved heat-pipe performance. The early models[17,18] used closed-form solutions to predict the heat transport capability of rectangular groove cross sections. Later computerized solutions of the applicable differential equations were used to predict the thermal conductance,[22,23] as well as transport capability.[19]

A study was conducted recently to develop a more extensive model for predicting the hydrodynamic behavior of axially grooved heat pipes.[20] The effects of fluid inventory, meniscus recession, liquid/vapor shear, and puddle flow all are taken into account. A Groove Analysis Program (GAP) was developed to solve the applicable differential equations.[21] In addition, a simplified closed-form solution was developed. Results of the study are summarized below.

Capillary Pumping Limit

The 0-g performance of a heat pipe is governed generally by the capillary pumping limit. The governing equations that determine the heat pipe's transport capability within this limit are as follows:[20]

For laminar vapor flow ($Re_v < 2000$),

$$\frac{\sigma}{R_X^2} \frac{dR_X}{dX} = \rho_\ell g \sin\beta + \left[\frac{8\mu_v}{\rho_v A_v R_v^2} + \frac{\mu_\ell}{K_X A_{\ell x} \rho_\ell} \left\{ 1 + \frac{\phi^2}{3} \psi \right\} \right] \frac{Q_X}{\lambda} \tag{1}$$

For turbulent vapor flow ($Re_v > 2000$),

$$\frac{\sigma}{R_X^2} \frac{dR_X}{dx} = \rho_\ell g \sin\beta + \frac{0.0656\mu_v^{0.25}}{\rho_v A_v^{1.75} R_v^{1.25}} \left(\frac{Q_X}{\lambda}\right)^{1.75} +$$

$$\frac{\mu_\ell}{K_X A_{\ell x} \rho_\ell} \left\{ 1 + \frac{\phi^2}{3} \psi \right\} \frac{Q_X}{\lambda} \tag{2}$$

The Hagen-Poiseuille equation is utilized in the foregoing for the laminar pressure drops, and the Blasius equation is used for turbulent vapor flow. The liquid/vapor shear interaction is a pressure drop in the liquid which results from a shearing effect at the liquid/vapor interface which results from the counterflow of the vapor. This term can be significant in uncovered liquid channels, as in the case of axial grooves. Hufschmidt[18] derived an empirical expression for rectangular grooves which is basically a modification of the Hagen-Poiseuille equation for the liquid to include the additional term

$$\frac{\phi^2 \psi}{3} = \frac{N}{3\pi\alpha} \; \frac{A_\ell'}{W^2} \; \left(\frac{R_i}{W} - \alpha \right)^{-3} \tag{3}$$

Solution of the appropriate differential equation leads to the maximum heat transport that can be provided by the pipe. The GAP program utilizes a fourth-order Runge-Kutta integration method with self-adjusting step sizes. It determines the meniscus recession at each axial location and the corresponding fluid inventory. The ideal charge and the maximum heat transport are determined with an option to calculate heat transport vs fluid inventory (e.g., Fig. 5). Single evaporator and condenser sections with uniform heat loads presently are used in the analysis.

Puddle Flow

Excess fluid that can be due to overcharge or to partial "dry-out" or drainage in the 1-g of the upper grooves will form a slug in the condenser in 0-g and a puddle in 1-g. Partial blockage of the condenser will cause a degradation in thermal conductance. In 1-g, the puddle also will affect the heat transport capability. The puddle, in addition to superimposing its own liquid flow, reduces the effective transport length and also the hydrostatic head, and a relatively small over-charge can result in substantial increases in measured 1-g transport. Since this additional performance will not be realized in 0-g, a model was developed to account for the contribution due to a puddle. The model consists of the three distinct flow regions shown in Fig. 3. The location of the zones is dependent on the amount of excess fluid and the tilt. Zone I consists of those grooves that are unaffected by the puddle (e.g., upper grooves). Zone II represents that length of grooves which extends out from the puddle, and zone III is the puddle with submerged grooves. In the grooves that extend beyond the puddle, the capillary pumping limit prevails; however, the effect of the puddle on these grooves is a reduction in both their transport

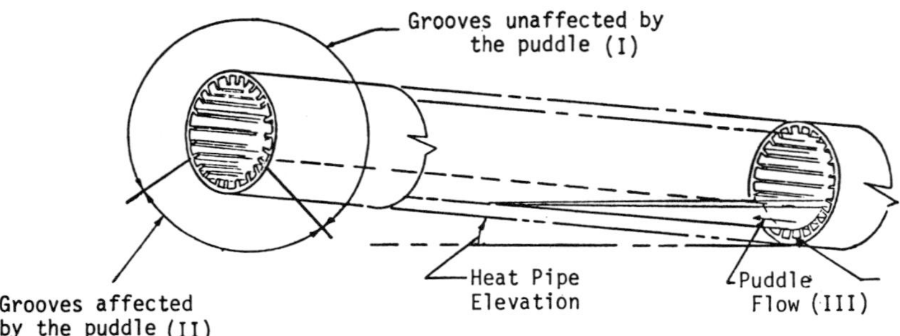

Fig. 3 Puddle flow schematic.

length and hydrostatic elevation. In the grooves submerged
under the puddle as well as in the puddle itself, liquid flow
is provided by the gravity head developed by the elevation
variation of the free surface of the puddle. The total
performance then is determined by the sum of the contributions
from each zone.

The mathematical model that was developed to predict the
performance with excess fluid is discussed in detail in
Ref. 20. The model is based on the following assumptions:
1) uniform heat addition and removal with single evaporator
and condenser sections; 2) one-dimensional laminar flow in the
puddle; and 3) the transport capability of the grooves
extending beyond the puddle (zone III) can be approximated
by a closed-form solution. The model consists of satisfying
the equation of motion for the puddle and mass continuity at
the puddle/groove interface. The governing equations are
as follows:

Equation of Motion for a Puddle

$$\rho_\ell g \cos \beta \frac{R_v}{2} \sin \frac{\theta_p}{2} \frac{d\theta_p}{dX} = \rho_\ell g \sin \beta + \frac{\mu_\ell Q_p(x)}{\lambda (KA)_p \rho_\ell} \tag{4}$$

The left-hand side of Eq. (4) denotes the pumping developed
by the free surface of the puddle. The hydrostatic and
dynamic losses within the puddle are on the right-hand side
of the equation. The first term is the hydrostatic head
developed as a result of the pipe inclination, and the second
is the liquid viscous loss.

Continuity Equation (Mass and Heat Flow Continuity)

$$\frac{dQ_p}{dx} = Q_{sp}\Omega_\ell + \frac{(QL)_{max}}{2\pi X_{eff}} \left(1 - \beta \frac{X}{h_{max}} \right) \Omega_r \frac{d\theta_p}{dx} \tag{5}$$

Table 3 Puddle flow functions

In the evaporator, $L_e \geq X \geq 0$:

$$X_{eff} = (1/2)X$$

$$\Omega_r = X/L_e$$

$$\Omega_\ell = 1/L_e$$

In the adiabatic section, $L_e + L_a \geq X \geq L_e$:

$$X_{eff} = X - (1/2)L_e$$

$$\Omega_r = 1$$

$$\Omega_\ell = 0$$

In the condenser, $L \geq X \geq L_e + L_a$:

$$X_{eff} = (X - L_e - L_a)\left[1 - (1/2)L_c (X - X_a - L_e)\right] + L_a + (1/2)L_e$$

$$\Omega_r = 1 - (1/L_c)(X - L_e - L_a)$$

$$\Omega_\ell = -1/L_c$$

where the parameters X_{eff}, Ω_r, and Ω_ℓ are listed in Table 3. The heat load (Q_{sp}) transported by the grooves that are affected by the puddle is

$$\frac{dQ_{sp}}{dx} = \frac{(QL)_{max}}{2\pi\, X_{eff}}\left(1 - \beta\,\frac{X}{h_{max}}\right)\frac{d\theta_p}{dx} \qquad (6)$$

The first term on the right-side of Eq. (5) is associated with the mass transfer (condensation/evaporation) at the free surface of the puddle element. The second term denotes the flow interchange from the puddle to the grooves. $(QL)_{max}$ in these equations is the predicted transport capability of an unaffected groove.

The puddle flow model also is included in the GAP program. Equations (4 - 6) are solved by numerical integration to yield Q_{sp} and θ_p. Specified conditions consist of 1) groove geometry and heat-pipe inclination; 2) evaporator, transport, and condenser lengths; and 3) the starting location of the puddle in the heat pipe. The total heat-pipe transport is calculated by summing the heat capacity Q_{sp} with the heat load carried by the unaffected grooves. The fluid inventory is calculated directly once the puddle angle θ_p has been established. In addition to capillary pumping limit and puddle flow, sonic and entrainments also are evaluated by the GAP program.

Closed-Form Solutions

In order to facilitate the groove optimization, the governing equations [Eqs. (1) and (2)] for capillary pumping were simplified to a closed-form solution[20]:

$$\frac{QL_{eff}}{N_L} = \frac{NN_g \left\{ 1 - (h/H)(W/2) \right\}}{1 + (\nu_v/\nu_\ell)(f_v + f_{\ell v})} \tag{7}$$

where f_v is the viscous vapor parameter,

$$f_v = \frac{4}{\pi} \frac{NN_g}{\left\{ (R_i/W) - \alpha \right\}^4 \, W^3} \tag{8}$$

$f_{\ell v}$ is the liquid/vapor shear parameter,

$$f_{\ell v} = \frac{N}{3\alpha\pi} \frac{A'_\ell}{W^2} \left(\frac{R_i}{W} - \alpha \right)^{-3} \tag{9}$$

N_g is the groove capillary flow factor,

$$N_g = \int_{(1/2)W}^{\infty} \frac{K_x A'_{\ell x}}{R_x^2} \, dR_x \tag{10}$$

and N_L is the liquid transport factor,

$$N_L = \lambda\sigma/\nu_\ell \tag{11}$$

In the derivation, the following assumptions were made: 1) laminar liquid and vapor flow; 2) effects of meniscus recession on capillary pumping and groove characteristics can be expressed by an integrated term N_g [i.e., Eq. (10)] ; and 3) a dimensional analysis indicates that the groove flow factor can be employed to approximate the liquid flow conductance KA.

With the aid of the GAP program, it was determined that the groove flow factor for a divergent groove with a sharp "land-tip" corner depends on the groove geometric parameters A_ℓ'/W, WP/W, and W. An empirical function was established for divergent grooves as follows:

$$N_g = 0.87 \left(A_\ell' / W^2 \right)^{3.1} \left(W/WP \right)^2 W^3 \qquad (12)$$

where A_ℓ' and WP are the groove area and the wetted perimeter, respectively, associated with a filled groove with a flat meniscus. A similar relationship also was developed for round land-tip corners and is presented in Ref. 20 for R_t/W less than 0.4.

The validity of Eqs. (7-12) has been established by comparison of predicted values with measured data for various fluids and groove geometries. Equation (7) is a useful tool in conducting parametric performance evaluations for various groove geometries to establish an improved groove design that meets the specified requirements.

Data Correlation

GAP was utilized to predict the heat-pipe performance for the ATS 6063 extruded groove tubing with methane, ethane, and ammonia. Complete comparisons between predictions

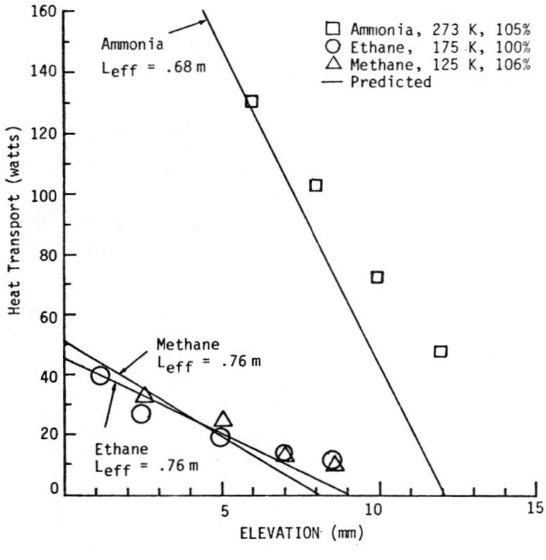

Fig. 4 Performance comparison between predictions and data.

and test data[1] are presented in Ref. 20. Figure 4 presents a comparison of measured and predicted transport capability vs elevation. The following observations can be made:

1) At low elevations, the predicted performance is higher than measured data. The lower the temperature, the higher this discrepancy will be. The effect seems to be related to vapor losses and is especially apparent in the turbulent vapor flow regime.

2) At intermediate elevations, good agreement with measured data was obtained.

3) At higher elevations, measured data are substantially higher than predicted. Drainage effects are believed to be causing this difference.

Figure 5 compares predicted performance as a function of cnarge at different elevations with experimental data reported in Ref. 19. The difference between predicted and measured performance for underfill conditions may be due to nonuniform liquid distribution within the heat pipe which causes some of the grooves to be partially empty while other grooves are filled completely. This would result in higher performance than predicted for uniformly filled grooves. For ideal charge, on the other hand, sufficient liquid is available to satisfy all grooves; maximum performance is achieved in each groove, and hence good agreement is obtained between predicted and measured performance. Similarly good correlation also is obtained for overcharge conditions.

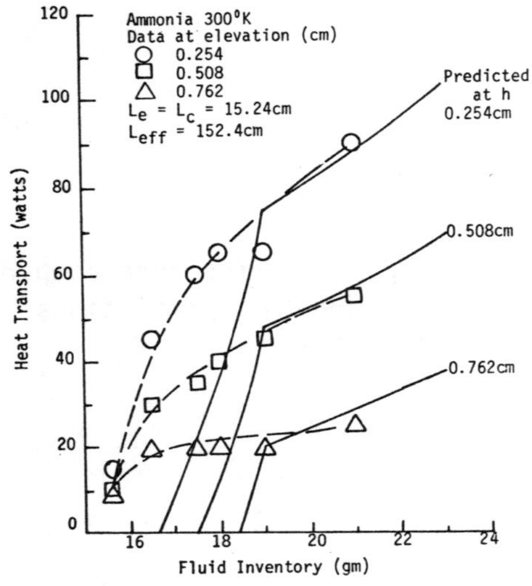

Fig. 5 Effects of fluid inventory on heat transport.

Groove Design Evaluation

The closed-form solution [Eq. (7)] shows that the heat transport capability QL_{eff} of an axially grooved heat pipe depends on various geometric parameters, as well as the fluid properties. However, Eq. (7) places no limit on the area of the groove which can be used in combination with the groove opening (W). It is known that, the larger the parameter A_{ℓ}'/W^2, the more pronounced the composite effect becomes. To account for this effect, a composite factor S is introduced which is defined as the ratio of the capillary pumping when the groove is filled to that obtained when the pipe is priming. The composite factor can be rewritten in terms of the groove's geometry as

$$S = 2 A_{\ell}'/WP\ W \tag{13}$$

Substitution of Eq. (13) into Eqs. (7) and (12) will lead to a relationship that shows that the heat transport capability increases rapidly with increasing values of the composite factor. The GSFC extrusion has a value of S = 0.85, whereas the Lewis covert groove has S = 1.45. There are no composite effects when S is less than or equal to one. However as in the case of the covert groove, when the composite factor is greater than one, the maximum transport during startup or self-priming is the fully primed transport capability divided by S. Groove geometries with composite factors greater than one will be required to meet future requirements. However, care must be taken to recognize the startup or

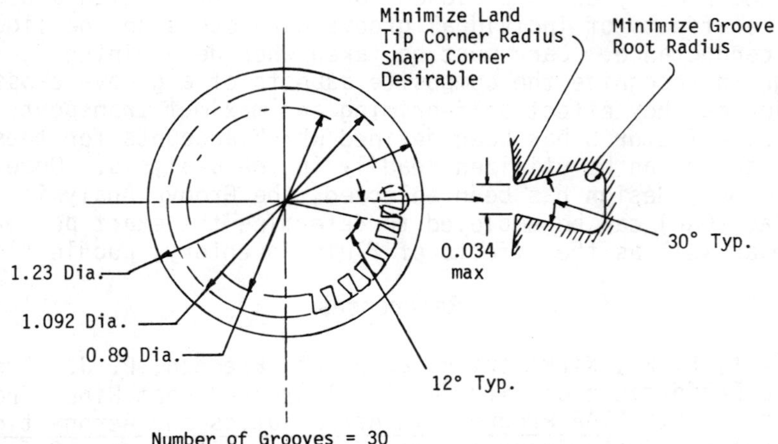

Fig. 6 Improved groove design.

self-priming requirements of the application, and in all cases the composite factor should be kept as low as possible within the bounds imposed by performance specifications.

A parametric analysis[20] was conducted for circular and trapezoidal grooves to evaluate their performance characteristics. It was concluded that trapezoidal grooves can yield higher performance than circular grooves at the same composite factor. Figure 6 illustrates a typical trapezoidal groove design with improved performance over the GSFC extrusion and Lewis covert grooves. The improved groove design can yield 75% higher heat transport with ammonia at ambient over GSFC and Lewis covert grooves. Its composite factor is 1.24, which is smaller than the Lewis covert groove, and its static height is twice that of the GSFC groove. Other geometries are defined readily depending on design goals.

Conclusions

The axially grooved heat pipe is being applied to a variety of aerospace systems. The extensive use and reliable performance has demonstrated the versatility of this design. Numerous design and performance requirements have been identified for future space missions utilizing both fixed-conductance and thermal control heat pipes. In addition to requiring new geometries with aluminum which will give increased transport, there is a well-defined need for axially grooved tubing in materials such as copper and stainless steel to accommodate high-temperature and/or low-conductance requirements.

Parametric analysis for preliminary designs are obtained readily using a closed-form solution, which accounts for the effects of individual groove parameters on the liquid flow conductance. Care must be taken when determining a design to recognize the composite aspects of a groove cross-section as they effect self-priming and maximum transport. A composite factor S has been defined which accounts for these effects and can be utilized readily in the analysis. Once a preliminary design has been selected, the Groove Analysis Program (GAP) can be employed to determine the exact performance as well as the effects of fluid inventory, puddle flow, etc.

References

[1]Schlitt, K. R., Kirkpatrick, J. P. and Brennan, P. J., "Parametric Performance of Extruded Axial Grooved Heat Pipes from 100 to 300 K," <u>AIAA Progress in Astronautics and Aeronautics: Heat Transfer with Thermal Control Applications</u>, Vol. 39, edited by M. Yovanovich, New York, 1975, pp. 215-234.

[2]Kroliczek, E. J. and Brennan, P. J., "Axial Groove Heat Pipes - Cryogenic Through Ambient," American Society of Mechanical Engineers, Paper 73-ENAS-48, July 1973.

[3]Harwell, W., Edelstein, F. and Ollendorf, S., "Orbiting Astronomical Observatory Heat Pipe Flight Performance Data," AIAA Progress in Astronautics and Aeronautics: Thermophysics and Spacecraft Thermal Control, Vol. 35, edited by R. G. Hering, New York, 1974, pp. 445-465.

[4]Berger, M. E. and Kelly, W. H., "Application of Heat Pipes to the ATS-F Spacecraft," American Society of Mechanical Engineers, Paper 73-ENAS-46, July 1973.

[5]McIntosh, R., Knowles, G., and Hemback, R. J., "Sounding Rocket Heat Pipe Experiment," AIAA Paper 72-259, 1972.

[6]Ollendorf, S., McIntosh, R., and Harwell, W., "Performance of Heat Pipe in Zero Gravity," International Conference on Heat Pipes, 1973.

[7]Edelstein, F., "Deployable Heat Pipe Radiator," Grumman Aerospace Corp., DHPR-75-13, April 1975.

[8]"Transient Thermal Response of A Thermal Control Canister," Grumman Aerospace Corp. Contract NAS5-2270, 1976.

[9]Harwell, W. and Ball, T., "Thermal Vacuum Tests on A Thermal Control Canister Breadboard," Contract NAS5-22980, 1976.

[10]Brennan, P. J. and Groll, M., "Application of Axial Grooves to Cryogenic Variable Conductance Heat Pipe Technology," 2nd International Heat Pipe Conference, April 1976.

[11]Jen, H. and Brennan, P. J., "Summary Report for Development of Copper/Water Axially Grooved Heat Pipes for Low Cost High Performance (LCHP) Applications," B & K Engineering, Inc., BK017-005, Sept. 1976.

[12]Bilenas, T. A. and Harwell, W., "Orbiting Astronomical Observatory Heat Pipes - Design, Analysis and Testing," American Society of Mechanical Engineers, Paper 70 HT/SPT-9, 1970.

[13]McIntosh, R., Ollendorf, S., and Harwell, W., "The International Heat Pipe Experiment," AIAA Paper 75-726, May 1975.

[14]Suelau, H. J., and Brennan, P. J., "Thermal Design of TIROS-N Heat Pipe Experiment Package (HEPP)," B & K Engineering, Inc., BK018-1024, Feb. 1977.

[15]Harwell, W., "Analysis and Tests of NASA Covert Groove
Heat Pipe," NASA CR-135156, Dec. 1976.

[16]"Summary Report for Effects of Non-Condensible Gas on
Axially Grooved Heat Pipe Performance," B & K Engineering, Inc.
June 1977.

[17]Frank, S., Smith, J. T., and Taylor, K. M., _Heat Pipe
Design Manual_, Martin Marietta Corp., Feb. 1967.

[18]Hufschmidt, W., Burck, E., DiCola, G. and Hoffman, H.,
"The Shearing Effect of Vapor Flow on Laminar Liquid Flow in
Capillaries of Heat Pipes," NAS TT-F-16601, Oct. 1975.

[19]Dynatherm Corp., "Summary Report for Fill Determination
Study of ATS-F & G Heat Pipes," Dece. 1972.

[20]Jen, H. and Kroliczek, E. J., "Axially Grooved Heat Pipe
Study," B & K Engineering, Inc., BK012-1009, June 1977.

[21]Jen, H. and Kroliczek, E. J., "User's Manual for Groove
Analysis Program (GAP)," B & K Engineering, Inc., BK012-1007,
June 1976.

[22]Chi, S. W., "Mathematical Modeling of Cryogenic Heat Pipes,"
NASA CR-166175, Sept. 1970.

[23]Schneider, G. E. and Yovanovich, M. M., "Thermal Analysis
of Trapezoidal Grooved Heat Pipe Walls," Univ. of Waterloo,
Aug. 1975.

EFFECTS OF ONE-SIDED HEAT INPUT AND REMOVAL ON
AXIALLY GROOVED HEAT-PIPE PERFORMANCE

Yasuhiro Kamotani*
NASA Goddard Space Flight Center, Greenbelt, Md.

Abstract

The performance of an axially grooved heat pipe with one-sided heat input and removal was investigated analytically. Under zero-g condition, the maximum heat transport of the pipe may decrease as much as 30%, depending on the liquid slug behavior in the condenser section. In one-g environment, the performance depends mainly on the fluid charge. The maximum heat transport, if overcharged, is almost equal to the value for uniform heating and cooling due to puddling effect. However, for some heater-cooler combinations, the temperature drop across the heat pipe becomes very large. Computed results for tilted heat pipes compare favorably with available experimental data.

Nomenclature

A_p	=	cross-sectional area of puddle
g	=	gravitational acceleration
H	=	puddle depth
h_c	=	condenser film coefficient
h_e	=	evaporator film coefficient
K_p	=	permeability
L_a	=	length of adiabatic section
L_c	=	length of condenser section
L_e	=	length of evaporator section
$\dot{M}_c, \dot{M}_g, \dot{M}_p$	=	defined in Fig. 9

Presented as Paper 77-191 at the AIAA 15th Aerospace Sciences Meeting, Los Angeles, Calif., Jan. 24-26, 1977. Copyright © American Institute of Aeronautics and Astronautics Inc., 1977. All rights reserved.

*Performed while the author was a NRC Resident Research Associate. Present address: Visiting Assistant Professor, Dept. of Mechanical and Aerospace Engineering, Case Western Reserve University, Cleveland, Ohio.

P_ℓ = liquid pressure
Q^ℓ = total heat transport
q = total heat transport for single groove
q_o = single-groove heat transport limit
R_g, R_{out}, R_v = defined in Fig. 1
R_{in} = $1/2\ (R_g + R_v)$
T = pipe wall temperature
T_v = vapor temperature
(r,θ,Z) = coordinate system defined in Fig. 2
α = defined in Fig. 8
β = tilt angle
ρ_ℓ = liquid density
μ_ℓ = liquid viscosity

Introduction

Axially grooved heat pipes have become useful tools for spacecraft thermal control. In recent years, their performances have been tested both on the ground and under zero-g covering a wide range of temperatures. Reference 1 presents a computer program to predict the maximum heat transport and the fluid inventory for a given grooved heat pipe and working fluid. The program currently is being improved under the contract of NASA Goddard Space Flight Center. In the preceding works, the predictions are based on the assumption that heat is applied uniformly to the whole evaporator section and removed uniformly from the whole condenser section. However, in practice this condition seldom is met, except in laboratory tests. When a heat pipe is attached to a certain object, the contact area often does not cover the whole circumference of the evaporator (condenser) section. The performance of an axially grooved heat pipe is more susceptible to such circumferentially nonuniform heat input and removal than that

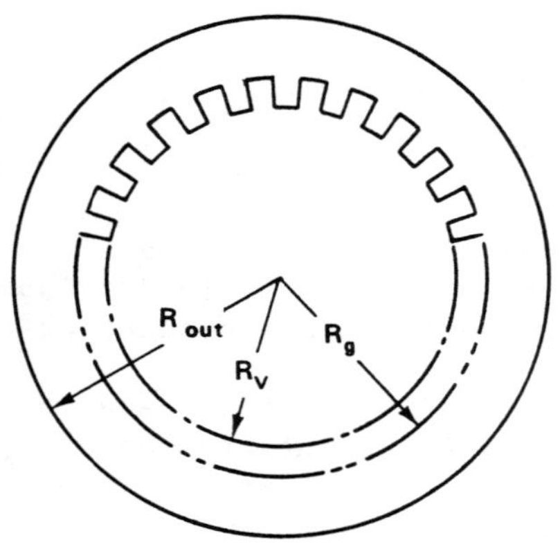

Fig. 1 Grooved heat-pipe cross section.

of other types of heat pipes, because the groove walls prevent the circumferential motion of liquid, resulting in premature dryout of some of the grooves and the degradation of the performance.

The preceding effect has been known, but very few systematic works on the subject are available. Reference 2 presents a simple analytical model to predice the evaporation and condensation rates for nonuniform heating and cooling. In Ref. 2 the pipe wall is considered as a one-dimensional conducting fin. The model does not consider axial conduction in the pipe wall, which becomes important when some grooves partially dryout in the evaporator and when a liquid pool is present in the condenser. The present analysis employs basically the same approach as that of Ref. 2; however, the present analysis is conducted numerically, which makes it possible to include the axial conduction term. The work is done for both zero-g and one-g environments. The maximum heat transport and the temperature drop across the heat pipe are predicted.

Analysis

Figure 1 shows a cross section of a typical axially grooved heat pipe. Each groove is designed to operate as an isolated liquid passage. Under steady-state conditions, the rates of condensation and evaporation must be equal for each groove. For uniform heating at the evaporator and uniform cooling at the condenser, all grooves are subject to identical boundary conditions, so that the performance of the heat pipe can be deduced from the behavior of any one groove (assuming no excess liquid). For nonuniform heating and cooling, however, the rates of condensation and evaporation of a groove may be different from those of other grooves, so that it is

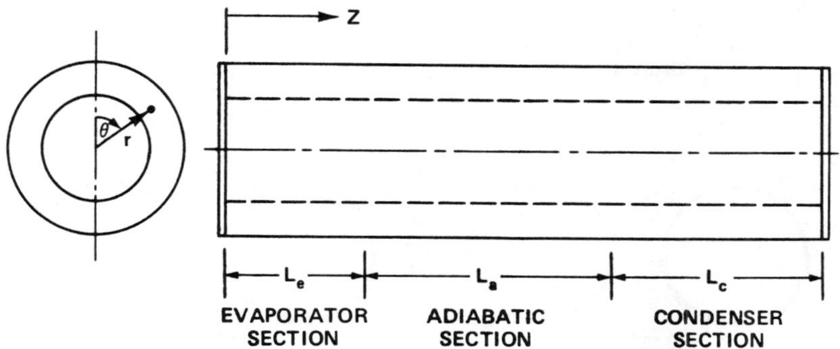

Fig. 2 Coordinate system.

necessary to study the behavior of all the grooves in order to predict the performance of the heat pipe.

The pipe wall temperature distribution is governed by a steady heat conduction equation, along with the exterior boundary condition (i.e., the way heat is applied to the evaporator and removed from the condenser) and the interior boundary condition (i.e., the rates of evaporation and condensation). Using cylindrical coordinates (Fig. 2), the governing equation is written as

$$\frac{\partial^2 T}{\partial r^2} + \frac{1}{r}\frac{\partial T}{\partial r} + \frac{1}{r^2}\frac{\partial^2 T}{\partial \theta^2} + \frac{\partial^2 T}{\partial z^2} = 0 \tag{1}$$

In the present analysis, four heater-cooler combinations are considered (Fig. 3). Both the heater and the cooler cover one-half of the evaporator and the condenser section, respectively, and a uniform heat flux is assumed in each region. The wall surface that is not covered by the heater or the cooler is assumed to be thermally insulated.

The boundary condition describing the thermal interaction between the groove and the liquid in it and between the liquid and the vapor is very complicated because of the complex geometry of the groove. To avoid an unnecessarily complicated solution in the present analysis, the inner surface thermal behavior is modeled as that shown in Fig. 4. An equivalent heat-transfer coefficient (h_e or h_c) is used to describe the

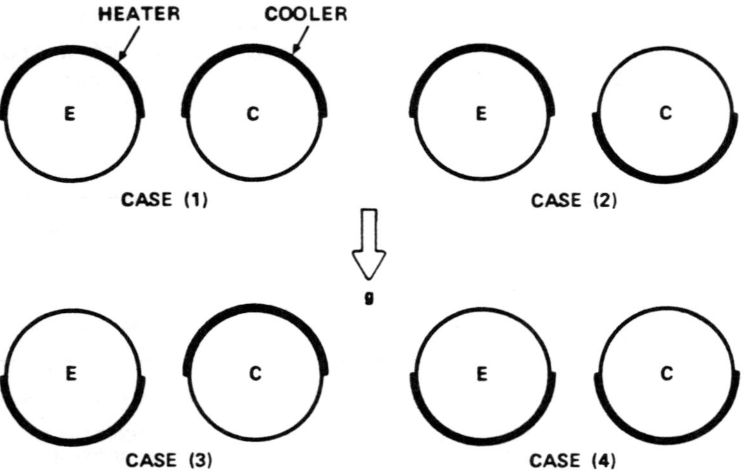

Fig. 3 Heater-cooler combination.

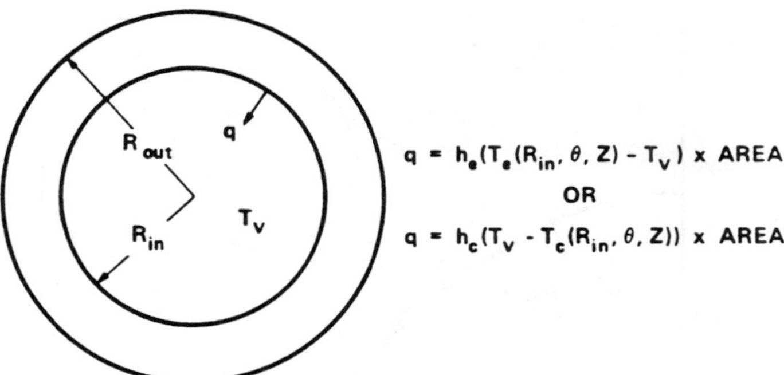

Fig. 4   Hypothetical inner wall surface.

thermal interaction between a hypothetical inner surface
(radius R_{in}) and the vapor. R_{in} is taken to be $1/2(R_v + R_g)$.
As has been shown previously (Kamotani [4,5]), h_e and h_c depend
on liquid meniscus shape, which means they may vary with heat
pipe operating conditions (e.g., heat flux, heat input and
removal mode). However, available experimental data for
symmetric heat transfer cases indicate that the mean values of
h_e and h_c (averaged over the entire evaporator and condenser
region, respectively) vary very little with heat fluxes.
Therefore, as a first approximation h_e and h_c are assumed to
be constant in the present analysis, namely they are assumed
to be independent of heat flux and heat input and removal
mode. T_v is assumed to be constant and given. The heat conduction through the pipe end caps and the conduction to the
pipe adiabatic section are neglected in the present analysis.

The results presented herein were obtained for a heat
pipe whose dimensions are as follows (in meters): $L_e = 0.15$;
$L_c = 0.3$; $L_a = 0.55$; $R_{out} = 0.795$ x 10^{-2}; $R_{in} = 0.486$ x 10^{-2};
aluminum extrusion pipe with 27 grooves; and working fluid,
ammonia (300 K). The numerical values of h_e and h_c are taken
from Schlitt et al.[3]: $h_e = 0.7$ W/cm^2-K, and $h_c = 1.36$ W/cm^2-K.
Both the evaporator and condenser walls are divided into five
sections radially, 27 sections circumferentially (corresponding
to the number of grooves), and 10 sections axially. An energy
balance for each node is formulated using a standard finite
difference approximation. The problem was solved numerically
using an iterative method. The detail is explained in Ref. 7.

In the evaporator section when the grooves are filled
with liquid, and if the heat-transfer coefficient (h_e) is
assumed to be constant everywhere, there is no axial variation
of the wall temperature, so that the problem becomes two-
dimensional. For a given total heat flux, the heat-transfer

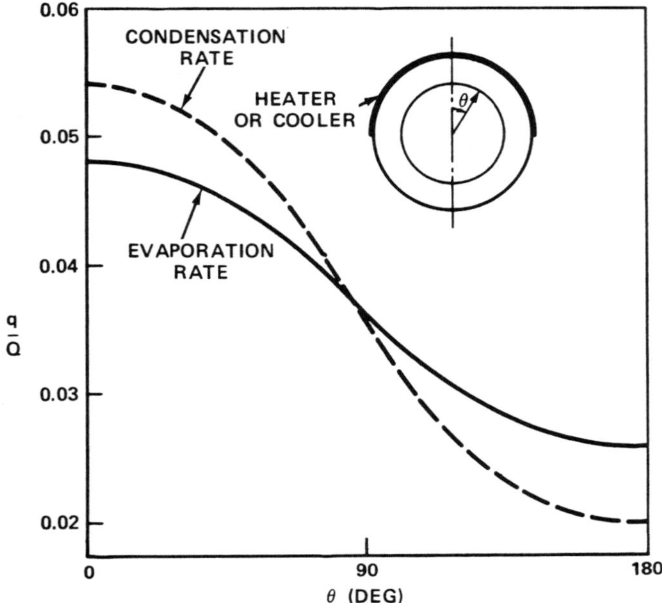

Fig. 5   Evaporation and condensation rates.

rate between the wall inner surface and the vapor depends
only on θ.   Figure 5 shows the relation between the local
heat-transfer rate (q) and θ from the numerical analysis.   The
same analysis was done for the condenser section.   The rates
of evaporation and condensation decrease progressively in
grooves away from the heater and cooler.   The condensation
rate decreases more rapidly with θ than the evaporation rate.
Therefore, even when the heater and the cooler are in the
same side of the pipe (cases 1 and 4 in Fig. 3), the condensation rate of each groove is different from the evaporation rate.   Such difference is more pronounced when the heater
and the cooler are on the opposite side of the pipe (cases 2
and 3). To maintain a steady-state operation, the preceding
mismatch has to be eliminated by assuming either groove-to-
groove liquid communication in the condenser or partial dryout
of some of the grooves in the evaporator, as explained below.
Both cases are discussed for both zero-g and one-g environments.

Zero-g Environments

In zero-g environment, if the condensation rates are
larger than the evaporation rates for some grooves, the excess
liquid tends to form a free liquid slug at the condenser end.
If it is assumed that the liquid slug communicates with all of

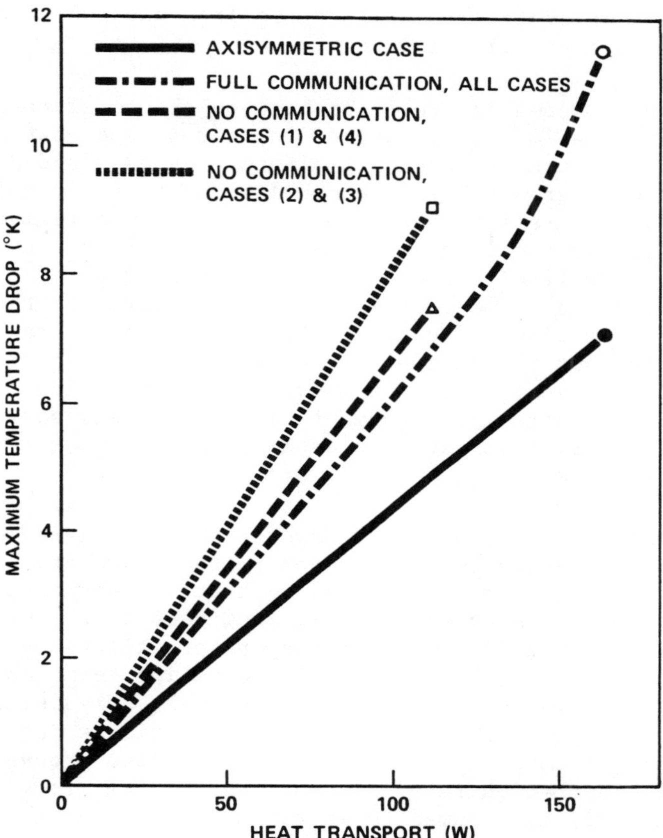

Fig. 6   Performance of heat pipe under zero g.

the starved grooves whose condensation rates are smaller than
the evaporation rates, the balance between the liquid input
and output is maintained for all grooves. If such is the case,
all of the grooves remain to be filled with liquid as the heat
transport increases until the heat transport of the groove
closest to the center of the heater (which transports the
largest heat) reaches the limiting value. The heat transport
limit for a single groove (q_o) is determined from a conventional heat-pipe analysis for uniform heating and cooling.
For the heat pipe and the working fluid discussed herein, it
was found to be q_o = 6.05 W. After the heat transport of the
groove closest to the center of the heater reaches the limiting value, the groove starts to dry out at the evaporator end,
although other grooves operate below the limit. The maximum
heat transport of the heat pipe is, then, obtained when the

heat transport of the groove furthest from the center of the
heater reaches the limiting value.

Figure 6 shows the relation between the total heat transport and the maximum temperature drop across the heat pipe.
All cases give the same result, although cases 2 and 3 require
more groove-to-groove liquid communication than cases 1 and 4.
Up to Q = 122 W, all of the grooves remain to be filled with
liquid. Beyond that, the temperature drop across the pipe
increases sharply with the heat transport because of the
partial dryout of some of the grooves. Compared to the values
for uniform heating and cooling, the maximum heat transport
is the same, but the maximum temperature drop is about 1.6
times.   Since very little is known about the motion of the
liquid slug in zero-g environment, it may be necessary to
obtain a conservative estimate assuming no groove-to-groove
liquid communication.

To maintain a steady-state operation when there is no
liquid communication in the condenser, the evaporation rate
of each groove has to be adjusted so that it becomes equal
to the condensation rate, thus assuming partial dryout in the
evaporation. In this case, the evaporator wall temperature
changes axially, so that the axial conduction term has to be
included in the analysis. In the numerical analysis, the
part of the groove which is filled with liquid is determined
by balancing the total condensation rate of the groove and
the total evaporation rate, and the rest of the groove (the
dried-out part) is assumed to be thermally insulated. Note
that since h_e is assumed to be independent of liquid meniscus
shape, a detailed study of the liquid meniscus variation near
the dryout region is not required in the present thermal
analysis. The maximum heat transport of the pipe is obtained
when the heat transport of the groove with the largest condensation rate (the groove closest to the center of the
cooler) becomes equal to the limiting value for a single
groove.

The result is shown in Fig. 6. Cases 1-4 have the same
maximum heat transport which is 68.1% of the value for uniform
heating and cooling. Cases 2 and 3 have larger evaporator
temperature drop and larger dryout region than cases 1 and 4.
One example of the pipe outside surface temperature variation
in the evaporator and the corresponding liquid distribution
for cases 2 and 3 are shown in Fig. 7. The wall temperature
and the circumferential temperature gradient increase toward
the evaporator end because of the increasingly large dried-
out region. The excess liquid from the dried-out region forms

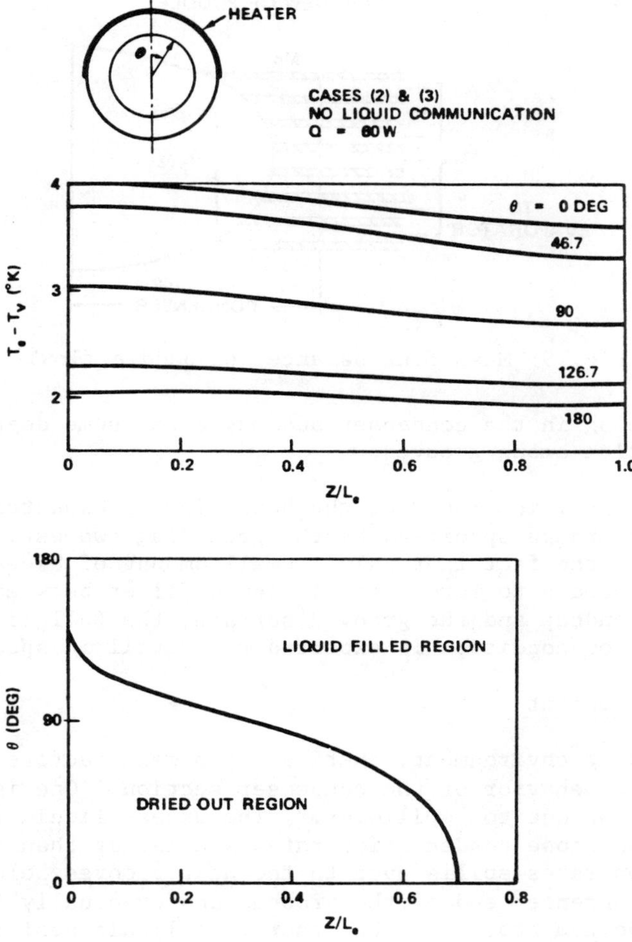

Fig. 7   Evaporator outer surface temperature distribution and liquid distribution.

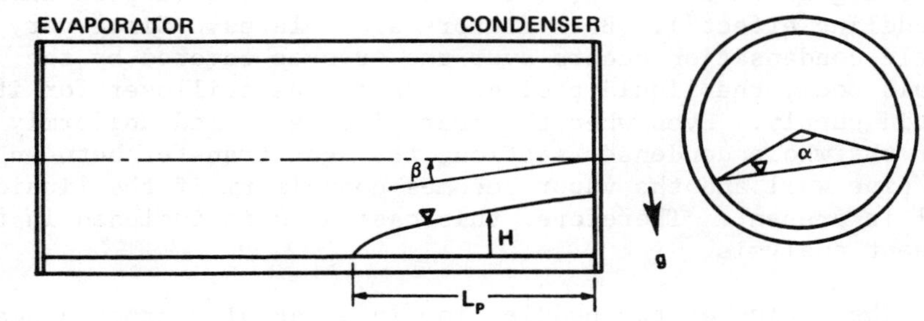

Fig. 8   Liquid puddle.

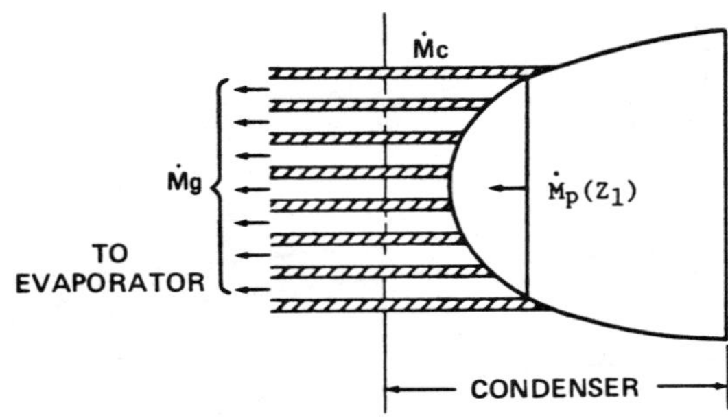

Fig. 9   Mass flux balance for puddle flow.

a liquid slug in the condenser and may cause some degree of communication among grooves.

Therefore, in practice, the heat pipe is expected to perform in the range specified by the preceding two estimates. Considering the fact that only a small amount of excess liquid is needed to form a small liquid fillet between the condenser endcap and the grooved surface, the full liquid communication condition is expected to prevail in space.

One-g Environment

In one-g environment, there are two main factors that influence the behavior of the condenser section. One is liquid communication due to "spillover." The excess liquid from the top grooves whose condensation rates are larger than the evaporation rates spills over to the next grooves below them near the condenser end of the grooves and eventually flows into the bottom grooves. The other is a liquid pool in the condenser bottom formed by overcharged fluid. The liquid pool reduces the transport length and the effective elevation (when the pipe is tilted, the evaporator up) of the bottom grooves, resulting in an increase of the bottom groove transport limit ("puddling effect"). Both factors are related. Since very little condensation occurs over the grooves covered by the liquid pool, the liquid pool depends on the spillover for its liquid supply. Even when the heat pipe is cooled uniformly over the whole condenser section, the heat transfer between the pipe wall and the vapor becomes nonuniform if the liquid pool is present. Therefore, this case also is included in the present analysis.

The motion of the puddle flow in an axially grooved heat pipe has been investigated by Kroliczek et al.[6] Thus only a brief description of the puddle motion is given herein. Referring to Fig. 8, the motion of the puddle is governed by the equation

$$\rho_\ell g \left(\frac{dH}{dZ} + \sin \beta \right) = \frac{dP_\ell}{dZ} = \frac{\mu_\ell \dot{M}_p}{\rho_\ell K_p A_p} \qquad (2)$$

where the vapor pressure term is assumed to be negligible. Equation (2) can be written in terms of α and $d\alpha/dZ$ and can be solved for α to determine the puddle shape once the mass flux in the puddle $[\dot{M}_p(Z)]$ is specified. Since the liquid in the puddle eventually flows into the grooves connected to the puddle, the total mass flux in the puddle at $Z = Z_1$ is given by (see Fig. 9)

$$\dot{M}_p(Z_1) = \dot{M}_g - \dot{M}_c$$

where $\dot{M}_g$ is the total mass flux of the grooves connected to the puddle, and $\dot{M}_c$ is the amount of liquid condensing over those grooves. Under a steady-state condition, $\dot{M}_g$ is equal to the total evaporation rates of the grooves. The liquid supply to the puddle is given by the upper grooves through spillover at the condenser end of the pipe. $\dot{M}_c$ and the amount of spillover depend on the condenser wall temperature distribution. Therefore, in order to determine the performance of the heat pipe under one g, it is necessary to solve the puddle shape and the distributions of temperature in the evaporator and condenser walls simultaneously. Since the analysis by Kroliczek et al.[6] does not include heat-transfer analysis, the puddle shape is determined by assuming a uniform mass flux change along each groove in the condenser. In the present analysis, the puddle shape determined according to the model by Kroliczek et al, is used as the first approximation in the iteration procedure.

The numerical procedure to determine the relation between the total heat transport and the mass charge under one g for the case of uniform heating and cooling is as follows:

1) Specify L_p (puddle length).

2) Using the method of Kroliczek et al., determine the puddle shape, and calculate the heat transport of each groove.

3) Determine the evaporator wall temperature distribution so that it satisfies the mass flux requirements of the grooves submerged in the puddle.

4) Determine the condenser wall temperature distribution so that the total condensation rate is equal to the total

evaporation rate, and calculate the amount of liquid condensing over the grooves influenced by the puddle (M_c).

5) Recalculate the shape of the puddle, and repeat steps 3-5.

6) Check the convergence of the iteration.

7) Calculate the liquid mass in the puddle and the total liquid mass in the grooves.

In the cases of nonuniform heating and cooling, the procedure is similar to the foregoing procedure; however, it is necessary to check whether the starved grooves get enough spillover liquid from the upper grooves; otherwise the grooves partially dry out in the evaporator, as in the case of no

liquid communication under zero g, as discussed previously. The computational results are given in Figs. 10 and 11 for zero tilt. In all cases, the maximum heat transport increases sharply with the total mass charge due to puddling effect. The

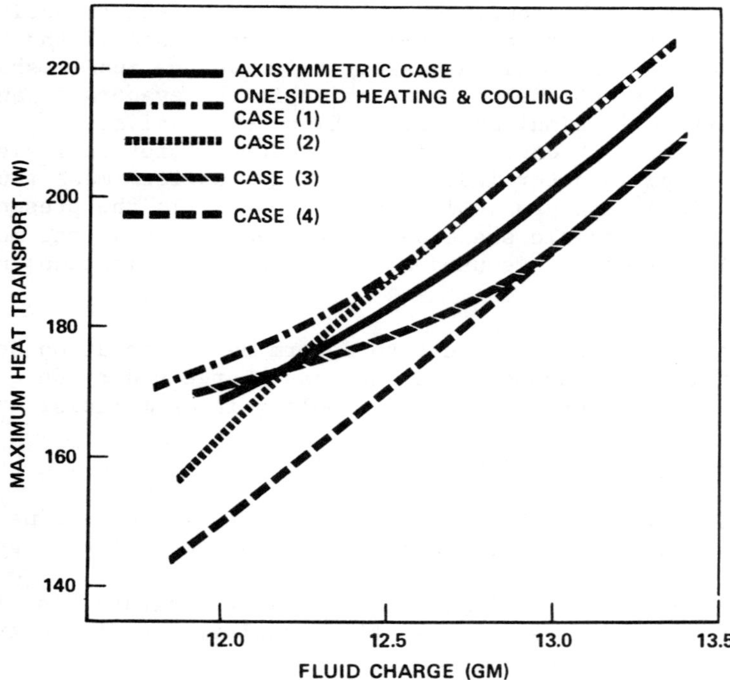

Fig. 10 Effect of fluid charge on maximum heat transport under one g.

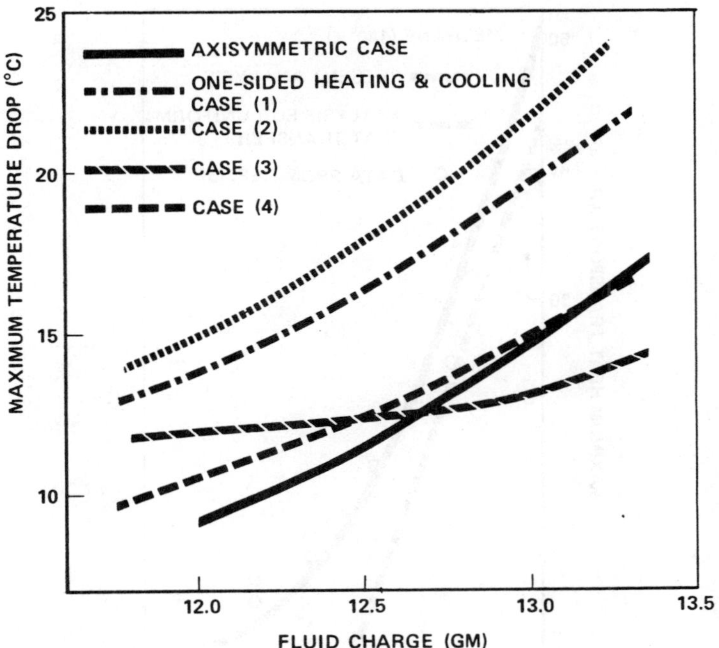

Fig. 11 Effect of fluid charge on maximum temperature drop under one g.

optimum charge for the case of uniform heating and cooling is
12.0 g. The heat pipe is expected to have a mass charge
that is equal to or larger than the optimum value. In that
range of mass charge, the differences in the maximum heat
transport among all cases are rather small (±6%), which may
make it difficult to measure the differences experimentally;
however, there are substantial differences in the maximum
temperature drop across the pipe, as can be seen in Fig. 11.

Cases 1 and 2 have slightly larger maximum heat transport
than other cases, but they produce very large temperature
drop across the pipe. When a liquid pool is present, a large
amount of heat is transported along the bottom grooves. Therefore, when the heater covers the upper half of the heat pipe,
it tends to increase the dryout region in the evaporator,
thereby increasing the amount of liquid in the puddle. As a
result, both the heat transport and the temperature drop increase. Conversely, when the heater is at the bottom (cases
3 and 4), it causes less temperature drop and smaller heat
transport. In the condenser section, the most efficient way
of cooling is to put the cooler at the top when a large puddle
is present and to put it at the bottom when the puddle is

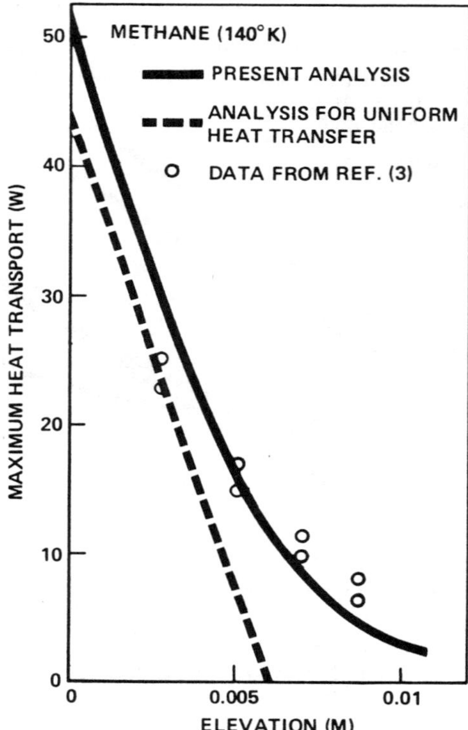

Fig. 12   Tilt vs   maximum heat transport.

small. Therefore, the best heater-cooler combination among
cases 1-4 is case 3 when the heat pipe is more than 5% overcharged, and case 4 when it is less than 5% overcharged. The
preceding also is true when the pipe is tilted.

When the pipe is tilted slightly beyond its static height
and operated at its capillary pumping limit, the grooves that
are not submerged in the puddle may dryout completely in the
evaporator, and only the bottom grooves can transport heat.
Such severe nonuniformity in heat transfer, which happens even
for the case of uniform heating and cooling, is considered to
give a good test for the present analysis. For that reason,
the results of the present analysis for the axisymmetric case
were compared to the data taken by Schlitt et al.[3] Figure 12
shows one comparison. A conventional analysis such as given
in Ref. 1, which does not take into account the nonuniformity
in heat transfer, predicts a linear change of the maximum heat
transport with tilt, as shown in Fig. 12. It is inadequate in
predicting the performance of the heat pipe near its static
height. On the other hand, the result of the present analysis

agrees well with the experimental data beyond the static
height. The discrepancy between the theory and the data for
small tilt angles is considered to be due to the loss of
capillary head of the top grooves caused by liquid communication (drainage effect). The drainage effect is not included
in the present analysis.

As for the effects of one-sided heating and cooling, the
present analysis needs experimental confirmation. Experimental works concerning the maximum heat transport, the distributions of temperature in the evaporator and condenser
walls, and the puddle shape are suggested.

Conclusions

1) In zero-g environment, the maximum heat transport of
the axially grooved heat pipe studied herein with one-sided
heating and cooling may differ from the value predicted for
uniform heating and cooling by as much as 30%. The temperature
gradient across the pipe increases as much as 80%. The performance of the heat pipe is influenced by the behavior of the
liquid slug in the condenser.

2) In one-g environment, the performance of the pipe depends very much on the fluid inventory. If overcharged, the
maximum heat transport is almost equal to the value for uniform heating and cooling. However, when the pipe is heated
at the top, the temperature drop across the pipe becomes
very large. The most efficient way of cooling is to cool
the pipe at the top when the pipe is more than 5% overcharged,
and to cool it at the bottom when it is less than 5% overcharged. The performance curves of tilt vs. heat transport
obtained by the present analysis agree well with available
experimental data.

References

[1] "Summary Report for Fill Determination Study of ATS-F & G
Heat Pipes," Dynatherm Corp., Dec. 1972.

[2] "Design, Fabrication and Testing of Four Hoop-shaped Heat
Pipes," Grumman Aerospace Corp., June 1975.

[3] Shlitt, K.R., Kirkpatrick, P.J., and Brennan, P.J., "Parametric Performance of Extruded Axial Grooved Heat Pipes from
100K to 300K," AIAA Paper 74-724, 1974.

[4] Kamotani, Y., "Analysis of Axially Grooved Heat Pipe Con-
densers," AIAA Progress in Astronautics and Aeronautics:
Thermophysics of Spacecraft and Outer Planet Entry Probes,
Vol. 56, edited by A.M. Smith, 1977, pp. 37-55.

[5] Kamotani, Y., "Thermal Analysis of Axially Grooved Heat Pipes," _Proceedings of 2nd International Heat Pipe Conference_, 1976.

[6] Kroliczek, E.J. and Jen, H.F., "Axial Grooved Heat Pipe Study," B & K Engineering, Summary Rept. BK012-1009, 1976.

[7] Kamotani, Y., _Handbook for Thermal Analysis Program of Axially Grooved Heat Pipes_, in preparation for NASA Goddard Space Flight Center, 1978.

A THERMAL DIODE HEAT PIPE FOR
CRYOGENIC APPLICATIONS

John A. Quadrini*
Grumman Aerospace Corporation, Bethpage, N.Y.

and

Craig R. McCreight†
NASA Ames Research Center, Moffett Field, Calif.

Abstract

The paper describes the development of a cryogenic thermal diode heat pipe for space-flight applications. The diode has ethane working fluid and uses the liquid blockage technique with an internal blocking orifice to accomplish shutoff in the reverse mode. The pipe is 0.635 cm o.d. x 75.82 cm long, including a 2.54-cm3 excess liquid reservoir. Experimental data are presented for forward-mode throughput vs tilt, film coefficients, and reverse-mode characteristics. Transport capacity is 1000 W-cm at 2.5-cm tilt. Evaporator and condenser film coefficients were 0.92 and 1.64 W/cm^2-K, respectively.

I. Introduction

In the time since heat pipes were recognized in the mid-1960's as efficient devices for spacecraft heat-transfer and thermal control applications, a wide variety of near-room-temperature heat-pipe thermal control techniques have been developed.[1-5] Some of these techniques already have been space-flight-tested successfully.[3-6]

During the 1976-1990 time frame, a large number of cryo-genic temperature payloads and/or secondary experiments will be flown. Heat pipes can be used to advantage with many of these

---

Presented as Paper 77-192 at the AIAA 15th Aerospace Sciences Meeting, Los Angeles, Calif., Jan. 24-26, 1977. Copyright © American Institute of Aeronautics and Astronautics, Inc., 1977. All rights reserved.

*Senior Engineer, Thermodynamics Section.
†Research Scientist, Project Technology Branch.

payloads. Also, some of these payloads may require a thermal
diode heat pipe. This device operates as a conventional heat
pipe with high conductance in one direction and very low thermal conductance in the reverse direction.

This paper covers two areas: cryogenic thermal diode
system characteristics, and the design, fabrication, and test
of an ethane thermal diode heat pipe. Under the first topic,
significant parameters affecting diode system performance
characteristics are described. Using typical Earth orbit environmental fluxes and detector/diode models, the amount of
energy absorbed and transmitted by the diode in making the
transition from the forward mode to the reverse mode and during
the reverse mode was found to depend upon amount of fluid
inventory, mass of diode system, and heat loads and temperatures imposed during transition. Discussion on the prediction
of the amount of energy absorbed is given, with particular
emphasis on how factors vary as a function of throughput,
radiator thermophysical properties, fluid properties, and
shutoff technique.

II. Selection of Diode Shutoff Technique

A variety of shutoff techniques are available for a cryogenic diode[2] including the liquid trap, liquid blockage, and
noncondensible gas blockage techniques. With the liquid trap
technique, a reservoir is provided at the evaporator end of the
pipe. The reservoir is empty during normal operation, but fills
with and retains liquid working fluid during reversal, starving
the wick to shut off the pipe. With the liquid blockage technique, a reservoir is provided at the condenser end of the
pipe, and extra working fluid is added to fill the reservoir
during normal operation. The extra fluid migrates to the
evaporator end during reversal, blocking the vapor space to
shut off the pipe. To retain liquid in 1-g tests, the evaporator vapor space must either be very narrow, or a plate with
a small orifice opening may be placed across the vapor space in
the transport section so that the meniscus for the blocking
liquid has sufficient capillary head to overcome gravity
effects. The noncondensible gas blockage technique is similar
in principle to that used in variable conductance heat pipes
(VCHP's) but the gas reservoir and pipe condenser regions get
much hotter than the evaporator during reversal, forcing all
the gas into the evaporator and transport sections and
compressing it to a relatively high partial pressure.

Prior to the selection of diode shutoff technique, an
evaluation illustrating the effect of pipe diameter and fluids

for the liquid trap, blocking orifice, and concentric artery liquid blockage techniques was performed. Brief calculations also were performed for noncondensible gas blockage, but the resulting reservoir volumes were too large for this technique to be considered further. Table 1 shows the pipe geometry and other pertinent information for these calculations.

The evaluations were performed for 1-g performance with the pipe in a level attitude for both forward- and reverse-mode operation. Calculations were made for a tunnel artery transport wick, with artery outside diameter 0.163 cm (0.064 in.) greater than tunnel diameter for all cases. Because of the many variables associated with diode selection, general comments are noted first, followed by some specific details.

For equal artery and pipe diameters, higher throughput is obtained with liquid trap compared with blocking orifice, with the difference representing the blocking orifice pressure loss.

From plots of reservoir volume and shutdown energy, the blocking orifice requires considerably less reservoir volume and shutdown energy. Shutdown energy is defined here as the energy required for transition from the forward mode to the reverse mode plus the reverse-mode heat leakage for 1 hr. Shutdown energy does not account for the transmittal of energy due to partial heat-pipe operation during this transition period.

Another important parameter of interest is the pipe pressure under ambient conditions. For a constant-outside-diameter envelope and given artery geometry, the highest specific volume is obtained with the liquid trap technique, result-

Table 1 Pipe geometry for parametrics

Pipe parameter	Dimension
Evaporator length	10.16 cm (4.0 in.)
Transport length	50.8 cm (20.0 in.)
Condenser length	30.48 cm (12.0 in.)
Effective length	71.12 cm (28.0 in.)
Web thickness	0.083 cm (0.0328 in.)
Web number	6.0 (evaporator and condenser only)
Pipe outside diameters	0.635, 0.953 cm (0.25, 0.375 in.)
Pipe wall thickness	0.071 cm (0.028 in.)
Wick pore size	38.1 μ (0.0015 in.)
Fluids	Methane, Freon 14, ethane
Pipe and wick material	Type 304 stainless steel

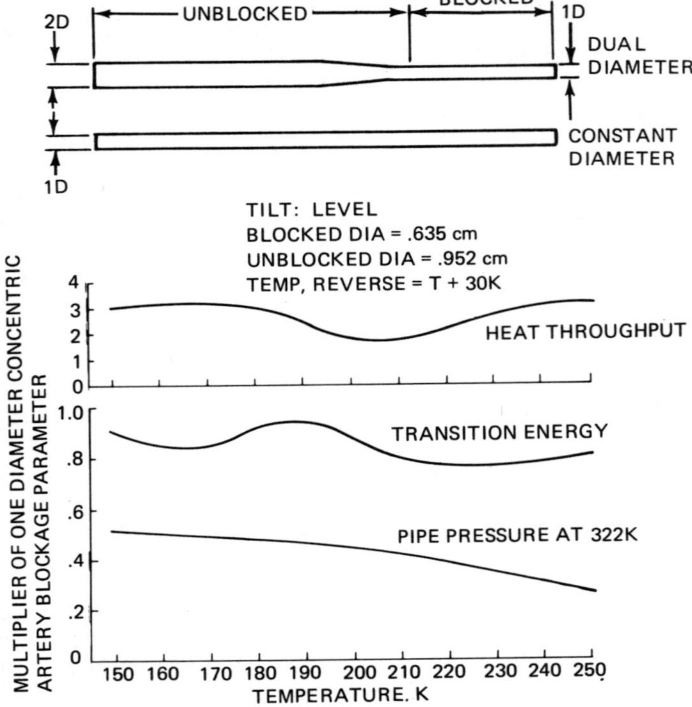

Fig. 1   Dual and constant diam concentric artery comparison, ethane.

ing in lower pipe pressures. Also, because of lower pressures, the pipe wall thickness can be reduced, with a corresponding reduction in shutdown energy. Although the amount of reverse-mode heat leakage under steady-state conditions is small for liquid trap blockage, the reduction in the shutdown energy will not approach the magnitudes for the blocking orifice technique. The liquid trap amount is primarily due to the energy required to evaporate all of the normal working fluid.

For liquid blockage diodes in 0-g, the liquid-blocked length increases with increasing reverse-mode temperature due to fluid expansion. This length is also inverse with the vapor space area. For all of the cases studied, the minimum blocked length was specified as the evaporator length plus 5.08 cm (2.0 in.) of the transport section. The amount of excess liquid required for shutoff is compared to the steady-state reverse-mode requirements. If this excess produces a blocked length greater than initially specified, the blocked length is adjusted accordingly up to a maximum equal to the total pipe length. Excess liquid beyond that required to fill the entire pipe length results in a system whereby the liquid is in a compressed state in the reversal mode.

The specific volumes associated with concentric artery, liquid blockage pipes having a constant outside diameter are the lowest of the techniques studies. Consequently, the pipe pressure under ambient conditions is the highest. Obviously, going to a design whereby a stepped outside diameter is employed as in the ATFE diode[5], retaining the smaller diameter associated with the blocked portions of the pipe, increases the pipe's specific volume, thereby reducing the pipe pressure under ambient conditions.

This design concept is illustrated in Fig. 1 for ethane and the geometry in Table 1, except where noted. The multiplier plotted on the ordinate is the ratio of the performance characteristic for a dual-diam concentric pipe to the corresponding performance characteristic for a constant-diam, concentric artery pipe. As shown, maximum heat throughput increases. Energy for transition is somewhat less for the

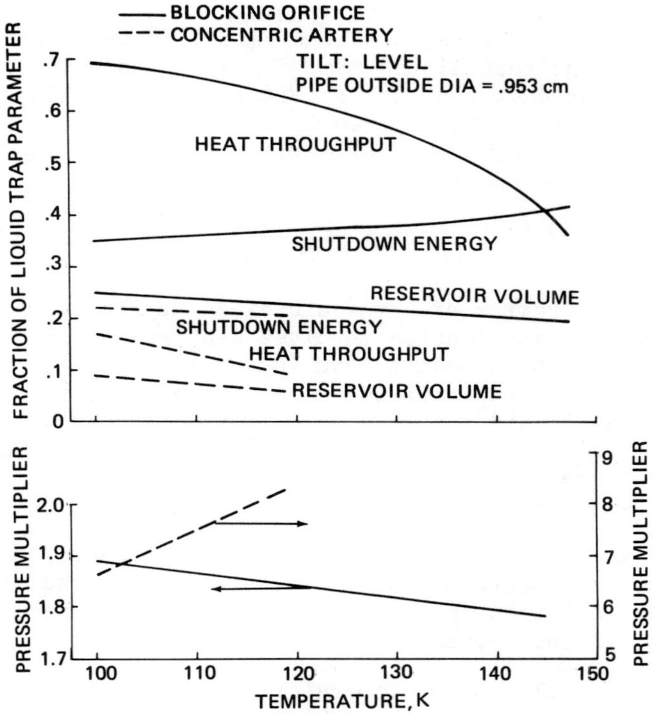

Fig. 2 Comparison of blocking techniques using methane.

dual-diam design because a slightly larger liquid charge is required, and expansion of this liquid reduces the excess liquid reservoir requirements for blockage. Excess mass requirements up to 170K (-154°F) are based on blockage at transition temperature, whereas above 170K (-154°F) steady-state requirements prevail because of the high vapor density under reverse-mode conditions. Another feature of the dual-diam design is the lower pipe pressures under room temperature (322K) conditions, as shown in Fig. 1. Reductions of 50% to 75% are obtained due to an increase in pipe total volume.

Comparison of the liquid blockage shutoff technique to the liquid trap technique for fluids such as methane, Freon-14, and ethane are shown in Figs. 2-4. The pressure multiplier is simply the ratio, at room temperature (322K), of the liquid blockage pipe pressure to the liquid trap pipe pressure. For maximum diode effect (i.e., turndown ratio), the shutdown energy curve must remain below the curve for maximum heat throughput. As indicated in the figures, the blocking orifice is better than liquid trap with methane as the working fluid up to 144K (-200°F), Freon-14 up to 160K (-170°F), and ethane up to 250K (-10°F). In making these comparisons, note again that the shutdown energy curve is based on a 1-hr reverse-mode heat leakage. The single-diameter concentric artery is worse than liquid trap for all temperatures and fluids considered. A closer examination of blocking technique formulations shows liquid trap to be attractive in terms of maximum diode effect for applications involving long evaporators and short condensers.

III. Shutoff Technique

From our review of cryogenic diode applications, the most common type of application involves heat rejection from a source package to a radiator during normal-mode operation, with shutoff when the radiator is exposed to a hot environment. The conventional liquid blockage technique[5] or the liquid blockage technique using an internal orifice[1] to shut the pipe off during reverse-mode operation is attractive in this case for several reasons:

1) The liquid reservoir can be heated easily by the hot environment to prevent liquid retention in the shutoff mode and cooled by radiation to a cold environment to promote liquid retention in the normal mode.

2) Generally, the evaporator is short compared with the condenser and total pipe length, minimizing blocking fluid requirements.

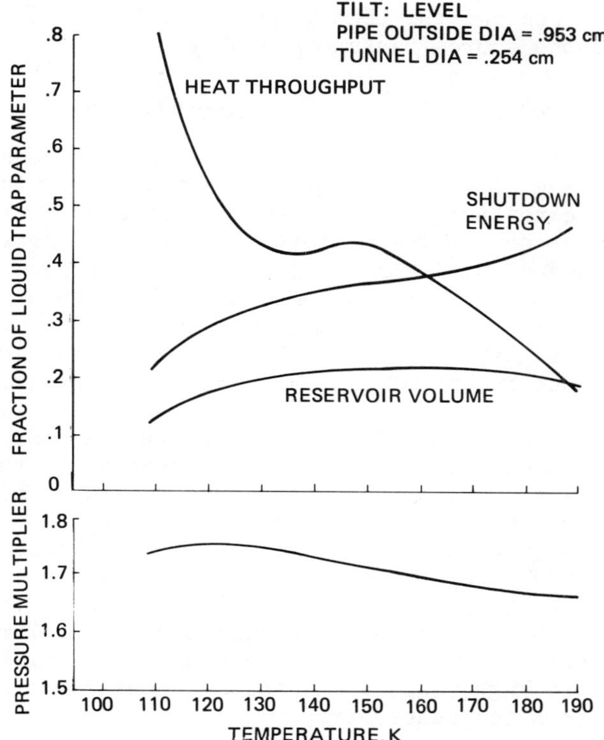

Fig. 3   Comparison of blocking orifice technique using Freon-14.

3)   The liquid reservoir volume may be reduced substantially or eliminated by the extent to which the blocked length liquid requirement is satisfied by expansion of the normal fluid inventory during the hotter reverse-mode condition.

4)   Liquid reservoir volumes are less sensitive to reservoir temperature when compared with reservoir volumes for the noncondensible gas blockage technique.

5)   Small liquid reservoir volumes are required compared with the noncondensible gas reservoir volumes (which range from 10 to 39 times the vapor space volume).

The liquid trap concept for blockage in this case is less attractive because of packaging considerations. For reverse-mode operation, the reservoir, to collect the normal inventory of working fluid, must become an integral part of the heat source package. Thus, there must be a thermal interface not only between the heat-pipe evaporator and source but also be-

tween the source and the heat-pipe reservoir. This arrangement
complicates the source package assembly.

IV. Diode Reversal Requirements

To initiate proper transient reverse-mode operation of
liquid blockage diodes, two conditions must be satisfied simultaneously: condition A, the internal surfaces of the diode
reservoir must have a temperature greater than the vapor
temperature; and condition B, the internal surfaces of the
diode evaporator must have a temperature less than its vapor
temperature. The time interval for transition from forward to
reverse mode is dependent upon the mass capacitances, absorbed
fluxes, and the amount of liquid required for blockage.

If only condition A is met, liquid will evaporate from the
reservoir and condense in the condenser region of the diode,
where it will accumulate as excess free liquid. The diode will
continue to operate in the forward mode, except as influenced
by the free liquid blocking a portion of the condenser surfaces. If condition B then should occur at any time after the
beginning of condition A, any free liquid present in the condenser would shift relatively quickly under capillary action to
the evaporator end of the diode, causing a partial blockage.
Additional blockage would occur as the remaining liquid
evaporated from the reservoir.

If only condition B is met, vapor will condense in the
reservoir until it is full, in addition to the reverse-mode
heat-pipe operation. After the reservoir is full, the diode
will operate as an ordinary heat pipe in the reverse mode with
no shutoff. If condition A then should occur at any time after
the beginning of condition B, liquid would begin to evaporate
from the reservoir and condense in the evaporator region of the
diode. The rate at which shutoff occurs would be controlled by
the rate of evaporation from the reservoir.

A number of computer runs were made to predict the transient thermal behavior of a diode and heat-pipe/phase-change
material system. The conceptual spacecraft system chosen for
analysis consists of a dissipator simulating a cryogenic sensor
package coupled to a radiator. The thermal coupling between
radiator and dissipator is comprised of a 51-cm (20-in.)-long
isothermalizer heat pipe and a 102-cm (40-in.)-long thermal
diode heat pipe. The isothermalizer heat pipe has a 1.59-kg
(3.5-lbm) phase-change material (PCM) canister clamped to it.
Both heat pipes have a 0.635-cm (0.25 in)-o.d. stainless-steel
envelope. For test purposes, the radiator also has a PCM canister attached to it which affects the heat capacity of the

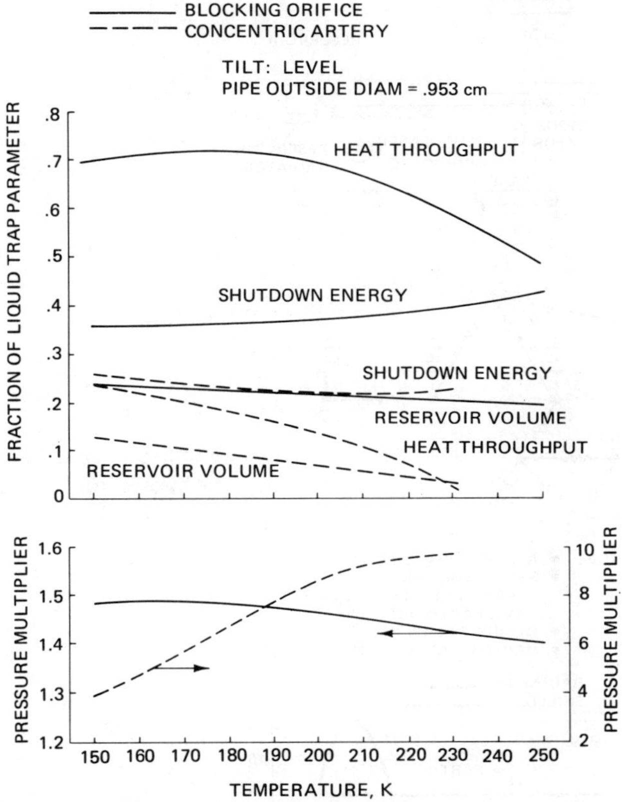

Fig. 4   Comparison of blocking techniques using ethane.

radiator node. The location of the radiator is assumed to be
on the antisun side of the spacecraft. The orbital analysis
accounted for a spacecraft launched into a 500-n. mi., sun
synchronous, gravity gradient orbit (inclination 99.1 deg).
It is assumed further that an Earth shield blocks Earth-
reflected solar radiation (albedo) and direct Earth radiation
(infrared emission) from the radiator. Typical incident fluxes
vs orbit position for the radiator and shield are shown in
Fig. 5. Also, Fig. 5 presents a schematic of the transient
thermal model.

The variables chosen which influence the diode characteristics were thermal capacitance of the main radiator, conductance between the heat pipe and diode, and absorbed flux on the
reservoir radiator. Each parammeter was varied independently
by choosing various values within a realistic range of values

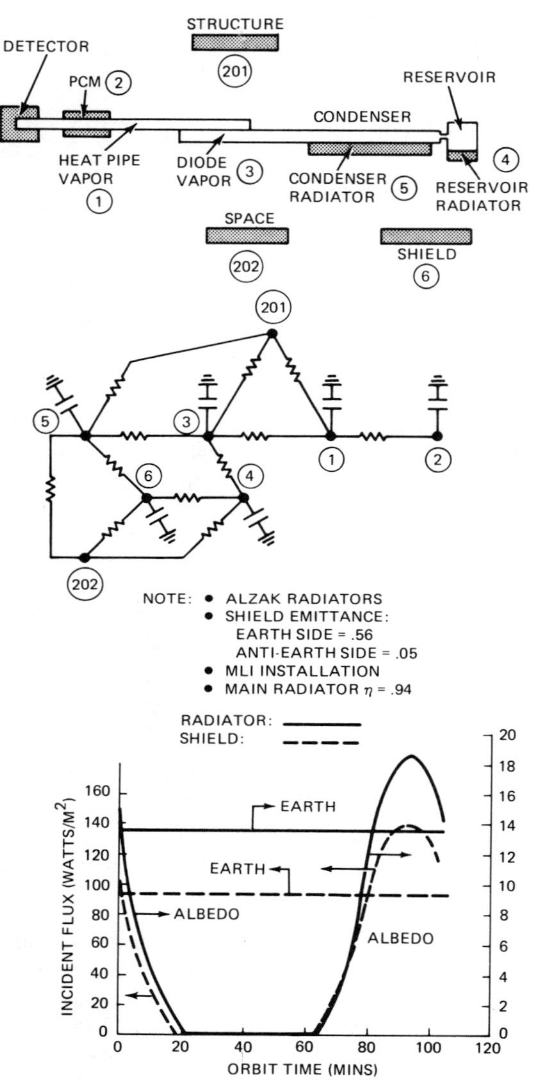

Fig. 5   Transient thermal network and incident fluxes.

associated with a 0.635-cm (0.25-in)-o.d. heat pipe and a blocking orifice diode with ethane as the working fluid. Heat-pipe throughput, independent of whether the diode was in the normal or reverse mode, was 3.0 W. The watt-hour capacity of the heat-pipe PCM was taken as 24, and initial conditions always assumed that the 170K (-154°F) PCM on the radiator had 0 W-hr capacity (i.e., the PCM is in a molten state).

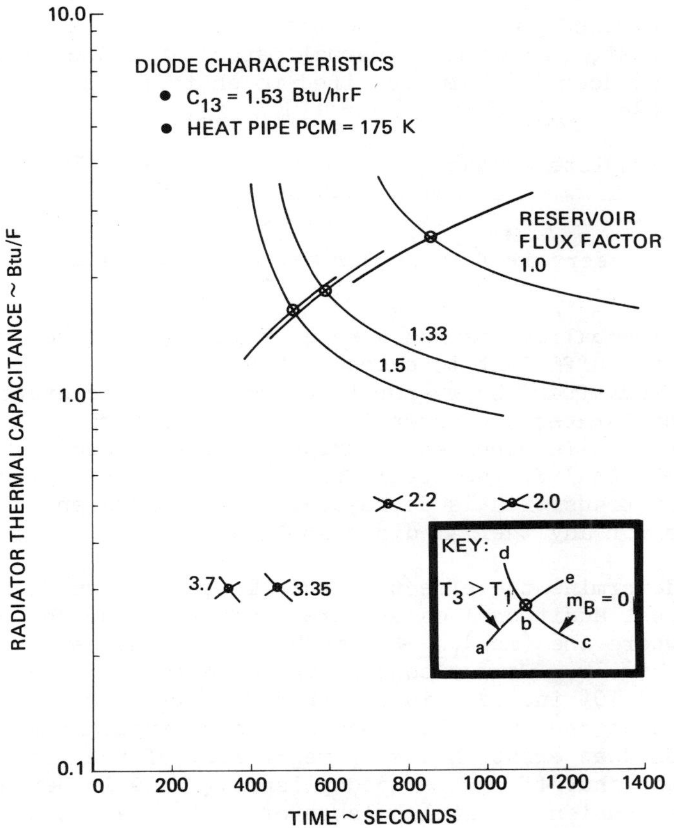

Fig. 6 Effects of $(mc_p)_{rad}$ on shutdown.

Fig. 6 illustrates the effect that radiator thermal capacitance has on complete shutdown of the diode for the baseline
design heat-pipe-to-diode conductance. This effect was considered for three reservoir-absorbed flux factors.

The key in Fig. 6 shows two curves: a-e, representing the
time (for each value of thermal capacitance) at which the diode
vapor temperature (T_3) first exceeds the heat-pipe vapor temperature (T_1); and d-c, representing the time (for each value of
thermal capacitance) at which all of the liquid is evaporated
from the reservoir $(m_B=0)$. For radiator thermal capacitance
values below point b (the intersection of the two curves), the
time interval between the two curves is the time required for
shutoff. For thermal capacitance values above point b, shutoff
is assumed to be essentially instantaneous, being limited only
by the time required for liquid to be wicked into the evapor-

ator and blocked portion of the transport sections. Note, however, that, as radiator thermal capacitance increases above point b, so does the time required to shutoff, as indicated by curve b-e.

For complete shutdown to occur in a reasonable period of time, the following criteria must be established:

$$\text{Reservoir flux factor} = \frac{(Q/mc_p)_{res}}{(Q/mc_p)_{rad}} > 1$$

The inequality assures a rate of increase of reservoir temperature sufficient to exceed the diode vapor temperature prior to and after the evaporation process. Once evaporation begins, both rates are essentially identical until all of the reservoir mass is expelled. Furthermore, provided that $(mc_p)_{rad} < 1$ Btu/°F, the total for shutoff and the reverse-mode energy for transition is decreased. This consequence is most desirable for any thermal diode application.

To determine the effect of this inequality on diode performance, additional cases were considered for the model network where the $(mc_p)_{rad} < 1$ Btu/°F. Performance results are shown in Fig. 6. Comparisons of the effects are given in Table 2. A 10% increase in absorbed flux illustrates the sensitivity that the inequality has on diode performance. A limitation does exist, however, regardless of the $(mc_p)_{res}$, as to how fast shutoff can be accomplished. This is dependent upon the transient thermal balance at individual nodes.

Fig. 7 illustrates the effect of approximately doubling the thermal conductance between the heat-pipe vapor and diode vapor. It shows that this change moderately increases the time between initiation of reversal and complete shutdown to take place, since condition B ($T_3 > T_1$) occurs sooner than before, whereas the time required to evaporate all of the reservoir mass remains essentially constant.

V.   Diode Constructed

The diode selected used ethane as the working fluid and a design goal requirement of 250 to 760 W-cm at 0.25-cm tilt. The design employed excess liquid to block the vapor space of the evaporator and part of the transport section during reverse-mode conditions. An orifice plate is positioned in the pipe at the blocking meniscus location, with the opening arranged to permit proper liquid distribution in both ground tests and zero-g operation.

Table 2   Comparisons of effects

$(mc_p)_{rad}$, Btu/°F	0.3	0.3	0.5	0.5
Reservoir flux factor	3.35	3.70	2.0	2.2
Shutoff, sec	466	343	1074	751
Energy for transition, W-hr	0.069	0.022	0.306	0.111

The fabricated diode consists of four sections: an
evaporator, a transport section, a condenser, and a reservoir.
The wick is a spiral artery formed of 250-mesh stainless-steel
screening. The pipe was made with a charge tube, permitting a
charge valve to be left on in order to accommodate the testing
of other working fluids. The envelope was made from type 304
stainless-steel tubing. This material was chosen to minimize
thermal conduction in the reverse-mode. The transition section, at the condenser end of the pipe, is integral with a
cylindrical reservoir having an inner core of aluminum channels.
No liquid communication was provided between the reservoir and
spiral artery. Detail design of the heat pipe is summarized in
Table 3. Because of the small inside diameter and narrow vapor
space for this design, 4.5 cm (1.77 in.) of vapor space length
can be blocked for every 5% overfill at 220K (-64°F). The
diode was charged with 4.8 g of preprocessed ethane, corresponding to a 7.1% overfill at 200K (-100°F).

This excess liquid at 200K (-100°F) will accumulate as a
puddle covering the bottom of the condenser during ground tests.

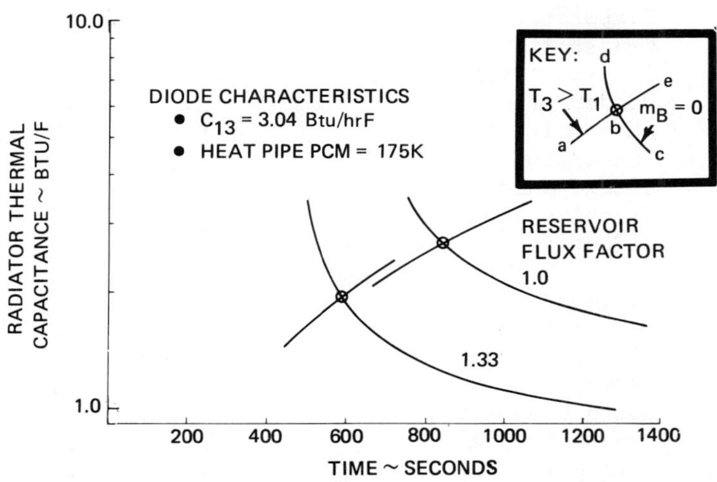

Fig. 7   Effects of $(mc_p)_{rad}$ on shutdown.

In zero g, the liquid distributes itself uniformly because of
the absence of body forces, forming a slug in the vapor
annulus, blocking that portion of the condenser closest to the
reservoir.

Prior to charging the diode, the pipe was pressure-tested
with GN_2 to 2.7 x 10^7 Nm^{-2} (4000 psi). Pipe pressure at 322K
(120°F) is estimated to be 1.1 x 10^7 Nm^{-2} (1600 psi), 5.5.
times below the calculated burst pressure of 6.07 x 10^7 Nm^{-2}
(8800 psi).

Other pertinent design information includes pipe, 304 -
1/8 HD stainless steel; screening, 250-mesh 304 stainless
steel; circumferential grooves, 63/cm (160/in.); reservoir,
6061 aluminum laminates, 0.239 cm (0.094 in.) thick with
0.127-cm wide x 0.127-cm deep (0.050-in x 0.050-in.) axial
machined grooves, core-machined to 1.448-cm (0.570-in.) o.d.
for press-fit into 304 - 1/8 HD stainless-steel cylindrical
shell; orifice height, 0.076 cm (0.030 in.); and orifice o.d.,
0.483 cm (0.190 in.). Circumferential grooves (63/cm) were
used throughout except for the transition and reservoir
sections. The orifice position separates the blocked and
unblocked transport sections, with the orifice opening positioned down so that gravity does not enhance diode reversal.
The orifice was located 10.16 cm (4 in.) from the transport
side of the evaporator.

Table 3 Detail design of heat pipe

Lengths, cm (in.)		
Evaporator	10.160	(4.0)
Transport (blocked)	10.160	(4.0)
Transport (unblocked)	24.130	(9.5)
Condenser	21.273	(8.375)
Reservoir	6.033	(2.375)
Transition	1.524	(0.60)
Charge tube	4.445	(1.75)
Inactive sections	2.540	(1.00)
Effective	50.010	(19.687)
Diameters, cm (in.)		
Pipe o.d.	0.635	(0.250)
Pipe i.d.	0.493	(0.194)
Artery o.d.	0.300	(0.118)
Solid tunnel o.d.	0.051	(0.020)
Reservoir o.d.	1.588	(0.625)
Charge tube o.d.	0.318	(0.125)

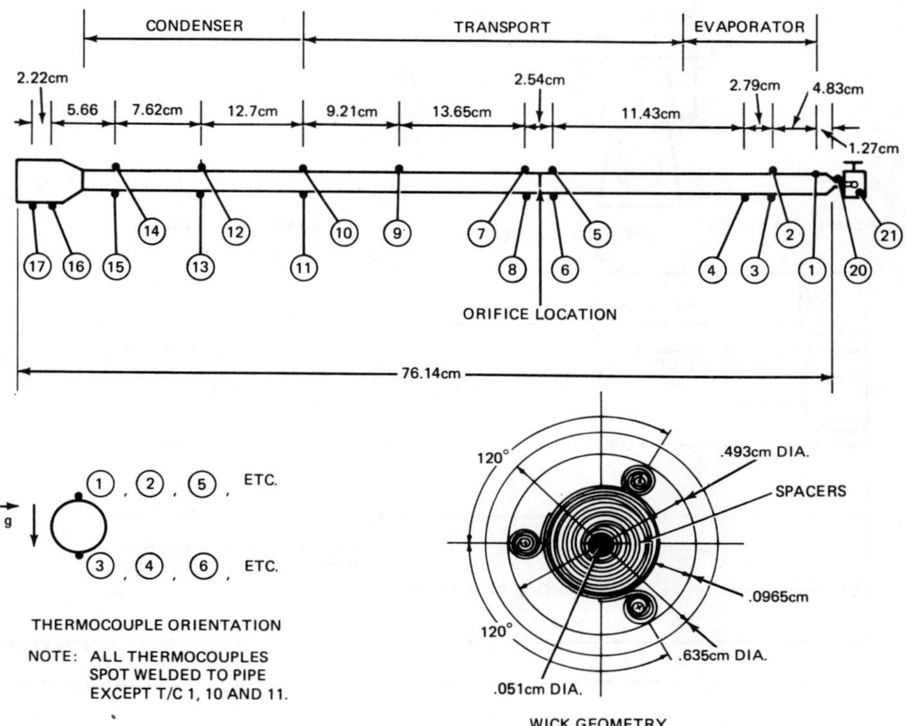

Fig. 8   Cryodiode instrumentation schematic.

Liquid communication between artery and pipe wall was achieved with three scroll-type webs equally spaced in the condenser and evaporator only. Standoffs were provided on the artery in the unblocked transport section. The fluid inventory for the nominal operating temperature (200K) is given in Table 4. Also shown is the dependence of the fluid inventory on temperature and the blocked length of the vapor annulus for an overfilled condition.

Table 4   Fluid inventory

Temperature, K	170	180	190	200	210	220	230
Relative fill, %	95.5	96.3	97.8	100	102.5	105.2	107.4
100% ethane charge, g	4.69	4.65	4.58	4.48	4.37	4.26	4.17
Length of blocked vapor annulus, cm	...	...	...	0	2.25	4.63	6.73

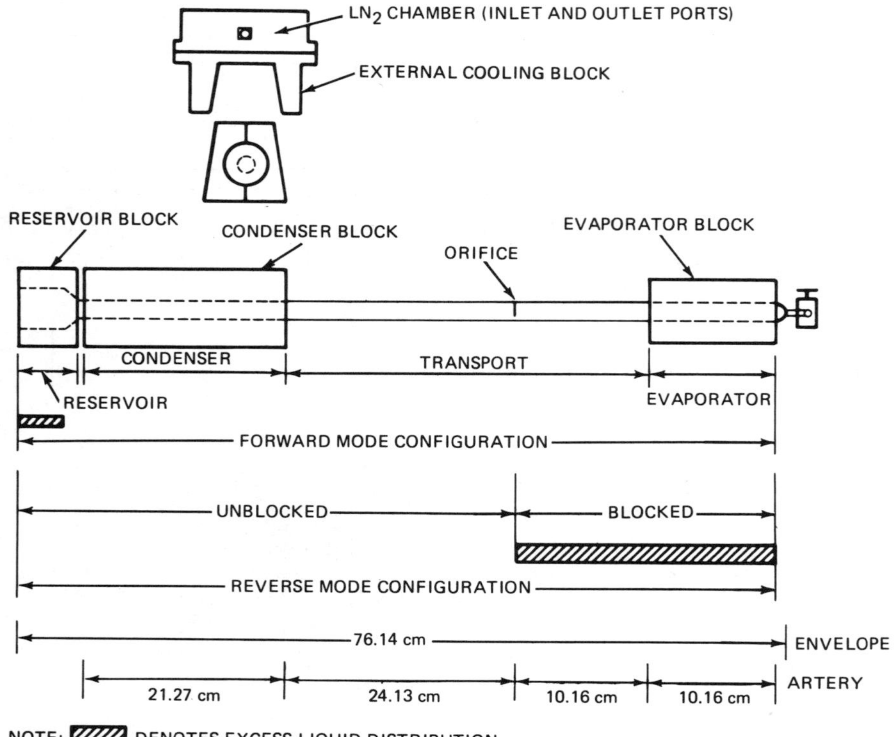

Fig. 9 Test configuration.

VI. Test Setup and Instrumentation

Low-temperature thermal performance tests were conducted
in both the normal and shutoff modes of operation. Normal-
mode tests were performed at various pipe attitudes (evaporator
above condenser) and a temperature of 200K (-100°F). Reverse-
mode operation was performed with the diode in a horizontal
attitude. Testing was performed in an insulated ambient enclosure where the environment surrounding the insulated pipe
was cooled by vaporizing LN_2. This enclosure provided a
controllable heat-pipe environment that was kept above the operational diode temperature. The pipe also was tested in a
thermal vacuum chamber.

Three aluminum masses were attached to the pipe: evaporator mass of 0.197 kg (0.44 lbm) to simulate a detector block;
condenser mass of 0.765 kg (1.69 lbm) for ease of mounting the
pipe assembly to an LN_2 sink; and a reservoir mass of 0.205 kg
(0.45 lbm), again for maintaining the reservoir in thermal

contact to the LN_2 sink. Affixed to the reservoir and evaporator masses were strip heaters. The evaporator heater simulates forward-mode heat loads, whereas the heater on the reservoir provides the reverse-mode environment to initiate diode reversal. A guard heater was installed on the charging valve to null the charge-tube/charge-valve heat leak during testing.

As mentioned, both the reservoir and condenser blocks were mounted to the LN_2 sink, referred to as the external condenser block and chamber. Its design permitted both reservoir and condenser blocks to be captured via a compression-type arrangement. An LN_2 chamber above the compression block provided the heat-pipe sink. Nichrome ribbon heaters attached to the outside surface of the external compression block controlled the forward-mode temperature and also provided an alternate means for diode reversal. The mass of the external condenser block and chamber was 4.654 kg (10.25 lbm).

The instrumentation used to monitor the diode temperature response is indicated in Fig. 8. All thermocouples were premium-grade type T and spot-welded to the pipe except where noted. Four additional thermocouples, not shown in Fig. 8, monitored the ambient environment. Most of the testing was performed with the test configuration shown in Fig. 9. The external condenser/chamber block was the mounting point of the pipe to the test fixture that supports the entire assembly. This assembly then can be tilted to obtain the desired pipe attitude. Some additional testing was performed with a very similar test setup in a small vacuum chamber.

VII. Performance Tests

Test results for the forward-mode transport capacity as a function of pipe attitude are shown in Fig. 10. Also shown on this figure is the diode theoretical performance. Experimental values are in good agreement with predictions, based on a 38.1-μ (0.0015-in.) wick effective pore diameter. This pore diameter is consistent with wick capillary rise measurements made prior to wick insertion, indicating no degradation of the effective capillary pore size of the spiral wick upon assembly.

Evaporator and condenser film coefficients were based on the average evaporator and condenser temperatures calculated from 18 separate test points. Thermocouples 1-4 were taken to represent the average evaporator temperature, whereas the average condenser temperature was represented by thermocouples 12-15, as shown in Fig. 9. Diode vapor temperature was

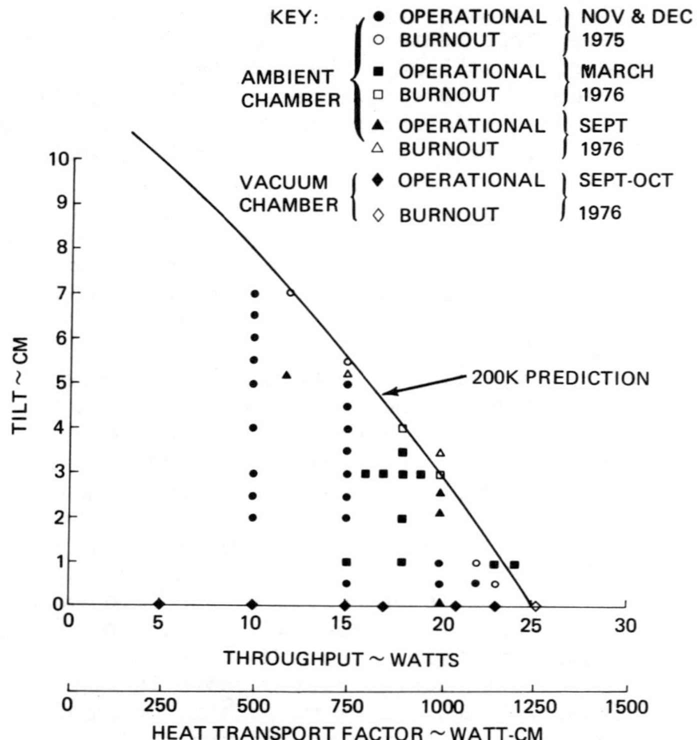

Fig. 10   Forward-mode throughput characteristics.

determined from thermocouple 9. Evaporator and condenser mean
film coefficients were found to be 0.92 and 1.64 W/cm^2-K (1620
and 2890 Btu/hr-ft^2-$^{\circ}$F), respectively. Total error involved in
calculating these coefficients from measured physical
quantities is ±5.1%. In general, these coefficients are more
representative of the higher heat flux runs, with lower values
obtained from low heat fluxes with a corresponding lower total
pipe temperature difference. The higher heat flux runs (>12W)
probably reflect a depressed meniscus and correspondingly
reduced internal surface areas blocked by liquid fillets,
thereby enhancing the heat-transfer process. These values are
well above film coefficients for ethane reported in Refs. 8
and 9 for axial grooved heat pipes and for circumferentially
grooved heat pipes with homogeneous and graded-porosity slab
wicks.

Reverse-mode test results permitted the energy absorbed by
the detector block (normal evaporator) to be determined. The
amount of energy absorbed, once reversal started, was a function of the time required for complete shutdown. The definition of the shutdown period for the reverse-mode tests, which

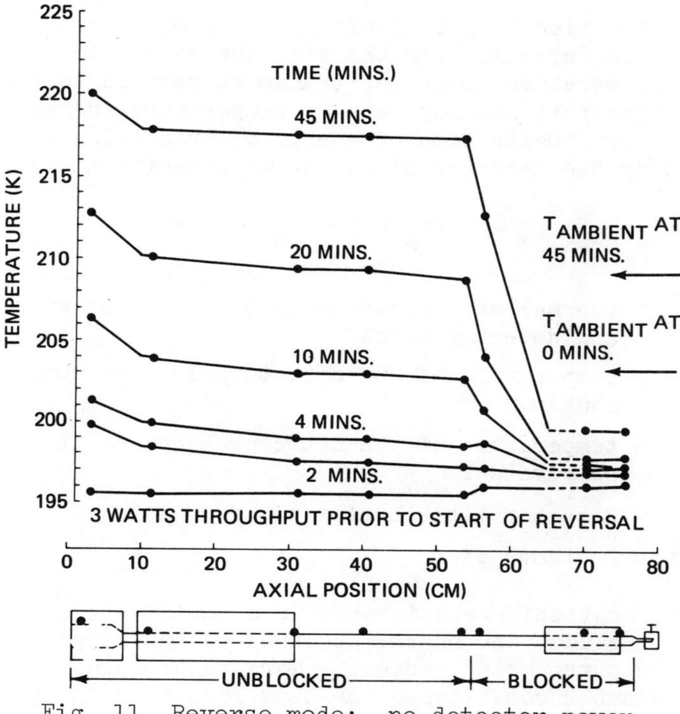

Fig. 11 Reverse mode: no detector power.

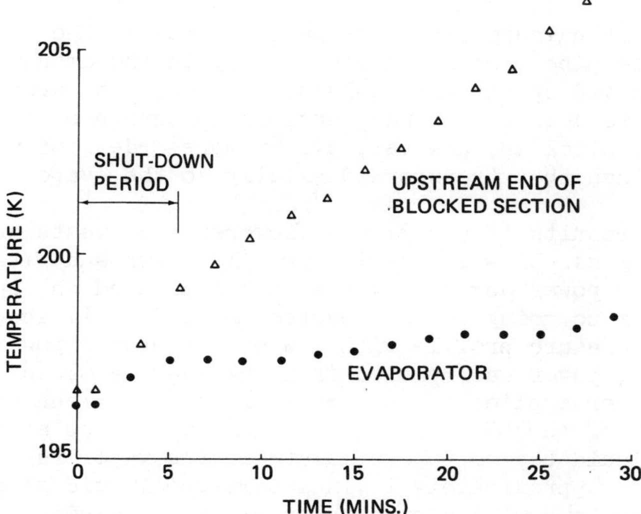

Fig. 12 Shutdown response of evaporator and upstream end of blocked transport sections.

neglects the time to get liquid out of the reservoir, was
taken on the interval from the time the reservoir begins to
rise in temperature above the transport section to the time the
blocked transport section rate of temperature increase drops
dramatically. Definition of energy of transition or the energy
absorbed by the detector block during transition was taken as

$$Q = mc_p \, (T_f - T_o) - Q_e \tau$$

where

mc_p = thermal mass capacitance of the diode evaporator
and detector block

T_f = temperature of the detector block at complete
shutdown

T_o = temperature of the detector block at the start
of reversal

Q_e = detector block heater power

τ = shutdown time

Note that this relationship is a conservative estimate,
since it includes the energy absorbed by the evaporator block
due to the surrounding ambient. Tare tests conducted in the
ambient chamber established the heat loss or gain by the
evaporator from the surrounding environment to be 0.058 W/K.
Throughout all testing, the ambient environment was maintained
2 to 10K above evaporator temperature.

A heat balance could not be performed at the condenser
end of the pipe because of uncertainty in the amount of heat
being removed by the LN_2 coolant. The net heat absorbtion
required to account for the sensible temperature rise of the
condenser block is, however, two or more orders of magnitude
greater than that transported axially to the evaporator.

The results of a series of reverse mode tests are indicated in Figs. 11 - 16. Axial profiles with and without
evaporator power during reversal are indicated in Figs. 11 and
14. Prior to going to the reverse mode, Fig. 11 shows the
pipe temperature profile with 3 W of evaporator power. At
time zero, power was removed from the evaporator and 91 W of
heater power applied to the external condenser/chamber, with no
reduction of the LN_2 flow to the chamber. After approximately
2 min, liquid begins to accumulate in the vapor region (see
Fig. 12). Approximately 1 min later, the liquid slug has
reached the end of the evaporator block. The pipe is shutdown
completely at 5.6 min, as indicated by the response of the
blocked transport section in Fig. 12 and the axial profile in

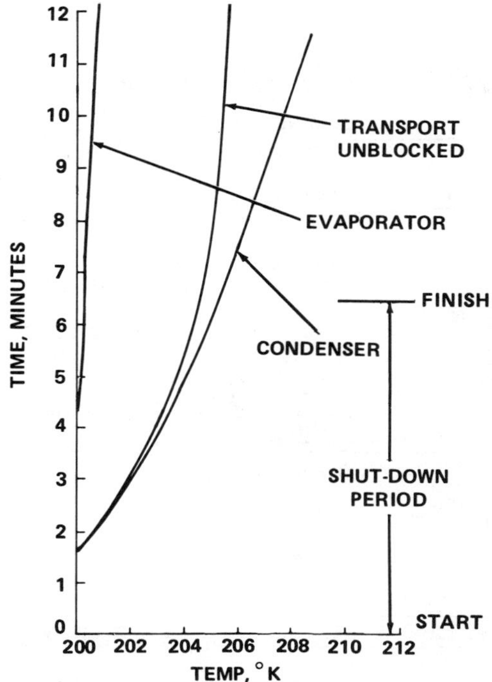

Fig. 13   Typical transient response.

Fig. 11 (approximately 4 min later, after shutdown). Fig. 13
shows the continuous response of the evaporator, unblocked
transport and condenser sections for similar conditions.

Several reverse-mode tests were performed without removing
evaporator block power during reversal. Power levels applied
ranged from 1 to 3 W, with 2 W on the evaporator for several
of the runs indicated in Figs. 15 and 16. In each reversal,
the shutdown was similar to that shown in Fig. 14. Shutdown
time and transition energy associated with shutdown increased
with a decreasing rate of condenser rise in temperature, as
shown in Figs. 15 and 16. Reservoir rate of temperature
increase was approximately 25% greater than the condenser rate
for these test conditions.

The amount of energy absorbed by the evaporator block
suggests that liquid removal from the reservoir is an expulsion
process rather than evaporation at the liquid-vapor interface
within the reservoir channels. If the reservoir were filled
completely with liquid, 0.176 W-hr of energy would be required
to evaporate its liquid volume. Since only 0.02 to 0.10 W-hr

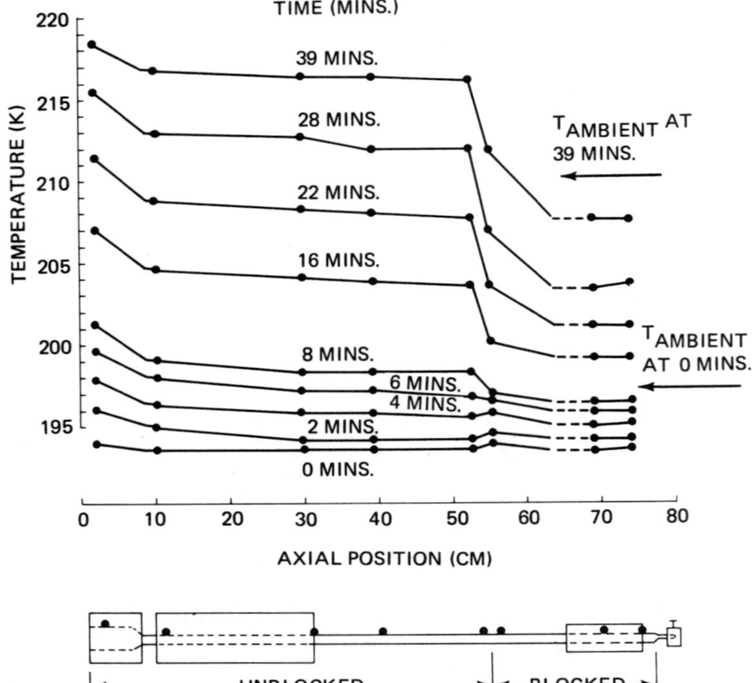

Fig. 14   Reverse mode: 1 W of detector power.

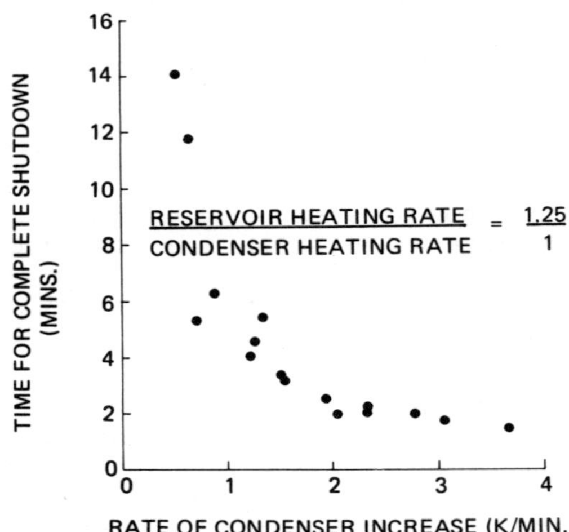

Fig. 15   Shutdown period vs condenser heating rate.

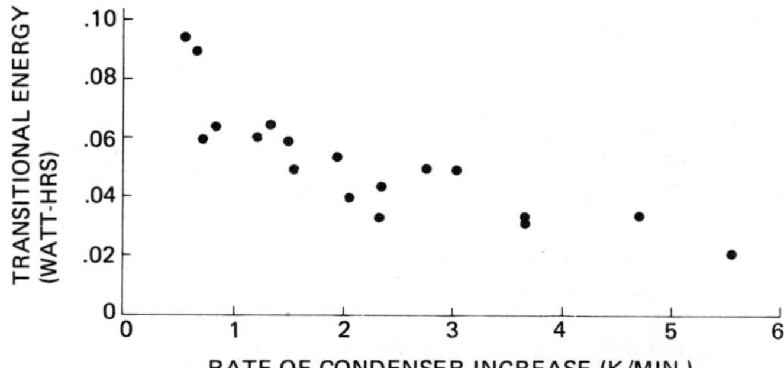

Fig. 16 Transitional energy vs condenser rate.

of transitional energy was absorbed by the evaporator block
(a conservative estimate, also) for the condenser heating
rates tested, it appears that vapor bubbles within each
reservoir channel must have expelled the liquid immediately
after reversal commenced. For this reservoir design, aluminum
channels were used in order to increase the heat-transfer rates
to the excess liquid during the forward to reverse mode. This
design provides an unobstructed passage for the expelled liquid
to the vapor space. It may be noted that the measured transitional energy values are an order of magnitude lower than the
corresponding values reported for a liquid trap diode[10].

Acknowledgements

This research was supported under Contract NAS 2-7492
from NASA Ames Research Center, Moffett Field, Calif. The
authors wish to express their thanks to W. Combs and P. Wagner
for help in manufacturing, F. Gorman for test activities, and
Georgia Institute of Technology co-op student D. Waller Jr.
for conducting most of the tests.

References

[1]Kosson, R., Quadrini, J., and Kirkpatrick, J., "Development
of a Blocking Orifice Thermal Diode Heat Pipe," AIAA Progress
in Astronautics and Aeronautics: Heat Transfer with Thermal
Control Applications, Vol. 39, edited by M.M. Yovanovich,
The MIT Press, Cambridge, Mass., 1975, pp. 245-258.

[2]Quadrini, J. and Kosson, R., "Design, Fabrication and Testing
of a Cryogenic Thermal Diode," NASA CR-137616, Dec. 1974.

[3]Kirkpatrick, J.P. and Groll, M., "Heat Pipes for Spacecraft
Temperature Control - An Assessment of the State-of-the-Art,"

Proceedings of the 2nd International Heat Pipe Conference, Bologna, Italy, March 31 - April 2, 1976, pp. 167-181.

[4] Quadrini, J.A. and Bocchicchio, R.L., "Application of Heat Pipes to Unmanned Space Power Systems," 5th Intersociety Energy Conversion Engineering Conference, Sept. 1970.

[5] Swerdling, B. and Kosson, R., "Design, Fabrication and Testing of a Thermal Diode," NASA CR-114526, Nov. 1972.

[6] Fine, H., Quadrini, J.A., and Ollendorf, S., "An Insight into the Features of the OAO-C Thermal Design, "Intersociety Conference on Environmental Systems, American Society of Mechanical Engineers Paper 73-ENAs4, San Diego, Calif., July 1973.

[7] Sherman, A. and Brennan, P., "Cryogenic and Low Temperature Heat Pipe/Cooler Studies for Spacecraft Application," Journal of Spacecraft and Rockets, Vol. 13, May 1976, pp. 288-293.

[8] Schlitt, K.R., Brennan, P.J., and Kirkpatrick, J.P., "Parametric Performance of Extruded Axial Grooved Heat Pipes from 100° to 300°K," AIAA Progress in Astronautics and Aeronautics: Heat Transfer with Thermal Control Applications, Vol. 39, edited by M.M. Yovanovich, The MIT Press, Cambridge, Mass., 1975, pp. 215-234.

[9] Groll, M., Pittman, R.B., and Eninger, J.E., "Parametric Performance of Circumferentially Grooved Heat Pipes with Homogeneous and Graded-Porosity Slab Wicks at Cryogenic Temperatures," Proceedings of the 2nd International Heat Pipe Conference, Bologna, Italy, March 31 - April 2, 1976, pp. 63 - 76.

[10] Brennan, P.J. and Groll, M., "Application of Axial Grooves to Cryogenic Variable Conductance Heat Pipe Technology," Proceedings of the 2nd International Heat Pipe Conference, Bologna, Italy, March 31 - April 2, 1976, pp. 183-195.

TWO-PHASE WORKING FLUIDS FOR THE TEMPERATURE RANGE 100°-350°C

E.W. Saaski*

Sigma Research, Inc., Richland, Wash.

and

L. Tower[+]

NASA Lewis Research Center, Cleveland, Ohio

Abstract

The decomposition and corrosion of two-phase heat-transfer liquids and metal envelopes have been investigated on the basis of molecular bond strengths and chemical thermodynamics. Potentially stable heat-transfer fluids for the temperature range $100°$ to $350°C$ have been identified, and reflux heat-pipe tests have been initiated with 10 fluids and carbon steel and aluminum envelopes to establish experimentally corrosion behavior and noncondensable gas-generation rates.

Nomenclature

C_a = concentration of solute species, cm^{-3}
C_{as} = saturation concentration of solute, cm^{-3}
D_{ab} = diffusion coefficient for solute in solvent, cm^2/sec
g = gravitational acceleration, cm/sec^2
h_{fg} = heat of vaporization, J/g
K_b = solvent thermal conductivity, W/cm-K
$Q_{a\ell}$ = average heat flux density over $0 \leq z \leq \ell$, W/cm^2
T_v = vapor temperature, K
T_w = wall temperature, K
$U_{y,z}$ = local film velocity in -y and -z directions, cm/sec
y = coordinate perpendicular to wall, cm
z = coordinate parallel to vertical wall, cm
δ = film thickness, cm

Presented as Paper 77-753 at the AIAA 12th Thermophysics Conference, Albuquerque, N. Mex., June 27-29, 1977. Copyright ©American Institute of Aeronautics and Astronautics, Inc., 1977. All rights reserved.
*Vice President and General Manager.
+Technical Monitor.

η =   similarity variable $= y/\delta$

θ =   defined by Eq. (11)

μ_b =   solvent viscosity, P

ρ_b =   solvent density, g/gm^3

B = $\exp[(3/8)\theta] \cdot \theta$

Introduction

The efficient transfer of heat in the temperature range 100°-$350^{\circ}C$ has been carried out historically by pressurized liquid loop systems. The heat pipe, however, presents an attractive alternate for both spacecraft and terrestrial applications, because of both its passive operating nature and its ability to transfer heat uniformly from complex structures. Delays in the application of heat pipes to this temperature range, however, have come, in part, from an inability to identify suitable working fluids. Over the past decade, heat-pipe development has concentrated on satellite thermal control applications over the temperature range -273° to $100^{\circ}C$. Work also has been done on liquid metal heat pipes for temperatures greater than about $350^{\circ}C$. For applications in the intermediate temperature range of 100° to 350°, there have been only limited tests described in the open literature to establish viable fluid/envelope combinations.[1-3] Conversely, there have been a number of studies up to about $100^{\circ}C$, and there has been a continued but generally unsuccessful effort to develop water/metal combinations over the 50° to $350^{\circ}C$ range; all common envelope materials, with the exception of copper, generate large amounts of hydrogen gas with water working fluid.

Fluid development in this range is relatively uncharted. Potential space applications include heat-transfer components of solar-powered heat engines, thermionic or thermoelectric converters, or high-power electronic equipment. In addition, the heat pipe and related two-phase systems now are being considered increasingly for terrestrial uses. Potential applications under active study are in the areas of solar power, geothermal power, and the thermal control of industrial processes.

In solar applications, heat pipes are being considered which must operate from 100° to $350^{\circ}C$. The lower end of the range is represented by heat-pipe heat sinks for solar cells using solar flux concentration and at the upper end by large solar power station designs. In the food-processing and other industries, heat pipes are being considered for waste heat reclamation in the range 50° to $300^{\circ}C$, whereas, in the paper industry, two-phase systems are being considered for roll and platen isothermalization.

The work summarized here was an initial analytical and experimental investigation into suitable working fluids for the range 100° to $350^{\circ}C$. Because the number of potential working fluids is almost unlimited, basic differences in chemical structure and bonding were used to categorize and rank large groups of compounds with similar chemical and stability characteristics. From this ranking, 10 fluids were selected for long-term reflux compatibility tests in aluminum and mild steel envelopes. The fluids selected were not necessarily the most stable compounds, but rather represented distinct chemical groups of interest. These groups were 1) aliphatic hydrocarbons, 2) aromatic hydrocarbons, 3) halogenated hydrocarbons, and 4) inorganic molecular fluids.

Definition of Stable Species

Hydrocarbons

Chemical bonding in organic molecules is chiefly covalent and dominated by the formation of hybrid electron orbitals from the standard s, p, and d electronic orbitals. One type of hybrid orbital is designated "sp" and is formed by combining an s orbital with a p orbital to form two equivalent sp orbitals. The orbitals are oriented 180 deg from each other. In sp^2 hybridization, a single s orbital and two p orbitals form three equivalent hybrid sp^2 orbitals, on a three-cornered arrangement. The sp^3 hybrid is formed from one s orbital and three p orbitals and is tetrahedral.

Because of their greater s character, sp orbitals are generally smaller than sp^2 orbitals, and sp^2 orbitals are smaller than sp^3 orbitals. A small orbital generally indicates a short, strong bond, and bond strength increases as $sp^3 \rightarrow sp^2 \rightarrow sp$. The hybrid orbitals are highly localized and are individually cylindrically symmetric. The cylindrically symmetric covalent bonds formed by these orbitals are commonly called σ bonds.

In some chemical compounds, these hybrid bonds do not accommodate all available electrons, and additional electrons are available in the remaining lobar p orbitals. If the p orbitals of the constituent atoms overlap, and the number of electrons available is 4N+2, where N is the number of atoms per molecule, then a hybrid orbital of "π electrons" is formed. The delocalized π electron clouds participate to some extent in all of the molecular bonds, enhancing overall molecular stability and, in particular, the stability of the various constituent σ bonds.

Thermal decomposition of an organic compound can be visualized as a stepwise process of bond dissociation, with the weakest bond essentially defining the compound's thermal stability. It is furthermore an experimental fact that substitutions at noncarbon σ bond positions do not influence significantly the stability of other σ bonds on a carbon skeleton, so that the search for highly stable classes of organic compounds can be, in part, performed by examining the various bonds by which atoms are attached to the carbon skeleton and identifying the weakest of these bonds. Table 1 is a summary of bond

Table 1 Bond dissociation energies for various working fluid vapors (kcal/mole)[a]

	H	CH_3	C_2H	C_2H_3	C_2H_5	CHO	OH	F	Cl	Br	I	NH_2	CN
H	104	104	<121	108	98	74	119	136	103	88	71	104	120
CH_3	104	88	~110	97	85	70-75	91	108	84	70	56	79	105-110
C_2H_3	108	97	...	...	94	~84	...	...	84	...	55	...	121
CH_3CO	89	81	...	...	83	~59	107	119	82	68	52	~98	...
C_2H_5	98	85	~109	94	85	~68	91	...	81	69	53	78	...
C_2H_5O	103	80	...	97	82	...	42	...	...	...	...	...	...
$n\text{-}C_3H_7$	98	85	~106	94	82	66	92	...	82	69	54	77	...
$i\text{-}C_3H_7$	95	84	~103	92	81	66	92	106	81	68	53	85	...
$n\text{-}C_4H_9$	94	78	...	...	78	...	...	...	...	...	...	...	...
$t\text{-}C_4H_9$	91	80	...	89	78	...	91	...	79	66	50	84	...
C_6H_5	112	102	~119	113	99	85	112	125	97	82	66	100	~124
$C_6H_5CH_2$	85	70	...	85	68	...	77	...	70	55	38	72	~95
$C_{10}H_7$	...	...	...	...	...	...	...	...	...	70	...	...	...
$C_{10}H_7\text{-}CH_2$	~76	...	...	...	...	...	...	...	...	...	...	...	...

[a]Variations of $\pm$ 10 kcal/mole are not uncommon among different investigators.

dissociation energies for a number of organic working fluids. Dissociation energies range from a low of about 40 to about 130 kcal/mole.

Because of the π bonding and hybridization of aromatic compounds, the entire class has superior stability. Benzene has an excellent C-H bond strength and would be a prime candidate working fluid, except for the fact that it has a rather low boiling point of 80.1°C. There are, however, other aromatic compounds that have more appropriate boiling points. These include napthalene (218°C), biphenyl (254.9°C), and the o-m-p terphenyls (332°-385°C). All can be expected to be rather stable species, and, in fact, these materials have been examined as potential reactor coolants.[4-6] A distinct disadvantage is the relatively high melting points in the range of 50° to 100°C.

Halogenated Hydrocarbons

With regard to halogenated hydrocarbons, a general rule is that the corresponding σ bond strengths are in the order F >Cl >Br >I. The σ bond resulting when fluorine is substituted for hydrogen in methane is about equal in strength to the original C-H bond, but there may be attractive heat-pipe working fluids resulting from fluoro substitutions in aromatic compounds; from Table 1, an increase of about 13 kcal/mole in bond strength is associated with substitution of fluorine for hydrogen in benzene. This is an area, unfortunately, that has not been explored in any detail.

Existing halogenated straight-chain hydrocarbons, in general, are not expected to be satisfactory heat-pipe working fluids above 100°-150°C. Existing data show that they are particularly susceptible to catalytic decomposition. Table 2, from Ref. 7, shows the 0.1 and 1.0% decomposition temperatures for

Table 2   Decomposition temperatures for halocarbon refrigerants in the presence of steel

Fluid	Decomposition Rate percent per year	
	0.1, °C	1.0, °C
R-11	89	129
R-12	100	145
R-113	91	137
R-114	97	146

Table 3   Time required for 0.1% decomposition of halocarbons R-12 and R-22 under various conditions[a]

Temperature, $^{\circ}C$	R-12, yr	R-12 + Fe_2O_3	R-22, yr	R-22 + OX.1[b]	R-22 + Fe_2O_3
100	$1.8(10^{16})$	730 yr	$4(10^8)$	1.2 yr	5.8 days
150	$4.1(10^{11})$	11 month	$1.6(10^5)$	4.9 days	2.2 hr
200	$9.1(10^7)$	1.7 days	340	. . .	. . .
250	$9.9(10^4)$	. . .	2.3	. . .	. . .

[a]From Norten.[8]

[b]Materials of similar activity: Fe powder, Fe_3O_4 powder, and CuO powder.

various halogenated alkanes in the presence of steel. These temperatures are generally in the range of $100°$ to $150°C$, and it is expected that other alkane-type halogenated refrigerants will be affected similarly by steel envelopes. The actual catalytic agent may be the metal oxides, as indicated by the work of Norten[8] summarized in Table 3; the oxides of iron and copper can accelerate the decomposition rates of R-12 and R-22 by as much as a factor of 10^6 to 10^{10}.

Dowtherm E (TM) is an example of a commercially available halogenated aromatic compound (O-dichlorobenzene) with a boiling point of $177.8°C$ and a maximum recommended working temperature of $260°C$. Available data on Dowtherm E indicate that aluminum catalyzes the decomposition of Dowtherm E to form hydrochloric acid, which, in turn, may accelerate chloride-activated corrosion. As will be discussed at a later point, chlorinated napthalene in carbon steel has been unsatisfactory as a working fluid and may represent a similar decomposition situation. The bond dissociation data of Table 1 indicate that the chloride bonds in both alkane and aromatic halocarbons are of comparable strength, and hence that the stability of chlorinated hydrocarbons may, in general, be unsatisfactory for high-temperature two-phase processes.

The halogenated aromatics display unusual properties. Hexafluorobenzene (C_6F_6) has a melting point of $5°C$ and a boiling point of $80.25°C$ and yet is considered thermally stable to $650°C$.[9] Although napthalene has a melting point of $80.1°C$, monochloronapthalene has a melting point of $-25°C$, and 1-fluoronapthalene has a melting point of $-9°C$. The boiling points are, respectively, $250°$ and $216°C$. Unfortunately, these potential heat-pipe fluids (with the exception of monochloronapthalene) are quite expensive because of research-quantity production methods now in use.

Inorganic Fluids

In addition to the well—known inorganic fluids ammonia, water, and sulfur dioxide, there are a large number of inorganic molecular fluids that are of potential interest, particularly metal halides. An inorganic molecular fluid differs from a molten salt in that the liquid phase is composed of molecules rather than ions, and the electrical conductivity is relatively low. Corrosion mechanisms that are accelerated by halide ions in aqueous solution, e.g., stress corrosion, may be inactive in the anhydrous molecular liquid.

Figure 1 shows the vapor pressure characteristics of a number of candidate inorganic working fluids. The inorganic fluids as a group are similar in thermal properties to the organic fluids, and their primary attractions are a potentially higher thermal stability and reduced generation of noncondensable gasses. However, some of the fluids are toxic or flammable and must be handled with care. Other potential problems relate to envelope/working-fluid chemical

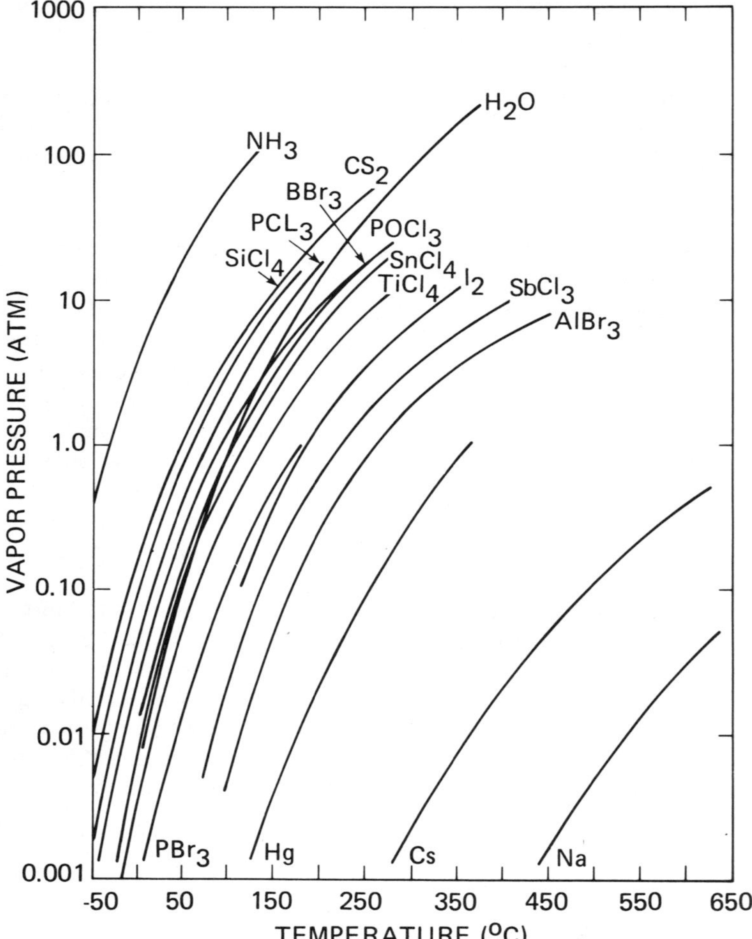

Fig. 1   Vapor pressure of various inorganic liquids of potential interest for heat-pipe applications.

reactions. For example, when the working fluid is the halide salt of a metal low on the electromotive series, and the heat-pipe envelope is made of a metal high on the electromotive series, there is the possibility that the envelope material will form a halide salt, and the less reactive metal will be deposited on the heat-pipe wall.

This tendency can be assessed by calculating free energy changes for postulated envelope/working fluid reactions or, in the case of halides, by comparing decomposition potentials. The double displacement reaction

$$fM_a + gM_b X_c \rightleftharpoons fM_a X_{cp} + gM_b \tag{1}$$

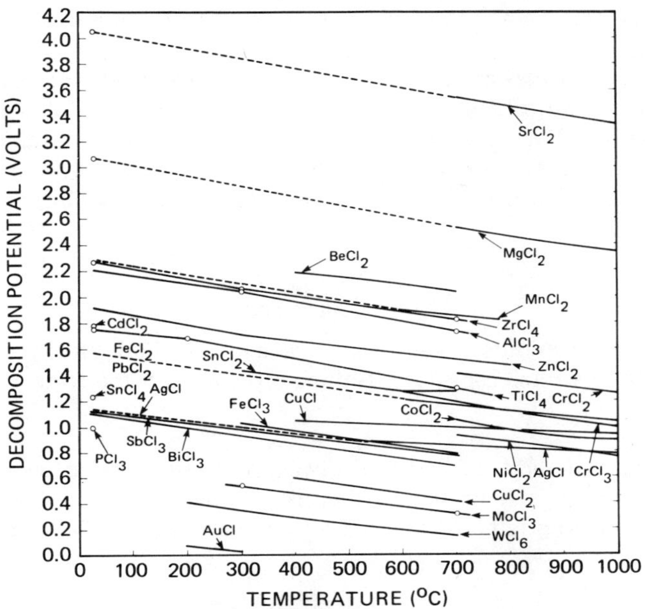

Fig. 2 Decomposition potentials for various inorganic chlorides.

represents the reaction of metal wall M_a with the halide salt M_bX_c to form a halide salt of the metal (M_aX_{cp}) and free metal M_b. This corresponds to two electrical cells of the type

$$M_a; M_aX_{cp}; X_{c2}, M_b; M_bX_b; X_{c2} \qquad (2)$$

where X_{c2} is the halide gas.

For each cell, there is a decomposition potential E_d, where E_d is the voltage that must be applied to inert electrodes in the halide to produce metal and halide gas. Neglecting polarization effects, at unit activities, the emf for a reaction of type (1) is

$$\Delta E^O = E_{da} - E_{db} \qquad (3)$$

where E_{da} and E_{db} are, respectively, the decomposition potentials for halides of metals "a" and "b."

If the standard emf, ΔE^O, is positive, then the reaction can proceed spontaneously, and the metal wall may corrode. Conversely, increasingly negative values of ΔE^O indicate unfavorable reactions. Figure 2 represents decomposition potentials for a large number of chlorides. Values at 25°C were calculated from data in Ref. 10 and were coupled to the values of Refs. 11 and 12 by a straight line extending to their lowest temperature estimates of E_d. On the basis of these decomposition potentials for chlorides, it can be generalized that working fluids are desired which have higher decomposition potentials than chlorides of the envelope material. For an aluminum envelope, most chloride fluids are seen to be of questionable usefulness. On the other hand, there are a number of fluids that may be acceptable with iron or stainless heat pipes, including $TiCl_4$, $SnCl_2$, $PbCl_2$, $AlCl_3$, etc.

Compatibility Tests

Ten working fluids have been installed in mild steel (A-178) and aluminum (6061) reflux capsules. Reflux capsules and the testing station are shown in Fig. 3.

A 1.6-cm-o.d. and 1.27-cm-i.d. heat pipe is surrounded over the lower 30.5 cm of length by an insulation shell consisting of multiple layers of high-temperature insulating felt within a 7.6-cm-i.d. metal shell. The lower 15.25 cm

Fig. 3　Reflux heat pipes and test stand.

of the insulated heat-pipe section is the evaporator, whereas the upper 15.25 cm of height is an adiabatic section. The heat-pipe condenser, which protrudes a maximum of 15.25 cm from the top of the insulation package, dissipates the heat load via free convection to laboratory ambient. The evaporator inside surface is covered with a single layer of screen. In the steel capsules the screen is 200-mesh stainless steel, and in the aluminum capsules it is 100-mesh (1100 alloy) aluminum. The reflux capsules are constructed of A-178 carbon steel and 6061 aluminum. This particular steel was selected on the basis of widespread national use in boilers, shell and tube heat exchangers, and other heat-transfer equipment. The aluminum alloy 6061 was selected because it is used extensively for aerospace heat-pipe applications, where low weight and high strength are desirable. The function of the screen is to distribute the fluid uniformly around the circumference to preclude thin-film evaporative instabilities. In addition, as with any heat pipe, evaporation will concentrate any dissolved species picked up from the condenser or adiabatic, and clogging of the screen can be evaluated through subsequent metallurgical examination. The adiabatic and condenser inner surfaces are not screen-covered to obtain as clear and unequivocal a picture as possible of erosion phenomena.

The heaters are tight-fitting cylindrical steel shells with a 0.10-cm-o.d. stainless sheathed heater wire brazed to the outer shell diameter. The sheathed heater wires have not been brazed or soldered directly to the heat pipes to simplify heater replacement in case of failure and to preclude diffusion of the braze or solder constituents through the heat-pipe wall. The heater shells incorporate an axial slot and a tensioning mechanism to maintain adequate contact with the heat pipe. In addition, the heat pipes are coated with a heat-transfer cement prior to heater attachment. The cement binder volatilizes at a low temperature, leaving a ceramic powder that aids heat transfer and allows the heater to be removed with minimal effort at a later time.

The testing station incorporates individual variable-voltage autotransformers for each heat pipe. Both the autotransformers and the $0.1°C$ resolution Doric Digitrend data-acquisition system are operated from isolation transformers to eliminate early experimental difficulties with ground currents. During operation, heat-pipe function is monitored using three special-limit type K thermocouples per heat pipe. These temperatures are measured on the outer wall at the centers of the evaporator and adiabatic sections and at the far end of the condenser section.

The evaporator-adiabatic temperature difference ΔT_{ea} provides an operational indicator of any changes in the evaporative surface or wicking which might influence evaporative heat transfer. Correspondingly, the adiabatic-condenser temperature difference ΔT_{ac} is an accurate measure of noncondensable gas buildup. A summary of the fluids used in these tests and pertinent operating data is given in Table 4.

Table 4 Corrosion summary

Working fluid (boiling point, °C)	Envelope[a]	Operating time, hr	Operating temperature, °C	Heat input, W	ΔT_{ea}[b] (init./final), °C	ΔT_{ac}[b] (init./final), °C
Carbon disulphide (46.5)	Al	2014	59	7.05	0.43/0.43	0.68/0.86
	C.S.	2014	60	6.4	0.33/0.33	0.10/0.10
Toluene (110.4)	Al	2014	130	26.2	2.53/2.80	2.53/2.94
	C.S.	672	119	26.0	1.72/1.56	3.36/4.86
Tin tetrachloride (114.1)	Al	. . .	159		Incompatible	
	C.S.	2014	159	26.0	1.24/3.32	1.54/10.6
Titanium tetrachloride (136.4)	Al	664	165	29.6	5.67/5.67	1.63/8.65
	C.S.	2014	152	38.8	2.60/4.25	1.70/1.68
1-fluoronapthalene (216.0)	C.S.	656	263	74.7	7.00/7.00	4.78/5.2
Napthalene (217.9)	Al	2014	218	52.5	10.7/11.7	4.50/4.1
	C.S.	2014	217	65.0	2.88/3.23	3.30/2.9
Monochloronapthalene (250)	C.S.	642	266→306	65.8	9.3/19.2	6.8/63.8
Biphenyl (255)	Al	2014	275	63.7	9.88/10.3	3.48/10.0
	C.S.	1675	267	89.3	5.80/5.37	3.72/3.80
Antimony trichloride (283.0)	Al	. . .	227		Incompatible	
	C.S.	1518	227	41.5	6.23/5.67	95.6/77.0
O-terphenyl (332)	Al	672	297→317	77.8	7.20/8.3	47.6/87.2
	C.S.	2014	295	71.7	9.17/9.50	52.8/55.8

[a] Al = 6061 aluminum, C.S. = A-178 carbon steel.

[b] ea = evaporator-adiabatic temperature difference; ac = adiabatic-condenser temperature difference.

The working fluids selected represent aromatic hydrocarbons (toluene, napthalene, biphenyl, and O-terphenyl), halogenated aromatic hydrocarbons (monochloronapthalene and 1-fluoronapthalene), and inorganic molecular fluids (carbon disulphide, tin tetrachloride, titanium tetrachloride, and antimony trichloride). The working fluids were all reagent-grade materials, with the exception of monochloronapthalene, which is of undetermined purity. The various fluids have been in test for varying periods of time up to 2000 hr. Operating temperatures have been selected so that the capsules are approximately at the normal boiling point or 1-2 atm above the boiling point. Table 4 shows the evaporator and condenser temperature drops ΔT_{ea} and ΔT_{ac} at test initiation and at the present time (April 1977). The evaporative temperature drops increased $1°$-$2°C$ for almost all heat-pipe and fluid combinations; most of this increase occurred shortly after test initiation. Several combinations also showed large initial gas-generation rates that did not continue, which may be attributable to decomposition of trace impurities. Brief discussions of each fluid group follow.

Aromatic Fluid Decomposition

Significant changes in performance were found for biphenyl/aluminum and O-terphenyl/aluminum. In both cases, the adiabatic-condenser temperature difference increased, indicating noncondensable gas generation. The evaporator-adiabatic temperature difference increased for all aromatic working fluids by about $1°C$ or less, with the exception of toluene, which showed a small decrease in evaporative temperature drop.

Halogenated Aromatic Fluid Decomposition

Monochloronapthalene/carbon steel exhibited large increases in both the evaporator and condenser temperature differences. Fluorinated napthalene/carbon steel, on the other hand, has not shown any degradation. The chlorinated napthalene was a commercial preparation used as a dielectric liquid in the electronics industry, and it is impossible to determine at this time whether the degradation is intrinsic to the compound or related to an impurity.

Inorganic Fluid Decomposition

Two inorganic tests failed within a short time of startup. Tin tetrachloride/aluminum and antimony trichloride/aluminum both produced gross corrosion of the evaporator and evaporator wick. This result might be expected because the decomposition potential for aluminum trichloride is on the order of 1 V above the decomposition potential for either tin tetrachloride or antimony trichloride; reaction of aluminum metal with either halide is favored thermodynamically. The remaining inorganic fluids have shown moderate changes in evaporator and condenser temperature difference. Antimony trichloride/carbon steel exhibited an initially large gas blockage that has not changed significantly.

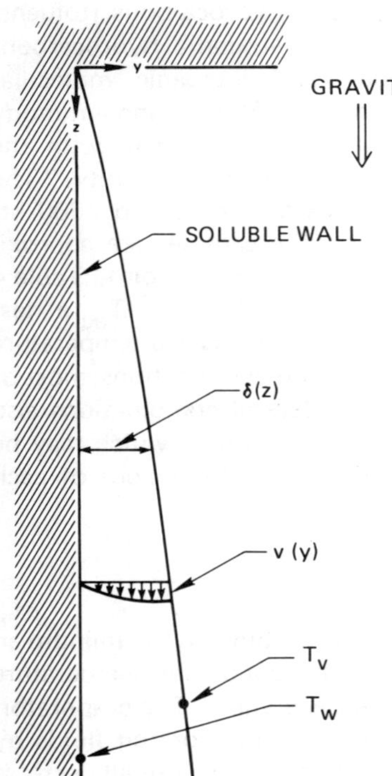

Fig. 4   Erosion of a vertical wall by a
condensing Nusselt film.

Erosion Modeling

To obtain an understanding of wall erosion behavior in refluxing systems
and its dependency on solubility, diffusivity, and heat-transfer rate, a simple
model for uniform wall erosion was developed. Although the model cannot
include the effects of galvanic corrosion, it does provide a benchmark where
erosion is rate-limited by mass transfer in the condensing film and in those
cases where the wall or wall deposits have a finite solubility in the working fluid.

Maximum nongalvanic erosion will occur in the condenser because of the
fresh condensate continually appearing upon the wall surface. Figure 4 presents
the physical model. The falling film produced by condensation is in contact
with a soluble wall. The saturated concentration of wall (species a) in the liquid
(species b) is C_{as} (g-moles/cm^3). The diffusivity of "a" into "b" is D_{ab}. For
most cases, diffusional mass transfer down the film will be much smaller than
convective mass transfer, and conservation of species "a" in the film yields

$$U_z \; \frac{\partial C_a}{\partial z} + U_y \; \frac{\partial C_a}{\partial y} = D_{ab} \; \frac{\partial^2 C_a}{\partial y^2} \qquad (4)$$

where velocities U_z and U_y are, respectively, parallel and normal to bulk film flow. Both the velocity distributions and the film thickness are assumed independent of species "a" concentration and are given by the the well-known Nusselt thin-film analysis as

$$U_y = -\frac{\rho_b g \delta^2}{8\mu_b} \cdot \frac{y}{z} \cdot \frac{y}{\delta}; \quad U_z = \frac{\rho_b g \delta^2}{\mu_b}\left[\frac{y}{\delta} - \frac{1}{2}\left(\frac{y}{\delta}\right)^2\right] \tag{5}$$

$$\delta(z) = [4K_b\mu_b(T_v - T_w)z/\rho_b^2 g h_{fg}]^{\frac{1}{4}} \tag{6}$$

The wall temperature T_w and vapor temperature T_v are assumed uniform, and the vapor and film surface temperatures are assumed equal. The gravitational constant is g, and z is measured from the maximum vertical position of condensation.

If Eq. (4) is assumed to have a similarity solution C_a (η), where $\eta = y/\delta$, then the boundary conditions take the form

$$\text{at } \eta = 0, \; C_a = C_{as} \tag{7}$$

$$\text{at } \eta = 1, \; \frac{dC_a}{d\eta} = -\theta C_a \tag{8}$$

The boundary condition (8) takes the form shown because of the nonconstant nature of the film thickness.

The transformed equation is found to have the solution

$$\frac{C_a(\eta)}{C_{as}} = \frac{1 + B \cdot [I(1) - I(\eta)]}{1 + B \cdot I(1)} \tag{9}$$

where

$$I(\eta) = \int_0^\eta \exp\left[-\frac{\theta x^3}{2}\left(1 - \frac{x}{4}\right)\right] dx \tag{10}$$

and the parameter

$$\theta = K_b(T_v - T_w)/\rho_b D_{ab} h_{fg} \tag{11}$$

If the solution (9) is used to calculate the erosion rate, the average diffusive flux density $\bar{\phi}_L$ at the surface $y = 0$ is found to be

$$C_s D_{ab} g(\theta)/\delta(L) \qquad\qquad (\text{cm}^{-2}\text{-sec}^{-1}) \tag{12}$$

Table 5 Dimensionless derivative factor for condensing film dissolution

θ	Derivative factor $g(\theta)$	$C_{a\delta}/C_{as}$
0.001	0.00133	0.9990
0.002	0.00266	0.9980
0.005	0.00665	0.9950
0.010	0.01325	0.9901
0.020	0.02634	0.9803
0.050	0.06465	0.9518
0.10	0.1255	0.9068
0.20	0.2373	0.8355
0.40	0.4283	0.6911
0.60	0.5862	0.5851
0.80	0.7197	0.4999
1.00	0.8347	0.4303
2.00	1.242	0.2201
4.00	1.703	0.07124
6.00	1.997	0.02630
8.00	2.221	0.01037
10.00	2.408	0.00425
20.00	3.077	0.00006
40.00	3.917	$<1.(10^{-5})$
60.00	4.505	
80.00	4.973	
100.00	5.368	

where the flux density has been averaged over the vertical range $0 \leq z \leq L$. The function $g(\theta)$ is given in Table 5 for $0.01 \leq \theta \leq 100$. This range covers most heat pipe fluids and heat flux densities. The dimensionless concentration of species "a" at the free surface of the condensing film also is tabulated.

When the various factors in Eq. (12) are related to an average heat flux density $Q_{a\ell}$, it is found for $\theta < 0.4$ that the erosion rate is approximately proportional to the solute saturated concentration C_{as} and the average condensing mass flux per unit area. At high heat flux densities ($\theta > 2$), the film is thick compared to the penetration depth of the solute, so that the dependence on heat flux density is small, and the erosion rate varies as the 0.64 power of D_{ab}. The importance of C_{as} is undiminished at large θ.

Acknowledgment

This work was performed under contract to NASA Lewis Research Center, Cleveland, Ohio.

References

[1] <u>Heat Pipe Design Handbook</u>, Dynatherm Corporation, August 1972.

[2] Kreeb, H., et. al., "Lifetest Investigations With Low Temperature Heat Pipes," <u>1st International Heat Pipe Conference</u>, October 15-17, 1973, Stuttgart, Germany.

[3] Basiulis, A., et. al., "Compatibility and Reliability of Heat Pipe Materials," <u>2nd International Heat Pipe Conference</u>, March 31-April 2, 1976, Bologna, Italy.

[4] Bolt, R. and Carrol, J. (eds.), <u>Radiation Effects on Organic Materials</u>, Academic Press, New York, 1963.

[5] <u>Organic Coolant Databook</u>, Monsanto Chemical Co., Tech. Publ. AT-1, July 1958.

[6] <u>Organic Liquids as Reactor Coolants and Moderators</u>, International Atomic Energy Agency, TR Ser. 70, Vienna, 1967.

[7] "Thermal Stability of the Freon Compounds," Dupont Product Information Booklet B-33.

[8] Norten, F. J., "Rates of Thermal Decomposition of $CHClF_2$ and CF_2Cl_2," <u>American Society of Refrigerating Engineers 53rd Annual Meeting</u>, Miami Beach, Fl., June 3, 1957.

[9] <u>Handbook of Aromatic Fluorine Compounds</u>, Olin Corp., Stamford, Conn.

[10] <u>Handbook of Chemistry and Physics</u>, Chemical Rubber Co., Cleveland, Ohio, 1973.

[11] Delimarskii, I.U.K. and Markov, B. B., <u>Electrochemistry of Fused Salts</u>, The Sigma Press, Washington, D. C., 1961.

[12] Janz, G. J., <u>Molten Salts Handbook</u>, Academic Press, New York, 1967.

EXCESS LIQUID IN HEAT-PIPE VAPOR SPACES

James E. Eninger[*] and D. K. Edwards[†]
TRW Defense and Space Systems Group, Redondo Beach, Calif.

Abstract

A mathematical model is developed of excess liquid in heat pipes which is used to calculate the parameters governing the axial flow of liquid in fillets and puddles that form in vapor spaces. In an acceleration field, the hydrostatic pressure variation is taken into account, which results in noncircular meniscus shapes. The two specific vapor-space geometries considered are circular and the "D shape" that is formed by a slab wick in a circular tube. Also presented are theoretical and experimental results for the conditions under which liquid slugs form at the ends of the vapor spaces. These results also apply to the priming of arteries.

I. Introduction

Heat pipes typically operate with more liquid than is necessary just to saturate the wick structure. For example, when a heat pipe is designed to operate over a wide temperature range, one calculates the fluid charge for the temperature that requires the most fluid. For other temperatures, excess liquid often is introduced intentionally to avoid an undercharge, which can degrade the transport capacity significantly. In arterial or tunnel-wick heat pipes, even a slight undercharge makes priming impossible.

As shown in Fig. 1, excess liquid forms fillets along corners and a puddle along the vapor-space bottom, forming low-permeability flow paths in parallel with the wick struc-

---

Presented as Paper 77-748 at the AIAA 12th Thermophysics Conference, Albuquerque, New Mexico, June 27-29, 1977.

*Member of the Technical Staff, Fluid Physics Dept.
†Professor, University of California at Los Angeles, Los Angeles, Calif.

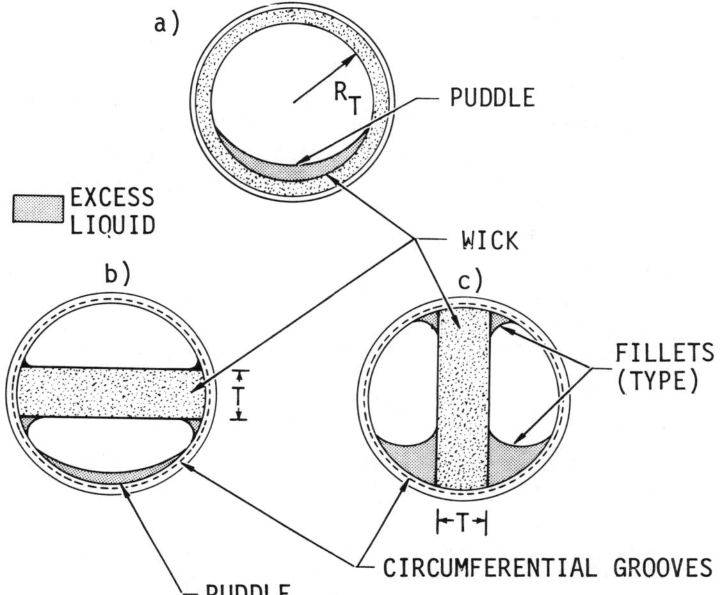

Fig. 1 The three geometries considered for the excess-liquid model: a) circular vapor space, b) horizontal slab wick, c) vertical slab wick.

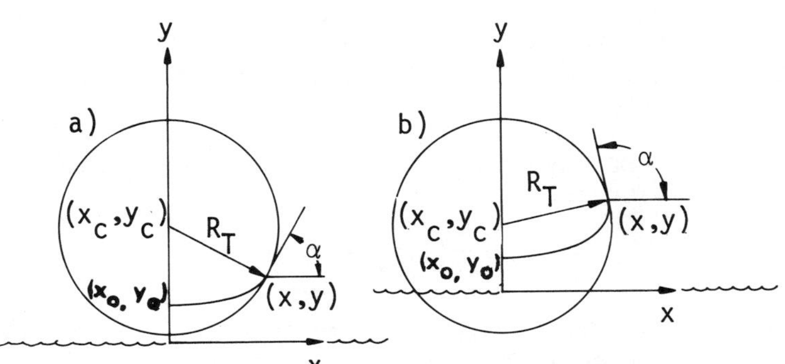

Fig. 2 Tangency condition: $x = R_T \sin \alpha$: a) $\alpha < \pi/2$, b) $\alpha > \pi/2$.

ture. Although excess liquid increases axial-transport capacity, it degrades heat transfer, because fillets and puddles block condensation surface. The condenser end of a vapor space even can fill completely with a liquid slug that blocks condensation heat transfer altogether. In this paper, we present methods for calculating the contribution of excess liquid to transport capacity and the conditions under which slugs form. For typical heat pipes, one cannot neglect variations of hydrostatic pressure with position on the free surface of the liquid. An earlier mathematical model in the computer program MULTIWICK[1] neglected these variations, and, although the resulting simplification allows application to quite general heat-pipe cross-sectional geometries, the results are only qualitatively correct. To retain the hydrostatic pressure variations in the present work, it is necessary to select specific types of cross-sectional geometries. We have selected the circular vapor space that occurs in axially grooved heat pipes and simple heat pipes with wick-lined walls, and horizontal and vertical "D-shaped" vapor spaces formed by slab wicks in circumferentially grooved tubes.

II. Fillets and Puddles in Vapor Spaces

The three geometries considered are shown in Fig. 1, where typical fillets and puddles are drawn as they might appear at a specific axial location of a heat pipe operating in Earth gravity. The key parameter that governs the meniscus configuration is the vapor-liquid pressure difference. Since it varies hydrostatically transverse to the heat pipe, to have a reference value we take its value at the center and define it there as the capillary stress. As the stress decreases, the size of the fillets and puddles increases. When the stress is sufficiently low, the vapor space will bridge and a liquid slug will form, as will be explained later. We now present methods for the calculation of the free-surface shape of a fillet or puddle, its cross-sectional area, and its free and wetted perimeter.

The equation governing the free surface of a liquid under the action of surface tension is

$$\Delta P = \sigma(1/R_1 + 1/R_2) \tag{1}$$

where ΔP is the local pressure difference between the vapor and liquid, σ is the surface tension, and R_1 and R_2 are the local radii of curvature in any two orthogonal planes that contain the normal to the surface. As the first of these, we take the x-y plane of the heat-pipe cross section. By assuming that the shape of the fillet or puddle changes gradually with distance along the heat pipe, we can neglect $1/R_2$ compared to $1/R_1 = 1/R$, and

$$\Delta P = \sigma/R \tag{2}$$

A differential equation is obtained for the free-surface shape by substituting for ΔP the hydrostatic pressure variation and by substituting for R its Cartesian formula. The result is

$$\rho g y = \frac{\sigma \, d^2 y/dx^2}{[1 + (dy/dx)^2]^{3/2}} \tag{3}$$

Here ρ is the difference in the liquid and vapor densities, and g is the gravitational acceleration. Also, the location of the coordinates is chosen such that the gas-liquid pressure difference is zero when y = 0. Equation (3), however, presents difficulties when dy/dx is infinite. To avoid these, we introduce

$$dy/dx = \tan \alpha \tag{4}$$

to transform Eq. (3) into two differential equations where the dependent variables are x and y, and the independent variable is the inclination α of the free-surface curve. The resulting equations are

$$2y \, dx/d\alpha = a^2 \cos \alpha \tag{5}$$

$$2y \, dy/d\alpha = a^2 \sin \alpha \tag{6}$$

where

$$a = \sqrt{2\sigma/\rho g} \tag{7}$$

is the capillary constant. Equation (6) is integrated readily, which results in

$$y = \sqrt{y_0^2 - a^2 (\cos \alpha - \cos \alpha_0)} \tag{8}$$

where the curve has an inclination α_0 at a point (x_0, y_0). With the solution (8), we can eliminate y from Eq. (5), which gives

$$\frac{dx}{d\alpha} = \frac{a^2 \cos \alpha}{2\sqrt{y_0 - a^2(\cos \alpha - \cos \alpha_0)}} \tag{9}$$

Although this equation can be integrated in terms of elliptic integrals,[2] a numerical integration is more convenient.

Equations (8) and (9) define a single free-surface curve that passes through the point (x_0, y_0) at an inclination α_0. The task now is to fit a segment of that curve inside the cross-sectional geometry of the heat pipe such that it represents a fillet or puddle.

Consider first a puddle. As shown in Fig. 2, the point at the bottom of the puddle is taken as the point (x_0, y_0), where $x_0 = 0$, $\alpha_0 = 0$, and y_0, which sets the vapor-liquid pressure difference at the surface of the puddle and is a selected positive value. Points (x,y) along the free-surface curve, which are calculated from Eqs. (8) and (9) for increasing values of α, are checked to see if the condition of tangency with the tube wall is met, which is

$$x - R_T \sin \alpha = 0 \qquad (10)$$

Figure 2 shows that there are two tangency conditions for which Eq. (10) is satisfied. Thus the same free-surface curve can represent two puddles. The capillary stress (vapor-liquid pressure difference at the center of the tube) is given by

$$\rho g y_c = \rho g (y + R_T \cos \alpha) \qquad (11)$$

where y and α are the values at the point of tangency. A series of puddles is calculated for different values of capillary stress by starting with different values for y_0.

The bottom fillet that forms in a heat pipe with a vertical slab wick (Fig. 1c) is calculated exactly the same as a puddle with Eqs. (10) and (11) applying, except that the initial condition is $x_0 = T/2$ and $\alpha_0 = -\pi/2$. Again two fillets are possible from a given free-surface curve. The other types of fillets depicted in Fig. 1 are calculated similarly except that only a single fillet is possible for a given free-surface curve.

We now turn to the calculation of liquid area, wetted perimeter, and free perimeter. The area and wetted perimeter are used in the calculation of the hydraulic diameter. Although the free perimeter is not used, it is an important parameter for future work on vapor-liquid interaction, which is neglected here. An elemental length dS_f of the free perimeter S_f is given by

$$dS_f = \sqrt{dx^2 + dy^2} \qquad (12)$$

With the use of Eqs. (5, 6, and 8), we rewrite Eq. (12) as

$$dS_f = \frac{a^2 \, d\alpha}{2\sqrt{y_0^2 - a^2(\cos \alpha - \cos \alpha_0)}} \qquad (13)$$

This equation can be integrated in terms of elliptic integral;[2] however, as in Eq. (9), it is more convenient to integrate numerically.

For the calculation of the area, refer to Fig. 3a. Although a particular fillet is shown, the expressions derived below apply directly to the puddle and other fillets as well. The area desired is $A_f = {}_{oa2b3}$, where the subscripts denote points and lines bounding the area. It is more convenient, however, to calculate A_{oa2}. Then, the desired area is

$$A_f = A_{023} + (A_{12b3} - A_{123}) - A_{oa2} \qquad (14)$$

The area of triangles A_{023} and A_{121} are calculated in terms of the coordinates for their vertices:

$$A_{123} = \frac{1}{2} \left| (x_2 y_3 - y_2 x_3) - (x_1 y_3 - y_1 x_3) + (x_1 y_2 - y_1 x_2) \right| \qquad (15)$$

$$A_{023} = \frac{1}{2} \left| (x_2 y_3 - y_2 x_3) - (x_0 y_3 - y_0 x_3) + (x_0 y_2 - y_0 x_2) \right| \qquad (16)$$

The area of the sector A_{12b3} is $\theta\, R_T^2/2$, where θ is the angle between vectors $\vec{P}_{21}$ and $\vec{P}_{31}$:

$$\theta = \cos^{-1} \frac{\vec{P}_{31} \cdot \vec{P}_{21}}{|\vec{P}_{31}|\ |\vec{P}_{21}|}$$

$$= \frac{(x_3 - x_1)(x_2 - x_1) + (y_3 - y_1)(y_2 - y_1)}{\sqrt{[(x_3 - x_1)^2 + (y_3 - y_1)^2][(x_2 - x_1)^2 + (y_2 - y_1)^2]}} \qquad (17)$$

It remains to calculate A_{oa2}. Consider Fig. 3b, where the vector $\vec{q}$ connects the point (x_0, y_0) to a general point along the free-surface curve. The area between $\vec{q}$ and the curve is A. We now calculate the increase in area dA when $\vec{q}$ is in incremented an amount corresponding to the increment $d\alpha$. The incremental area dA is the area of the triangle bounded by $\vec{q}$ and $d\vec{q}$; thus,

$$dA = \frac{1}{2} |\vec{q} \times d\vec{q}| \qquad (18)$$

with

$$\vec{q} = (x - x_0)\,\vec{i} + (y - y_0)\vec{j}$$

$$d\vec{q} = dx\,\vec{i} + dy\,\vec{j}$$

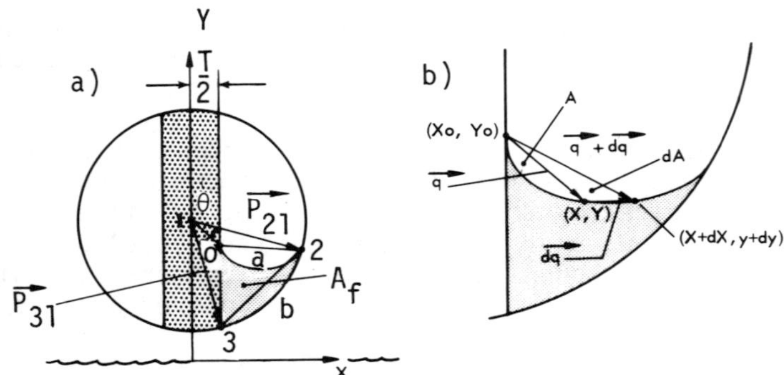

Fig. 3 Calculation of the fillet area A_f: a) diagram defining points, lines, and vectors; b) a blowup of Fig. 3a showing the vector $\vec{q}$ that connects (x_0, y_0) to a point (x,y) on the free-surface curve.

we find

$$dA = \frac{1}{2} \left| dy(x - x_0) - dx(y - y_0) \right| \tag{19}$$

Equations (5) and (6) are used to obtain

$$\frac{dA}{d\alpha} = \frac{a^3}{4y} \left| (x - x_0) \sin \alpha - (y - y_0) \cos \alpha \right| \tag{20}$$

Since x is given in terms of elliptic integrals, there is no hope of obtaining an analytic integral for A. Therefore, the integration is carried out numerically from α_0 to the point where the boundary condition on the tube wall is met. Referring to Fig. 3a, we see that the wetter perimeter is given simply by

$$S_w = y_0 - y_3 + R_T \theta \tag{21}$$

The computer program FILLET[3] was written to carry out the calculations outlined in this section. The output from FILLET is a table of hydraulic diameter and the area of the excess liquid for the range of capillary stress which can exist in a heat pipe.

To illustrate the calculations, we focus attention on the puddle in a circular vapor space. Two parameters are required to fix the puddle configuration. The first is the Bond number

$$B = \rho g R_T^2 / \sigma \tag{22}$$

which is the ratio of hydrostatic forces to surface tension. The second parameter is the dimensionless capillary stress

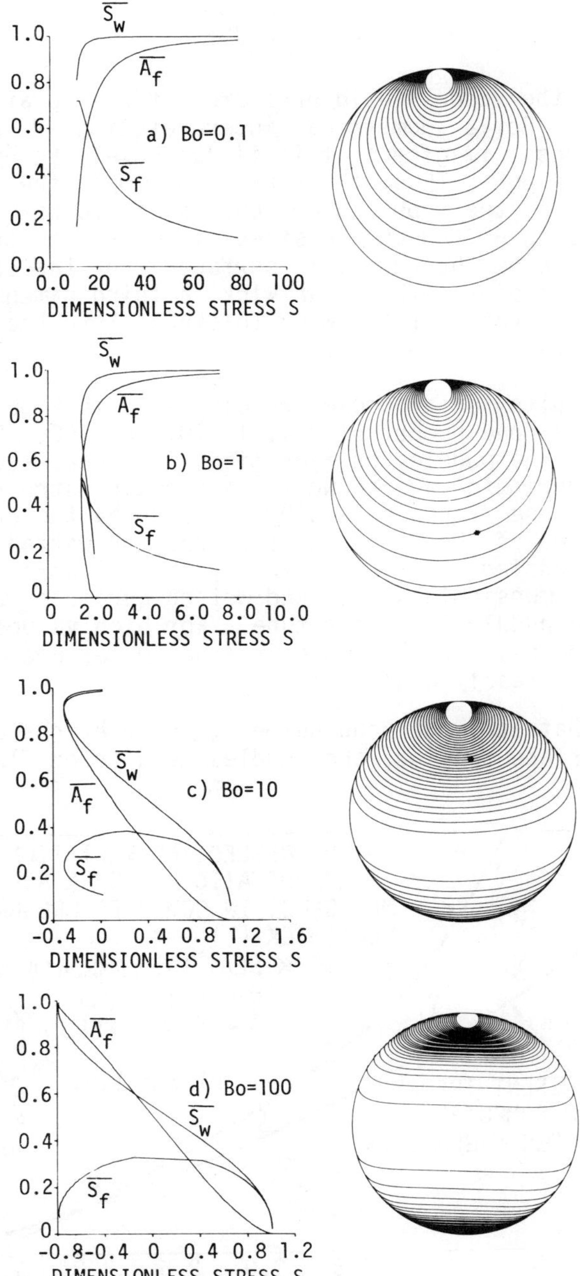

Fig. 4 Fillet configurations in horizontal circular cylinders. Dimensionless fillet area $\overline{A}$, wetter perimeter $\overline{S}_w$, and free perimeter $\overline{S}_f$ are plotted as a function of dimensionless stress S. An asterisk denotes the neutrally stable meniscus.

$$S = P_o/\rho g R_T \tag{23}$$

where P_O is the vapor-liquid pressure difference at the center
of the tube. For a geometrical interpretation of the dimensionless stress, imagine that it is due solely to hydrostatic
pressure $P_O = \rho g h$. Thus, the stress is simply the ratio of
the height h of the tube center above a liquid reservoir to
the tube radius R_T. When the stress is negative, the tube
center lies below the reservoir surface. For the case of fillets in a heat pipe with a slab wick, a third dimensionless
parameter, the ratio of the wick thickness T to the tube radius
R_T, is required.

The results for a puddle are displayed in Figs. 4a-4d for
values of the Bond number of 0.1, 1, 10, and 100. For each
Bond number, the puddle configuration, its area, free perimeter,
and wetted perimeter are shown for a range of stress. The puddle area is made dimensionless by dividing by the cross-
sectional area of the tube, and the free and wetted perimeter
are made dimensionless by dividing by the circumference, thus
making the dimensionless area and wetted perimeter approach
unity as the puddle fills the tube. For high values of stress
when the puddle is shallow, the free perimeter and wetter
perimeter are nearly equal.

Note that at large Bond numbers, where hydrostatic forces
dominate surface tension, the puddles tend to be flat, whereas

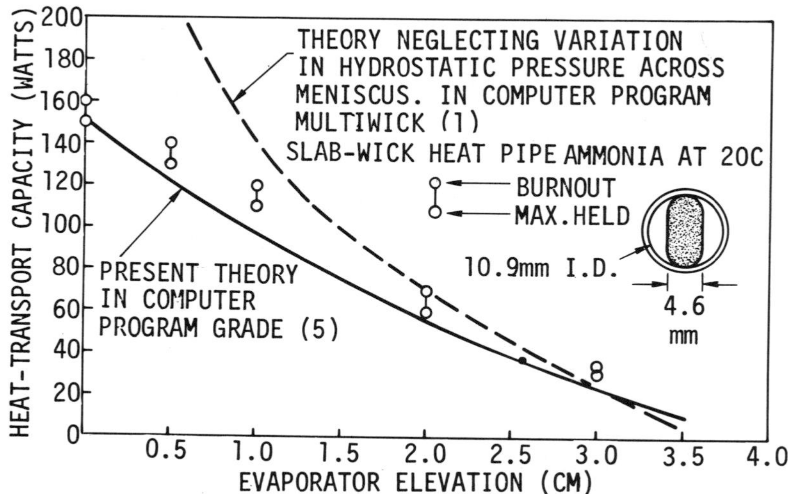

Fig. 5 Comparison of theory to data for homogeneous slab wick
heat pipe (wick porosity = 0.86, fiber diameter = 0.127 mm;
evaporator, adiabatic, and condenser lengths of 60 cm).

for small Bond numbers, the puddle menisci are nearly circular. As will be discussed in the next section, those liquid configurations for which the fillet area increases with increasing stress are unstable and cannot exist. This is the case for all puddle configurations when B = 0.1. The critical free surfaces, which are neutrally stable, are denoted in the figure by an asterisk. For stresses below the critical stress, a liquid slug forms.

The computer programs FILLET and GRADE[3] were used to predict the performance of a 180-cm homogeneous slab-wick heat pipe with ammonia. The comparison with experiment is shown in Fig. 5. Also shown is the prediction of the computer program MULTIWICK, which models excess liquid without taking into account the hydrostatic variation of excess liquid across the meniscus.

III. Excess-Liquid Slugging of Vapor Spaces and the Priming of Arteries

In this section, we focus on the conditions under which a liquid slug forms in a vapor space or under which an artery primes. There are two distinct situations: vapor spaces with and without solid boundaries. The former occurs in most simple heat pipes, such as axially-grooved or slab-wick heat pipes with end caps. The latter occurs primarily in gas-loaded heat pipes where the condenser end protrudes into a gas

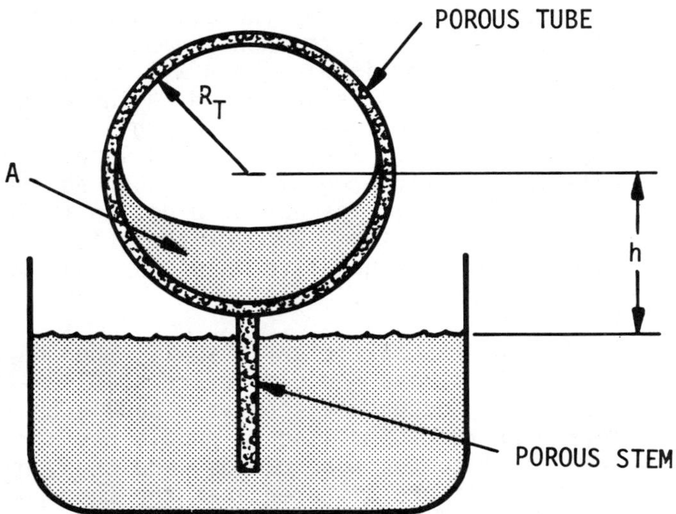

Fig. 6 As the cylindrical tube is lowered at a critical value of h, a liquid slug will form abruptly.

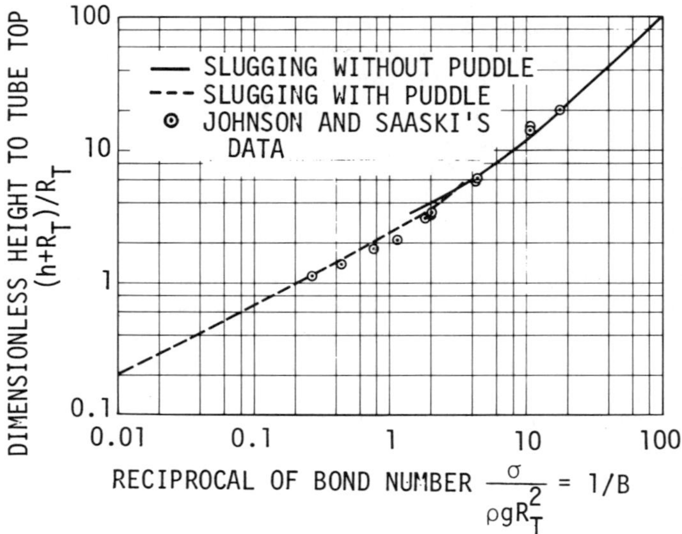

Fig. 7 Slugging criterion for open-ended cylindrical tubes.

reservoir and is otherwise uncapped. We consider this latter situation first, because it is a good introduction to the former.

Slugging of Open Vapor Spaces

Consider an experiment, as shown in Fig. 6, where a horizontal open-ended porous tube is lowered into a liquid pool. The capillary stress is measured by the elevation h of the tube center above the pool. If the Bond number is sufficiently large, a puddle of cross-sectional area A will form before the tube slugs with liquid. As h is decreased, A increases. The tube will slug abruptly with liquid when dA/dh = ∞. This slugging criterion, shown in Fig. 7 by the dashed line, is found from the calculations of the program FILLET such as those displayed in Fig. 4. In Fig. 7, instead of plotting the critical height to the center of the tube, we plot the critical height to the top, which eliminates negative values and allows the use of logarithmic graph paper.

If the Bond number is sufficiently small, then dA/dh is positive for all puddles, and hence none are stable. Therefore, slugging abruptly occurs as soon as a puddle begins to form on the bottom of the tube. Such a puddle has a radius of curvature nearly equal to the tube radius, and thus, by Eq. (1), the gas-liquid pressure difference at the tube bottom is σ/R_T:

$$\rho g(h - R_T) = \sigma/R_T$$

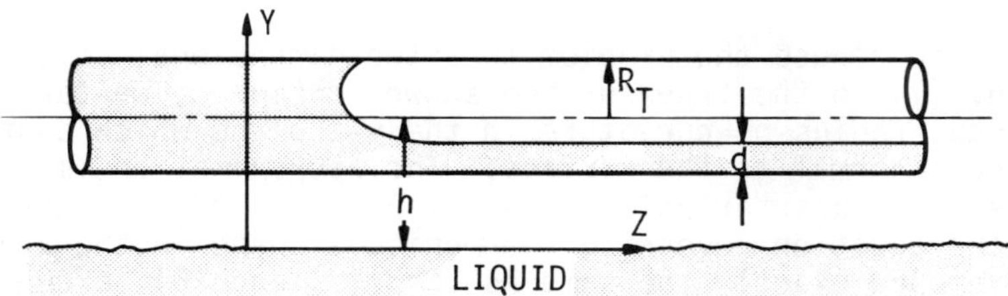

Fig. 8 Liquid slug with a puddle of depth d.

or, in dimensionless form,

$$h/R_T - 1 = 1/B \qquad (24)$$

This criterion for slugging is displayed in Fig. 7 by the solid line. The intersection of the dashed and solid lines at a Bond number of 0.33 represents the point at which a stable puddle is just possible. The analytical calculation of this critical Bond number poses an intriguing unsolved academic problem. Also displayed in Fig. 6 are data taken by Johnson and Saaski[4] on the priming of open-ended glass and screen-walled tubes ranging from 0.51 to 3.2 mm with 2-propanol and Freon 113 as the liquid.

Slugging in a Closed Vapor Space

In a vapor space with a closed end, a slug first begins to form at the end and then grows progressively longer. Such a slug is depicted in Fig. 8. The closed-vapor-space slugging criterion has a wider application than the open criterion; therefore, in this section we present detailed results for the "D" shape as well as the circular vapor space.

The critical capillary stress for the formation of a slug is higher for the closed vapor space because the free surface develops curvature in two orthogonal planes rather than in just one. Thus, if the stress is sufficiently low for a liquid slug to form in an open vapor space, to clear it the stress must be increased to the higher critical stress for slugging of the closed vapor space. Similarly, although the maximum stress to initiate priming of an empty open-ended artery is set by the slugging criterion for open vapor spaces, the capillary pumping, that is, the stress necessary to initiate emptying, is set by the slugging criterion for closed vapor spaces.

The mathematical modeling of the slug that forms in the closed vapor space is difficult because the free surface varies in three dimensions. We consider first the circular vapor space and employ an approximation suggested by Johnson and

Saaski[4] to reduce the problem to a two-dimensional one. Applying Eq. (1) to the free surface shown in Fig. 8, we take R_1 as the local radius of curvature in the y-z plane which contains the local normal to the surface. The approximation is to use a constant characteristic value for R_2 equal to the tube radius R_T. The equation governing the meniscus shape in the z-y plane is developed exactly the same as Eqs. (5) and (6), except that the constant term $1/R_T$ is carried along. Thus, instead of Eq. (6), we obtain

$$(2y - 1/R_T)\ \frac{dy}{d\alpha} = a^2 \sin \alpha \qquad (25)$$

Referring to Fig. 8, we see that the boundary conditions for the free-surface curve are

$$\alpha = 0 \text{ at } y = h - R_T + d \qquad (26a)$$

$$\alpha = -\pi \text{ at } y = h + R_T \qquad (26b)$$

In case slugging occurs without a puddle forming first, we take d = 0. Integration of Eq. (26) between the preceding boundary conditions results in

$$d^2 + d(2h - 2R_T - a^2/R_T) + 4(a^2 - R_T h) = 0 \qquad (27)$$

from which the no-puddle slugging condition is (d = 0)

$$h/R_T = 2/B \qquad (28)$$

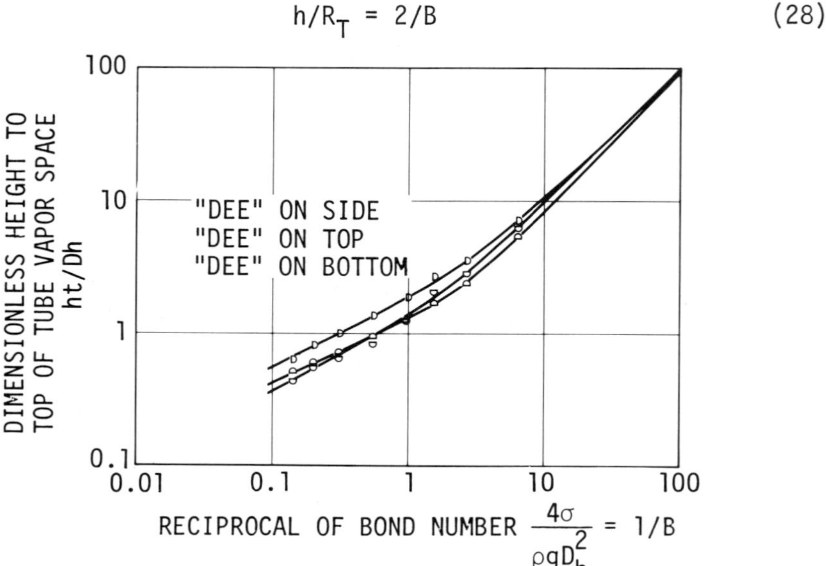

$$\text{RECIPROCAL OF BOND NUMBER } \frac{4\sigma}{\rho g D_h^2} = 1/B$$

Fig. 9 Slugging criterion for closed-ended cylindrical tubes.

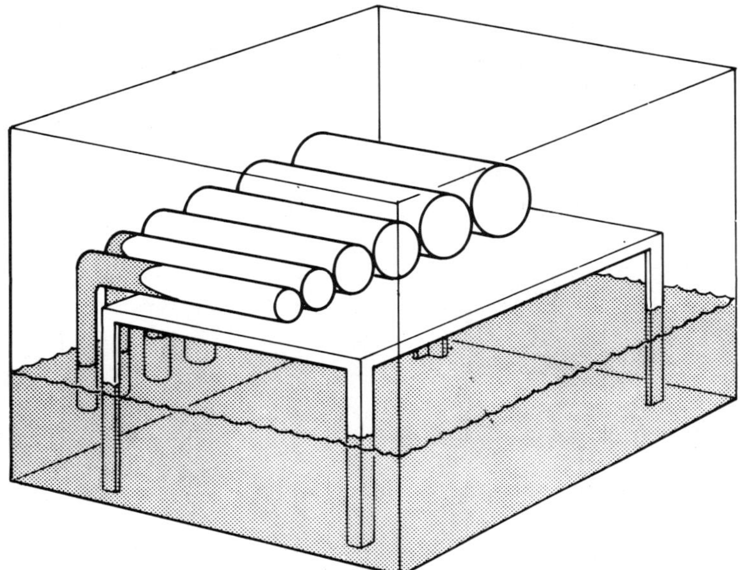

Fig. 10   Schematic diagram of apparatus to measure
the slugging criterion for closed-ended cylinders.
Slug is shown formed in the smallest tube.

Equation (24) gives the condition for a puddle just about to
form. The two curves given by Eqs. (24) and (28) intersect at
B = 1. Thus, for B < 1, the slug forms without a puddle, and
Eq. (28) is the appropriate criterion. For B > 1, we use a
modification of the program FILLET to calculate the puddle
depth as a function of h/R_T and B:

$$d/R_T = f(h/R_T, B) \tag{29}$$

This relation is used to eliminate d from Eq. (27). The resulting slugging criterion for B > 1 and that for B < 1 are displayed in Fig. 9.

The validity of the approximation of Johnson and Saaski
can be assessed by comparing the results either to experiment
or to an exact numerical solution. We carried out an experiment with a series of glass tubes that varied in diameter from
2 to 30 mm. As shown in Fig. 10, one end of each tube was
drawn down to half the original diameter, and the tapered section was bent 90 deg. The set of tubes was placed on a level
platform inside a glass jar with the tapered sections down.
The jar was slowly filled with acetone. At a critical level,
a liquid slug, which initiated in the tapered section, moves
rapidly through the smallest tube. The height of the critical

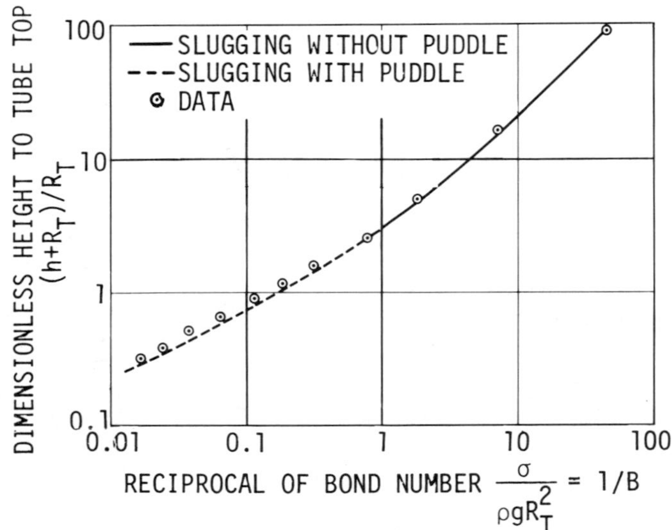

$$\text{RECIPROCAL OF BOND NUMBER} \quad \frac{\sigma}{\rho g R_T^2} = 1/B$$

Fig. 11 Empirical slugging criterion for "D-shaped" vapor spaces.

level is measured with a cathetometer. The jar again is filled slowly until the slug moves through the second tube, and the height is measured, and so forth. The results are displayed in Figure 9. There is good agreement except at large values of B, where the theory underpredicts the critical stress. One expects this, because at these large values of B the puddle itself nearly fills the tube before slugging occurs. As seen in Fig. 4d, when the puddle level is near the top, the transverse radius of curvature R_2 is on the average much smaller than the tube radius R_T that is used to approximate it in the theory.

The slugging criterion for "D-shaped" vapor spaces was measured experimentally for shapes corresponding to a typical ratio of the wick thickness to tube diameter of 0.465. Long

Table 1 Constant for the empirical curve fit for the slugging condition for "D- shaped" vapor spaces $[h_T/D_h = 1/B + (\alpha/B^\beta)(1 - \eta e^{-\gamma\sqrt{B}})]$.

	α	β	γ	η
D on top	0.820	0.40	1.5	2.9
D on bottom	0.68	0.40	1.5	1.9
D on side	1.5	0.5	1.0	1.2

stainless-steel strips of the appropriate width were inserted
into each glass tube and held in place with a spring. Slugging
of the resulting vapor space was measured with the "D" on the
top and the bottom, which corresponds to a horizontal wick,
and on the side, which corresponds to a vertical wick. The
results are displayed in Fig. 11. The Bond number is based on
one-half of the hydraulic diameter D_h instead of the tube
radius. The faired lines through the data are given by the
empirical fit

$$h_T/D_h = \frac{1}{B} + \frac{\alpha}{B^\beta} (1 - ne^{-\gamma\sqrt{B}})$$

where the constants are given in Table 1.

Acknowledgment

This work was supported by NASA Ames Research Center under
Contract NAS 2-8310.

References

[1]Edwards, D. K., Eninger, J. E., and Marcus, B. D., "User's
Manual for the TRW MULTIWICK Program," TRW Rept. 20509-6027-
RU-00, prepared under NASA Contract NAS-11476, 1974.

[2]Eninger, J. E., Edwards, D. K., and Luedke, E. E., "Flight
Data Analysis and Further Development of Variable-Conductance
Heat Pipes, Research Report No. 2," NASA CR-137953, 1976.

[3]Eninger, J. E., "User's Manual for GRADE II," NASA CR 137954,
1976.

[4]Johnson, G. D. and Saaski, E. W., "Arterial Wick Heat Pipes,"
American Society of Mechanical Engineers, Publ. 72-WA/HT-36,
1972.

PERFORMANCE OF GRAVITY-ASSISTED HEAT PIPES OPERATED AT SMALL TILT ANGLES

Yasuhiro Kamotani*

Case Western Reserve University, Cleveland, Ohio

Abstract

Performance of a gravity-assisted heat pipe operated at small tilt angles is investigated theoretically, including the effect of vapor shear. The vapor shear creates a backflow region at the surface of the puddle and thereby degrades the pipe performance. A formula is obtained for the liquid pressure drop which accounts for the vapor shear effect. The performance of the pipe is strongly influenced by the fluid inventory and the tilt. It is found that, if the fluid charge exceeds a certain amount, the pipe operation becomes unstable at a relatively small heat transport.

Nomenclature

D_h	=	hydraulic diameter
f	=	resistance coefficient
g	=	gravitational acceleration
H	=	puddle depth
H_0	=	puddle depth at $Z=0$
h	=	H/R
h_0	=	H_0/R
ΔH_v	=	defined in Fig. 7
Δh_v	=	$\Delta H_v/R$
L_c	=	length of condenser section
L_e	=	length of evaporator section
P	=	pressure
$\dot{Q}$	=	heat flux
Q_T	=	total heat transport
$\dot{Q}_{T,B}$	=	heat transport at burn out limit

Presented as Paper 77-750 at the AIAA 12th Thermophysics Conference, Albuquerque, N. Mex., June 27-29, 1977. Copyright © American Institute of Aeronautics and Astronautics, Inc., 1977. All rights reserved.

*Visiting Assistant Professor, Department of Mechanical and Aerospace Engineering.

$\dot{Q}_{T,V}$	=	heat transport at vapor-pressure limit
$\dot{Q}_{sonic}$	=	heat flux at sonic limit
R	=	pipe inner radius
R_e	=	Reynolds number
S	=	cross-sectional area
U	=	velocity
$\bar{U}$	=	average velocity
(X,Y,Z)	=	coordinate system defined in Fig. 1
β	=	tilt angle
θ	=	puddle angle defined in Fig. 1
μ	=	dynamic viscosity
ρ	=	density
λ	=	latent heat of evaporation
ϕ	=	dimensionless liquid velocity
σ	=	surface tension
(ξ,η)	=	nondimensionalized coordinate system

Subscripts

ℓ	=	liquid phase
v	=	vapor phase

Introduction

It is known that relatively high heat transport capability can be achieved when a heat pipe is operated with the condenser end elevated so that the condensate return to the evaporator occurs with the help of gravity. Mainly because of this advantage, considerable attention has been given to terrestrial applications of heat pipes in recent years, particularly in areas involving energy conservation[1], energy recovery[2,3], solar energy collection[4], and permafrost stabilization[5].

For a given heat pipe, the highest heat transport is obtained when it is operated vertically (condenser up). However, when the heat pipe (or heat pipes) is a part of an integrated system, it sometimes is required to operate at small tilt angles, including horizontal position. Under those conditions, the condensate flows along the bottom surface of the pipe ("puddle pumping") instead of being distributed uniformly over the wall surface, as in the case of vertical and high-tilt operation. In order to predict the performance of heat pipes operated at small tilt angles, it is necessary to study the behavior of the puddle as well as that of the counterflowing vapor. Kaser[6] developed a simple mathematical model for the liquid flow (vapor phase not included) and compared his performance predictions with some test results with Freon 21.

Bienert et al[7] modified the foregoing model by including the frictional and dynamic pressure gradient in the vapor. Later Bienert et al[8] found that the preceding model is inadequate for working fluids having low vapor pressures (e.g., water), which led to the investigation of the shear action of the vapor at the surface of the puddle. Bienert[8] estimated the vapor shear effect for a one-dimensional configuration (liquid and vapor flowing over a flat wide plane) and then superposed the results to account for variable liquid depth at a given section. However, the performance prediction based on the foregoing model was found to differ from the test results with water as much as 50%. The present work was initiated to investigate such a large discrepancy. In the present analysis, the motion of the puddle is studied numerically using a two-dimensional model, including the effect of the vapor shear. Velocity distributions in the puddle and the resistance co-efficient of the flow are computed. Then, using the results, the performance of gravity-assisted heat pipes operated at small tilt angles is investigated. The predicted heat transport capabilities are found to be in good agreement with available experimental data. The present analysis also shows that the vapor-pressure limit plays an important role when heat pipes are operated at small tilt angles; especially, it explains pipe wall temperature fluctuations, which have been observed in some heat pipe tests.

Analysis

Basic Equations

Figure 1 shows a cross section of a typical gravity-assisted heat pipe with a puddle flow and the coordinate system adopted herein. In the present analysis, the free liquid surface is assumed to be a plane in the x direction. In reality, the free surface shape may be more complex, depending on the Bond number = $\rho \ell g R^2 / \sigma$ (ratio of gravitational to surface tension forces), contact angle, and the pipe inner wall surface condition (when it is threaded or lined with wire-mesh screens). For water at 100°C and R = 0.012 m, the Bond number is about 230. For such a large Bond number, the liquid surface is expected to be very flat in the x direction, except in the region very close to the pipe wall.

The change of liquid pressure P_ℓ (actual pressure minus hydrostatic pressure) across a differential element dZ of liquid puddle due to depth variation, tilt and surface pressure change for a small tilt angle (cos β ≐ 1) is given by

$$\frac{dP_\ell}{dZ} = \frac{dP_v}{dZ} + \rho_\ell g \left(\frac{dH}{dZ} + \sin \beta \right) \tag{1}$$

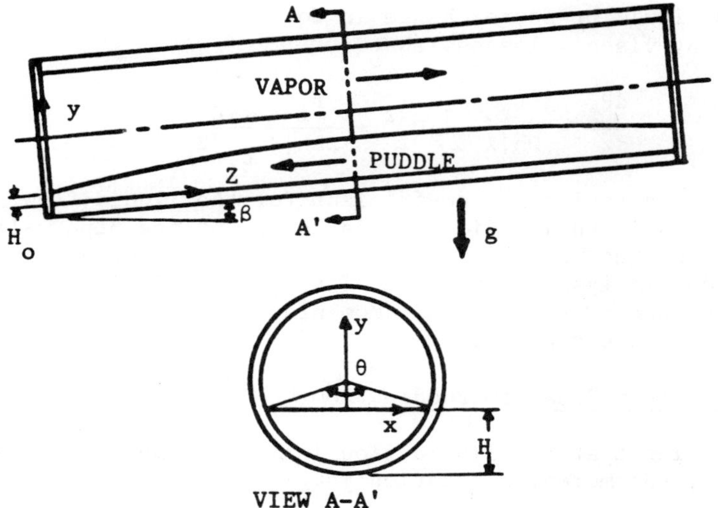

Fig. 1   Cross section of heat pipe.

The liquid surface pressure is taken to be equal to the vapor
pressure because of the aforementioned flat liquid surface
assumption. The vapor flow is assumed to be one-dimensional;
that is, at a given section the vapor flow has a very uniform
velocity distribution, and the vapor shear stress is nearly
uniform in the x direction. Because of the one-dimensional
vapor flow assumption, the present analysis applies mainly to
a heat pipe with a turbulent vapor flow. The momentum equation
of the vapor flow is, then,

$$\rho_v \bar{U}_v \frac{d\bar{U}_v}{dZ} = -\frac{dP_v}{dZ} - \frac{f_v}{D_{h,v}} \frac{\rho_v}{2} \bar{U}_v^2 \tag{2}$$

where f_v is a resistance coefficient for the vapor flow. By
analogy with a turbulent pipe flow, f_v is assumed to be given
by

$$f_v = 0.316/\mathrm{Re},v^{0.25} \tag{3}$$

The vapor mass flux is related to heat flux as

$$\rho_v \bar{U}_v S_v = \dot{Q}/\lambda \tag{4}$$

Substituting Eqs. (3) and (4) into Eq. (2), after some rearrangements one obtains

$$\frac{dP_v}{dz} = \frac{\dot{Q}^2}{\rho_v \lambda^2 S_v^2} \left(\frac{1}{\dot{Q}} \frac{d\dot{Q}}{dz} - \frac{1}{S_v} \frac{dS_v}{dz} + \frac{f_v}{2D_{h,v}} \right) \tag{5}$$

The liquid pressure change in Eq. (1) is assumed to be balanced by viscous losses, that is

$$\frac{dP_\ell}{dz} = \frac{f_\ell}{Dh,\ell} \frac{\rho_\ell}{2} \bar{U}_\ell^2 = \frac{f_\ell}{2Dh,\ell} \frac{\dot{Q}^2}{\rho_\ell \lambda^2 S_\ell^2} \tag{6}$$

where f_ℓ is a resistance coefficient for the liquid flow. The viscous loss in the liquid flow is caused by the pipe wall friction and the shear force of the counterflowing vapor. In the present analysis, the value of f_ℓ in Eq. (6) is determined by solving the two-dimensional momentum equation of the liquid flow, as discussed below.

Effect of Vapor Shear on Puddle Flow

Assuming that the liquid flow is laminar and fully developed, the momentum equation is given by

$$\frac{\partial^2 U_\ell}{\partial x^2} + \frac{\partial^2 U_\ell}{\partial y^2} = \frac{1}{\mu_\ell} \frac{dP_\ell}{dz} \tag{7}$$

Introducing dimensionless quantities

$$\phi = U_\ell/\bar{U}_\ell, \quad \xi = x/R, \quad \eta = y/R$$

Eq. (7) can be written as

$$\frac{\partial^2 \phi}{\partial \xi^2} + \frac{\partial^2 \phi}{\partial \eta^2} = \frac{R^2}{\bar{U}_\ell \mu_\ell} \frac{dP_\ell}{dz} = -\frac{f_\ell}{2} Re,\ell \left(\frac{R}{Dh,\ell}\right)^2 \tag{8}$$

The liquid velocity has to satisfy no-slip conditions at the pipe wall surface, and at the liquid-vapor interface one has

$$\frac{\partial U_\ell}{\partial y} = -\frac{1}{\mu_\ell} \frac{f_v}{8} \rho_v \bar{U}_v^2 \tag{9}$$

Under steady-state conditions, the liquid mass flux must be equal to the vapor mass flux at given z, so that

$$\rho_\ell \bar{U}_\ell S_\ell = \rho_v \bar{U}_v S_v \tag{10}$$

Substituting Eq. (10) into Eq. (9), after some rearrangements one obtains

$$\frac{\partial \phi}{\partial \eta} = -\frac{f_v}{8} \left(\frac{\rho_\ell}{\rho_v}\right)\left(\frac{S_\ell}{S_v}\right)^2 \left(\frac{R}{Dh,\ell}\right) Re,\ell = -C \tag{11}$$

Also, by definition,

$$\bar{\phi} = 1 \tag{12}$$

Geometrical variables Dh, ℓ, S_ℓ, S_v can be written in terms of θ. Therefore, from Eq. (8) together with the boundary conditions, one can see that for given values of Re, v, Re,ℓ, ρ_ℓ/ρ_v, and θ the liquid velocity distribution depends on f_ℓ, but, since the velocity distribution has to satisfy Eq. (12), the value of f_ℓ can be fixed. In the present analysis, the velocity distributions were computed numerically using a successive iteration method, and the values of f_ℓ were calculated for various values of Re, v, Re, ℓ, ρ_ℓ/ρ_v, and θ. For the purpose of heat-pipe analysis, the preceding results are correlated to give a reasonably simple expression for f_ℓ. It was found that the equation

$$f_\ell = (64.0/Re,\ell) + 8.2 \times 10^{-4} \theta^6 f_v (\rho_\ell/\rho_v) \tag{13}$$

expresses the computed values of f_ℓ within ±8% error (see Fig. 2), which was considered to be satisfactory for the present heat-pipe analysis. The first term of the right-hand side of Eq. (13) is due to the pipe wall shear, and the second term is due to the vapor shear.

In addition to the value of f_ℓ, the effect of the vapor shear on the liquid velocity distribution is important in understanding the behavior of the heat pipe. Figure 3 presents some examples of the velocity distribution in the puddle. In

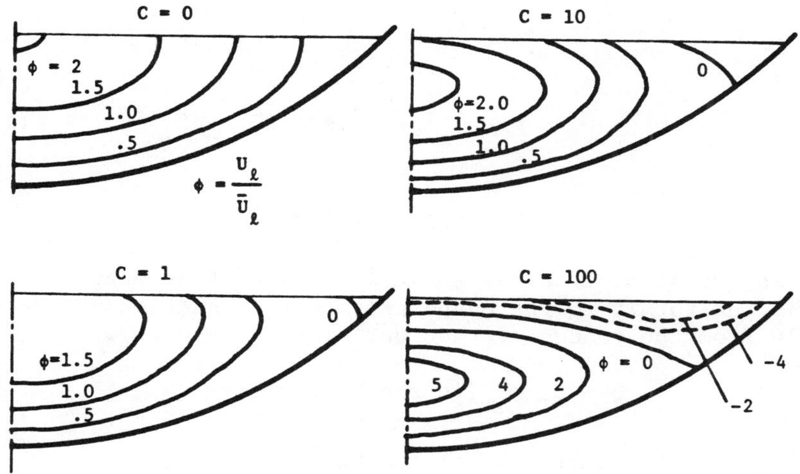

Fig. 2 Contours of constant speed in puddle (θ = 90 deg.).

the figure, C is a parameter to describe the shear stress at
the liquid-vapor interface. C is defined in Eq. (11).

As the value of C increases, a backflow region appears
in the corner and eventually covers the entire top surface
of the puddle. In the backflow region, the liquid is transported to the direction of the condenser. Under steady operating conditions, the backflow causes a circulatory flow in the
puddle. To take into account the effect of such circulatory
flow, it is necessary to extend the present analysis to a
three-dimensional one, but the task is left for future work.

Performance Prediction

Combining Eqs. (5, 6, 13 and 1), after some rearrangements the following equation for the liquid depth variation is
obtained:

$$F(\dot{Q},\theta,\zeta) \frac{dh}{d\zeta} = G(\dot{Q},\theta,\zeta,\beta) \tag{14}$$

where

$$F(\dot{Q},\theta,\zeta) = \frac{\rho_\ell^2 g \lambda^2 R^6}{L} - 8 \frac{\rho_\ell}{\rho_v} \frac{R}{L} \frac{(1-\cos\theta)\dot{Q}^2}{(2\pi-\theta+\sin\theta)^3 \sin\theta/2} \tag{14a}$$

$$\begin{aligned}
G(\dot{Q},\theta,\zeta,\beta) = &\ f_\ell \frac{\dot{Q}^2}{\theta^2[1-(\sin\theta/\theta)]^3} \\
&+ 4 \frac{\rho_\ell}{\rho} \frac{R}{L} \frac{\dot{Q}^2}{(2\pi-\theta+\sin\theta)^2} [\frac{1}{\dot{Q}} \frac{d\dot{Q}}{d\zeta} \\
&+ \frac{f_v}{4} \frac{L}{R} \frac{2\pi-\theta+2\sin\theta/2}{2\pi-\theta+\sin\theta}] \\
&- \rho_\ell^2 g \lambda^2 R^5 \sin\beta
\end{aligned} \tag{14b}$$

and

$$h = H/R, \quad \zeta = Z/L$$

In the present analysis, it is assumed that heat is
applied uniformly to the evaporator section and removed uniformly from the condenser section. Thus, $\dot{Q}$ can be written as

$$\dot{Q} = \dot{Q}_T \frac{Z}{Le} \quad \text{(in the evaporator)}$$

$$\dot{Q} = \dot{Q}_T (1 - \frac{Z-Le}{Lc}) \quad \text{(in the condenser)}$$

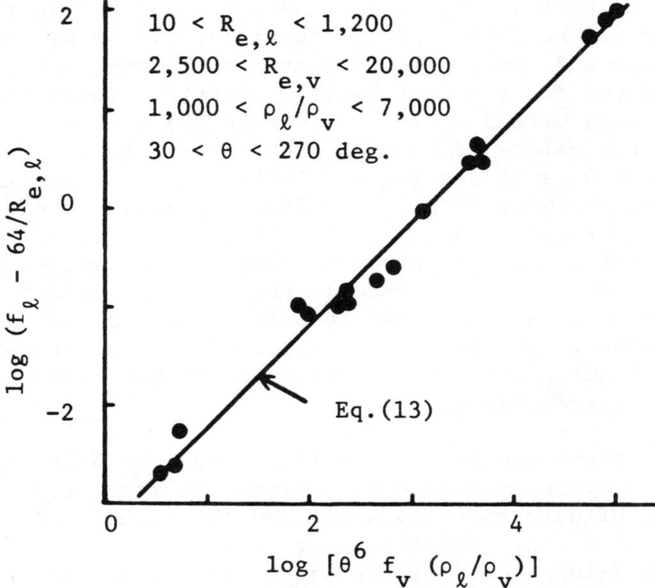

Fig. 3   Resistance coefficient of liquid flow.

Equation (14) determines the variation of h (or θ) with
ζ, if $\dot{Q}_T$ and the initial liquid depth h_o (at ζ = 0) are
specified. In the present analysis, Eq. (14) was solved
numerically using the fourth-order Runge-Kutta method.

Most of the present results are obtained for a heat pipe
whose dimensions are as follows: Le = 1.22 m, Lc = 1.22 m,
R = 0.0118 m, and working fluid is water. Also, the inner
surface of the pipe evaporator section is threaded to distribute the liquid circumferentially. The preceding heat pipe
has been studied experimentally by Dynatherm Corporation under
NASA Goddard Space Flight Center contract as part of a heat-
pipe heat-recovery system study[8].

Performance at Zero Tilt

When β = 0 (horizontal position), the maximum heat transport of the heat pipe is determined by the following procedure.
As explained previously, for given $\dot{Q}_T$ and h_0, Eq. (14) determines the variation of h with ζ. By integrating h over the
entire pipe length, the total mass charge is computed. By
repeating the preceding calculation for various values of $\dot{Q}_T$
and h_0, one can determine the relation between $\dot{Q}_T$ and h_0 for a
given total mass charge.

When no heat is applied to the pipe, the liquid stays at
the bottom of the pipe, and h = const (=h_0). As $\dot{Q}_T$ increases,
h_0 decreases and eventually reaches its minimum, beyond which
the pipe burns out (burnout limit). Figure 4 shows the re-
lation between $\dot{Q}_T$ and θ_0 ($h_0 = 1 - \cos\theta_0/2$) near the burnout
limit for two values of mass charge. As can be seen in the
figure, when $\theta_0 \doteq 15$ deg ($h_0 \doteq 0.0086$), $\dot{Q}_T$ becomes maximum. If
$\dot{Q}_T$ is increased beyond that, the liquid quickly recedes from
the evaporator end, and the pipe wall temperature in the region
starts to rise. In the previous studies of gravity-assisted
heat pipes (e.g., Refs. 6 and 9), there was an uncertainty as
to what value of h_0 to choose in order to calculate the maximum
heat transport. The present analysis shows that at burnout
limit the liquid depth at the beginning of the evaporator has
a certain nonzero value.

Since there are no data available for the heat pipe
mentioned previously at $\beta = 0$, the present results are compared
with a set of data taken by Abhat and Nguyenchi[9]. The heat
pipe is 0.51 m long, and the inner diameter is 1.14 cm. It
has a 0.19-m-long adiabatic section. The inner surface of the
pipe is covered by 2-1/2 layers of 160-mesh screen. Figure 5
shows the comparison. The predicted $\dot{Q}_{T,B}$ is slightly smaller
than the data, mainly because the present model does not include heat transport through the screen wick. The extra-
polation of the experimental data to zero mass charge (no
liquid for the puddle, but the wick is saturated) shows that
the wick can transport 10∿15 W of heat, which is close to the
difference between the prediction and the data. Since at 45°C
water vapor has relatively low density, the vapor shear effect
is expected to be large. Figure 5 shows that the present prediction, without taking into account the liquid pressure drop
due to the vapor shear stress, is about 1.6 times as large
compared to the prediction with the vapor shear effect.

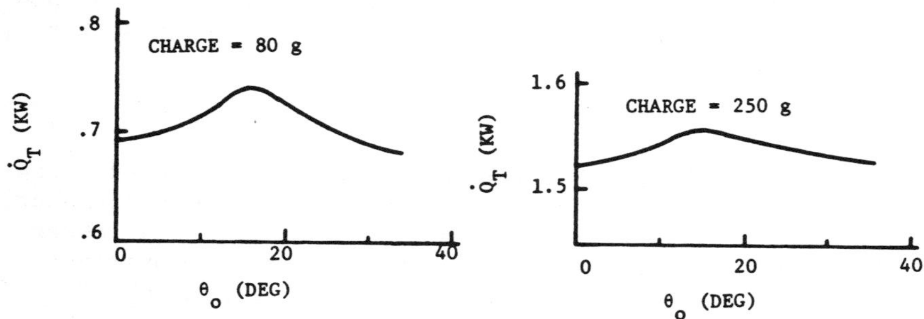

Fig. 4 $\dot{Q}_T$ vs θ_0 near burnout limit [water 60°C , zero tilt].

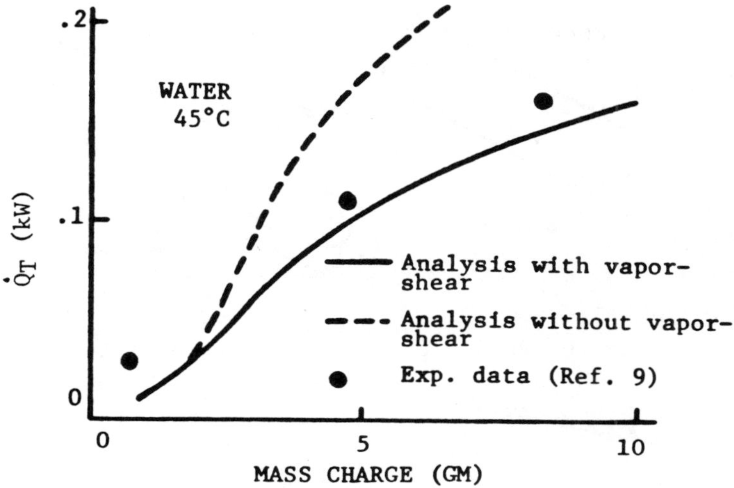

Fig. 5   Heat transport vs mass charge at zero tilt.

As can be seen in Fig. 5, maximum heat transport increases
with mass charge. Ultimately the heat-pipe operation is
limited by two major factors. One is sonic limit, and the
other is vapor-pressure limit. Heat-pipe operation is limited
when vapor flow reaches sonic velocity. According to Busse,[10]
the sonic limit is given by

$$\dot{Q}_{sonic} = 0.5 \; S_v \lambda (\rho_v p_v)^{1/2}$$

The vapor-pressure limit arises from the fact that under
certain conditions $F(\dot{Q},\theta,\zeta)$ in Eq. (14) becomes zero, and thus
$dh/d\zeta$ becomes infinite. Physically this means that beyond a
certain vapor speed, no matter how one changes the puddle
shape, the vapor-pressure gradient due to vapor area change
is always larger than the liquid pressure head, and thus one
no longer can sustain a steady liquid flow to the evaporator.
When heat pipes are operated at $\beta = 0$, the vapor-pressure limit
is not very restrictive in practice, because it occurs only
when the mass charge is very large.

Performance at Small Tilt Angles

Compared to the previous case for $\beta = 0$, the procedure to
predict maximum heat transport when $\beta > 0$ is more complicated.
When no heat is applied to the heat pipe, the liquid stays at
the bottom of the evaporator section (Fig. 6). As $\dot{Q}_T$ increases, the amount of liquid in the pool starts to decrease.
Unlike the case for $\beta = 0$ where a large amount of liquid

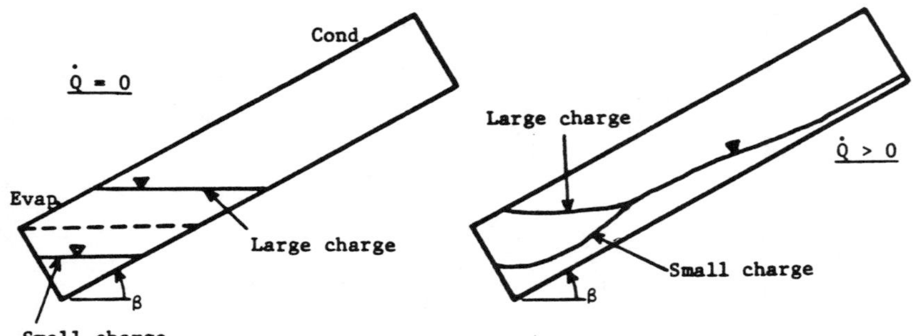

Fig. 6   Puddle profile of nonzero tilt.

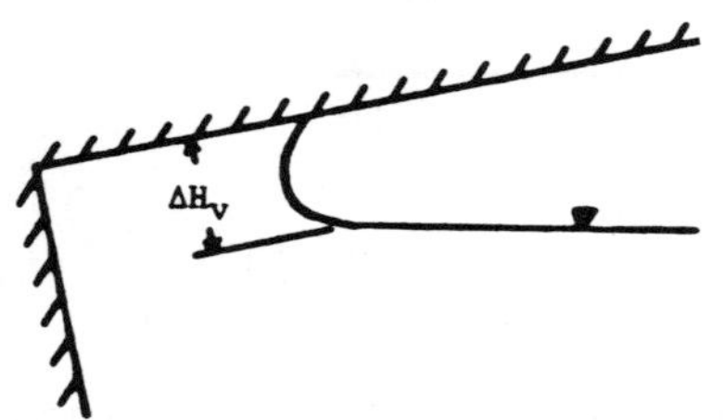

Fig. 7   Detail of puddle profile
near pipe wall.

stays in the condenser section, in the case for $\beta > 0$ the
puddle depth is much smaller in the condenser region so that a
large amount of liquid stays in the evaporator, especially
when the mass charge is large (Fig. 6). A large liquid pool
in the evaporator restricts the vapor passage, causing the
aforementioned vapor-pressure limit to occur before the heat
pipe reaches the burnout limit.

In order to calculate the vapor-pressure limit when a
large liquid pool is present in the evaporator, it is necessary
to know the shape of the liquid surface near the region where
it touches the pipe wall (Fig. 7). The liquid surface shape
near the liquid-pipe wall interface depends mainly on the
surface tension of the liquid, contact angle, tilt, and the
inner wall surface condition. To make the problem
mathematically tractable, it is assumed in the present analysis
that, at the beginning of the region where Eq. (14) is applicable, the vapor space has a finite area specified in Fig. 7
by ΔH_v or $\Delta h_v = \Delta H_v/R$. Since the value of ΔH_v is, in general,
not known a priori, the vapor-pressure limits are calculated
for various values of Δh_v. In the numerical analysis, the
solution blows up at and beyond the vapor-pressure limit, so
that the exact value of $\dot{Q}_T$ at the limit cannot be calculated.
In the present analysis, $\dot{Q}_T$ is varied by 10-W increments
until the solution blows up. When it happens, $\dot{Q}_{T,v}$ is taken
to be equal to the value at the previous step. Therefore,

$\dot{Q}_{T,v}$ contains a maximum of 10-W error, but it is a small fraction of $\dot{Q}_T$ in the present analysis. The results are given in Fig. 8.

According to the visual observation study of the behavior of the puddle conducted by Dynatherm Corporation, it is rather difficult to determine the vapor-pressure limit experimentally, because the phenomenon usually is coupled with boiling within the liquid pool. Since threads or screens on the evaporator inner wall surface supply good nucleation sites, nucleate boiling in the evaporator generally starts at a heat flux below the vapor-pressure limit. As a result, the liquid surface is very much disturbed by the time the heat flux approaches the vapor-pressure limit. When we have such finite disturbances, the liquid surface may become unstable at a slightly smaller heat flux than the value calculated previously, since a finite-size hump in the liquid surface could cause enough vapor flow blockage to make $F(\dot{Q},\theta,\zeta)$ in the region zero or negative temporarily. If it happens, the hump is amplified, and, as the vapor pressure builds up behind the hump, it is carried to the condenser direction. Since the liquid surface almost everywhere in the evaporator is unstable for such large disturbance, by the time the hump leaves the evaporator its size becomes very large, and it carries a large amount of liquid with it all the way to the condenser end, where it loses its forward momentum and comes back to the evaporator, and it starts the foregoing process again. As

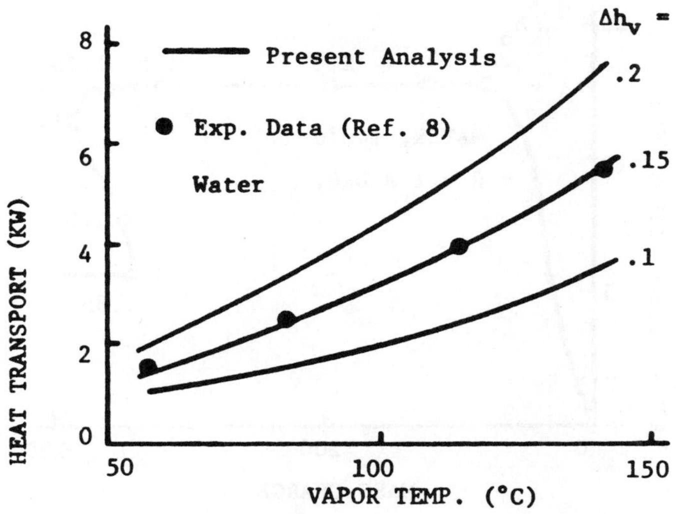

Fig. 8 Onset of unstable operation.

we increase heat flux, the preceding process becomes very
violent. At large heat fluxes, one can hear knocking sounds
at the condenser end. As the hump moves back and forth, the
pipe wall temperature fluctuates because, while a large amount
of liquid is in the condenser region, it temporarily changes
the condensation rate. Bienert et al.[8] measured heat fluxes
at the onset of unstable operations, which are shown in Fig. 8.
As can be seen in the figure, the present calculation of the
vapor-pressure limit for Δh_v = 0.15 agrees well with the data.
Although the vapor-pressure limit does not cause immediate
failure of heat-pipe operation, it is desirable to limit heat
fluxes below the limit for safe long-time operation.

In the preceding computations, the unstable point is
located near the beginning of the evaporator section, where
the vapor space is very small. Therefore, if the mass charge
is small and the liquid does not touch the pipe upper wall,
one expects to obtain larger $\dot{Q}_T$ at the vapor-pressure limit
because of larger vapor space at the beginning ot the evaporator. To check this, values of $\dot{Q}_T$ at vapor-pressure limits
are calculated for various values of h_0 (or Δh_v, since
$\Delta h_v = 2 - h_0$). In each case, the total mass charge also is
computed. Figure 9 shows the relation between $\dot{Q}_{T,v}$ and mass
charge at vapor-pressure limit. The maximum value of h_0 is
taken to be 1.85 (Δh_v = 0.15), as discussed previously. In
the range $1.65 \le h_0 \le 1.85$, $\dot{Q}_{T,v}$ increases with decreasing h_0,
and the unstable point shifts from the beginning of the eva-
porator to the region near the end of the evaporator. In

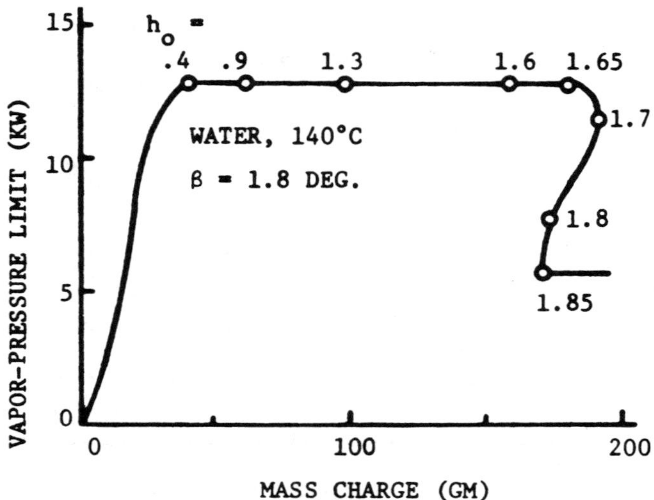

Fig. 9 Vapor-pressure limit vs mass charge.

terms of mass charge, $\dot{Q}_{T,v}$ is multivalued. What is expected
to happen in practice when the mass charge is, for example,
180 g is that the wall temperature starts to fluctuate at
$\dot{Q}_T$ = 5.8 kW, but the fluctuation level does not increase with
$\dot{Q}_T$ up to $\dot{Q}_T$ = 12.8 kW, beyond which the fluctuation level
starts to increase. In the range $0.4 \leq h_0 < 1.65$, the unstable
point stays in the region near the end of the evaporator.
For a long heat pipe such as the one studied herein, the
liquid surface profile in most of the evaporator region and
the condenser is independent of the initial liquid depth h_0
and depends only on $\dot{Q}_T$. As a result, the value of $\dot{Q}_{T,v}$ is
independent of h_0 (and mass charge), as seen in Fig. 9. In
the range $h_0 < 0.4$, the burnout limit becomes dominant. The
heat pipe burns out without experiencing the pipe wall temperature fluctuation.

To see the effect of tilt on the vapor-pressure limit,
two values of $\dot{Q}_{T,v}$, one for large mass charge and one for
small mass charge (the dividing charge is approximately equal
to the one represented by a broken line in Fig. 6), are calculated at several tilts and are shown in Fig. 10. $\dot{Q}_{T,v}$ for
large mass charge increases sharply with tilt, partly because
of increased gravity head and partly because of steeper liquid
depth variation in the region near the beginning of the evaporator, increasing the vapor space as a result. Beyond 25 cm
tilt, the vapor-pressure limit for large mass charge no longer
exists; that is, the liquid surface instability at the vapor-

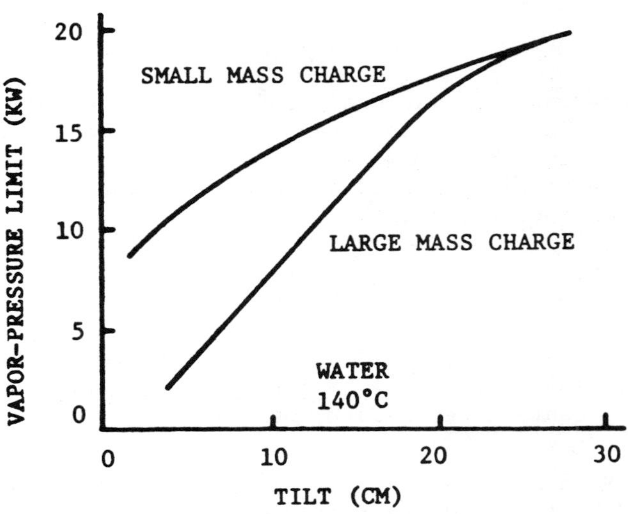

Fig. 10 Vapor-pressure limit vs tilt.

pressure limit occurs at the end of the evaporator for any
mass charge.

Based on the present analysis, the visual observation
study currently is being conducted by Dynatherm Corporation.
Together with the present result, it is expected to give a
better understanding of the performance of gravity-assisted
heat pipes operated at small tilt angles.

Conclusions

1) For gravity-assisted heat pipes with working fluids
having low vapor pressures, the vapor shear effect at the
liquid surface is very important. It causes a backflow in the
puddle and reduces the maximum heat transport. A formula is
obtained for the liquid pressure drop which accounts for the
vapor shear effect.

2) An equation is derived which predicts the performance
of gravity-assisted heat pipes operated at small tilt angles
including the effect of vapor shear. The predicted heat transport capabilities are found to be in good agreement with available experimental data.

3) When the fluid charge of the pipe exceeds a certain
value, the operation at small tilt angles becomes unstable
because of the liquid surface instability and the resultant
blockage of the vapor flow passage.

Acknowledgment

The present work was conducted while the author was a
NRC Resident Research Associate at NASA Goddard Space Flight
Center. The author acknowledges useful discussions with
S. Ollendorf of NASA Goddard and with W. Bienert and D. Trimmer
of Dynatherm Corporation.

References

[1] Basiulis, A. and Johnson, J.H., "High Temperature Heat Pipes
for Energy Conservation," _10th Intersociety Energy Conversion
Engineering Conference_, Newark, Del., 1975.

[2] Ernst, D.M. and Rhodes, R.D., "Energy Recovery with Heat
Pipes," _10th Intersociety Energy Conversion Engineering Conversion Engineering Conference_, Newark, Del., 1975.

[3] Ruch, M.A., "Thermal Recovery Units," _10th Intersociety Energy Conversion Engineering Conference_, Newark, Del., 1975.

[4] "Research Applied to Solar Thermal Power Systems," Univ. of Minnesota and Honeywell Systems and Research Center, Semi-Annual Rept., NSF/RANN/SE/GE-34871/PR/74/4, 1975.

[5] Water, E.D., Johnson, C.L., and Wheeler, J.A., "The Application of Heat Pipes to the Trans-Alaska Pipeline," _10th Intersociety Energy Conversion Engineering Conference_, Newark, Del., 1975.

[6] Kaser, R.V., "Heat Pipe Operation in a Gravity Field with Liquid Pool Pumping," unpublished paper, 1972.

[7] Bienert, W.B., Trimmer, D.S., and Wolf, D.A., "Application of Heat Pipes to Solar Collectors," _10th Intersociety Energy Conversion Engineering Conference_, Newark, Del., 1975.

[8] "Final Report for a Heat Pipe Heat Recovery System Study," Dynatherm Corp., Rept. DTM-76-3, 1976.

[9] Abhat, A. and Nguyenchi, H., "Investigation of Performance of Gravity Assisted Copper-Water Heat Pipes," _Proceedings of 2nd International Heat Pipe Conference_, Bologna, Italy, 1976.

[10] Busse, C.A., "Theory of the Ultimate Heat Transfer Limit of Cylindrical Heat Pipes," _International Journal of Mass and Heat Transfer_, Vol. 16, pp. 169-177, 1973.

A ZERO-G VARIABLE-CONDUCTANCE HEAT PIPE
USING BUBBLE PUMP INJECTION

C.C. Roberts Jr.*

Packer Engineering Associates, Naperville, Ill.

Abstract

Methods for varying the thermal conductance of a heat
pipe include liquid flow control, vapor flow control, and
condenser blockage. A new way to vary heat-pipe thermal conductance has been achieved by using a bubble pump to control
return liquid flow. This has been proven in a 1-g environment.
In an attempt to apply this concept to the 0-g environment,
various heat-pipe fluid models were tested on a 0-g simulator.
The design appears feasible over a range of Bond numbers from
0 to 16.

Nomenclature

Δa = change in wetted area of bubble pump, cm^2
ΔA = change in wetted area in condenser, cm^2
B = Bond number, dimensionless
D_b = bubble pump diameter, cm
D_c = condenser diameter, cm (see Fig. 5)
D_e = diameter of the evaporator, cm (see Fig. 5)
D_{fb}= diameter of flow baffle, cm (see Fig. 5)
D_i = injector diameter, cm (see Fig. 5)
D_r = flow baffle reservoir channel diameter, cm
       (see Fig. 5)
g = gravitational acceleration constant, m/sec^2
L = length of heat pipe, cm
ΔH = displacement (see Fig. 5), cm
Δh = displacement (see Fig. 5), cm
ϕ = simulator tilt angle, rad
ρ = density, g/cm^2
σ_{vs}= surface tension, vapor solid, N/m
σ_{lv}= surface tension, liquid vapor, N/m
σ_{ls}= surface tension, liquid solid, N/m

Presented as Paper 77-752 at the AIAA 12th Thermophysics
Conference, Albuquerque, N. Mex., June 27-29, 1977.

*Engineering Consultant.

Θ = wetting angle, rad
ΔS, Δs = energy change, Nm

Introduction

Variable thermal conductance can be achieved in a heat pipe by controlling liquid flow as shown in Fig. 1. This design requires a 1-g environment to operate[1]. In Fig. 1a, there is no thermal conductance from the heat source to the heat sink, since there is no liquid in the wick. All of the liquid is in the liquid reservoir. When heat is applied to the bubble pump injector, liquid is injected into the heat-pipe wick, which immediately evaporates and flows to the condenser in vapor form (Fig. 1b). This pumping action of the bubble pump controls the thermal impedance of the heat pipe. It should be noted that this device acts unlike the typical buoyancy driven bubble pump[2,3], where bubbles are formed in the tube and are allowed to rise due to buoyancy forces (see Fig. 2). The type of bubble pump discussed in this paper is driven by the bubble expansion near a heating

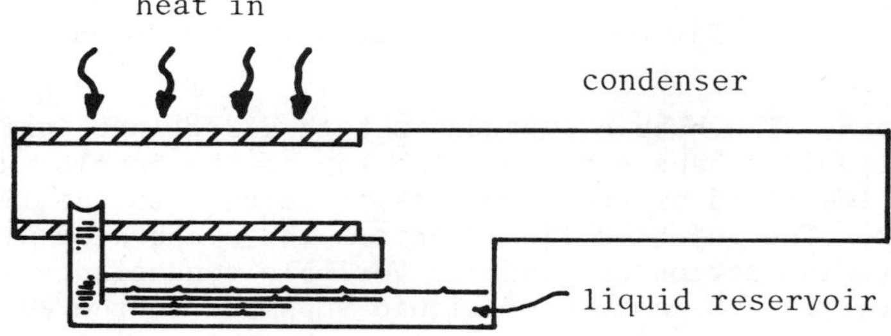

a) Low thermal conductance

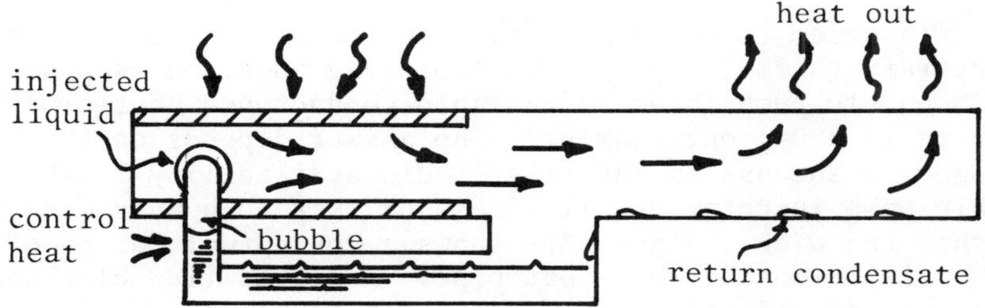

b) High thermal conductance

Fig. 1 1-g bubble-pump-injected variable-conductance heat pipe.

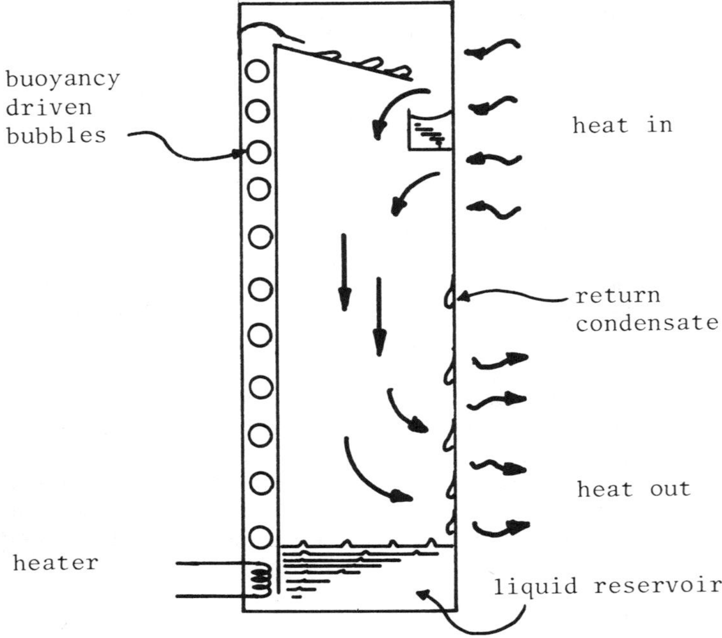

Fig. 2   Two-phase flow thermosyphon[2].

point[4]. The bubble expands in both directions, and a slug
of liquid exhausts through the top of the bubble pump before
the lower end of the bubble is injected into the liquid reservoir. The injected liquid enters the evaporator wick, and
heat-pipe action continues. Variable conductance is achieved
by varying the amount of liquid supplied to the evaporator.
This type of variable-conductance heat pipe is characterized
by a wide range of thermal impedance, insensitivity to source
or sink temperature, and no moving parts.

In order to extend the application of this device to
spacecraft thermal design, the following question is posed:
can a bubble-pump-injected variable-conductance heat pipe
operate in a 0-g environment? The answer depends on the
designer's success in insuring liquid availability in the
bubble pump injector at all times and that no bypass flow
reaches the wick. Figure 3a shows a conceptual design of the
0-g variable-conductance heat pipe. A liquid flow blockage
device is placed on the walls of the tube to prevent liquid
from bypassing the heat-pipe injector. The injector tube now
is placed at the extreme end (condenser) of the heat pipe. At
the condenser end of the heat pipe, excess liquid collects,
having been driven there by the vapor pressure difference

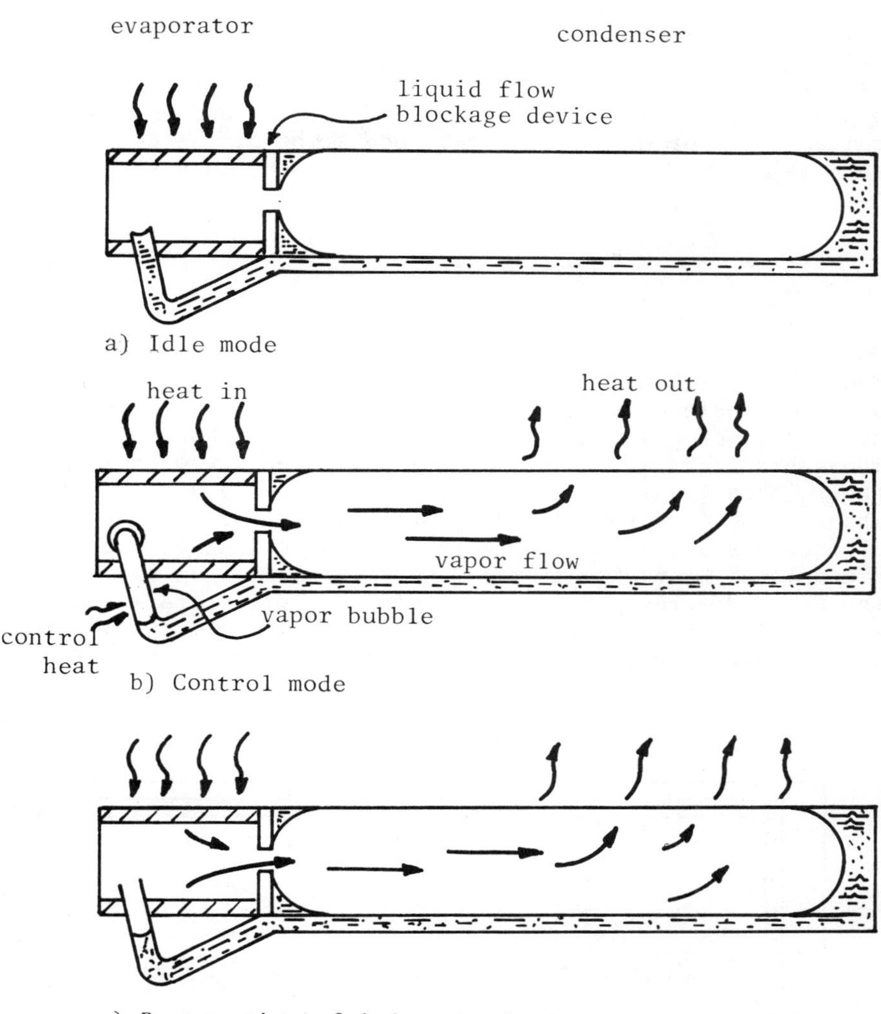

Fig. 3 0-g bubble-pump-injected variable-conductance
heat-pipe concept.

between evaporator and condenser. This helps to insure liquid
at the entrance to the injector tube. The injector tube exhaust is placed close to the heat source to insure that the
injected liquid is directed to the heat source wick. The
liquid injector tube is also quite long to provide sufficient
back-pressure resistance, so that when the bubble forms it
will not be driven back into the condenser end. Consequently,
as the bubble expands, it will exhaust first out through the

injector and into the wick (Fig. 3b). Once the pressure is
relieved and the control heat removed, the liquid in the
injector will be restored by capillary action (Fig. 3c). This
paper presents conceptual design information and an experimental analysis of the hydrodynamic operation of the 0-g bubble-
pump-injected variable-conductance heat pipe in a simulated
0-g environment.

Method of Analysis

     A detailed analysis of the operation of the 0-g bubble
pump heat-pipe requires operation in the 0-g environment.
However, a preliminary analysis can be accomplished by using
a two-dimensional simulator, as shown in Fig. 4. In Fig. 4,
two pieces of glass are separated by a thin rubber gasket.
The gasket acts as the two-dimensional boundary of the heat
pipe. When liquid is placed between the two glass plates,
whose plane is normal to the gravitational forces, the liquid
surface assumes a shape that is similar to that of a three-
dimensional heat pipe in the 0-g environment when viewed from
the top. Reference 5 reviews a wide variety of research
applied to the simulation of three-dimensional liquid systems
in 0-g environments, with two-dimensional simulators. In

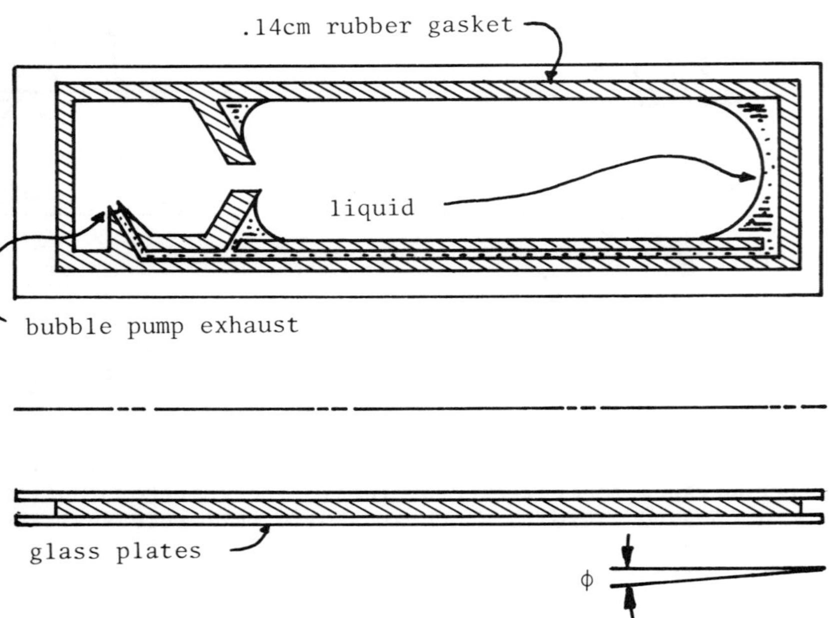

Fig. 4 Two-dimensional 0-g simulator.

order to have a valid two-dimensional simulator, three conditions must be met:

1) The liquid solid contact angles must be the same for the two-dimensional simulator and the three-dimensional prototype.

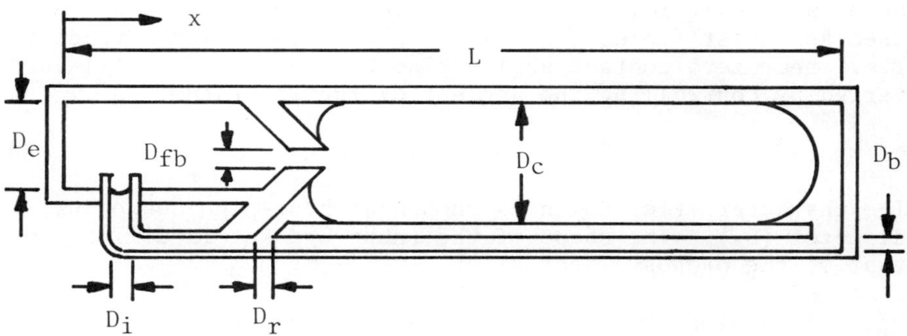

a) Critical heat-pipe dimension

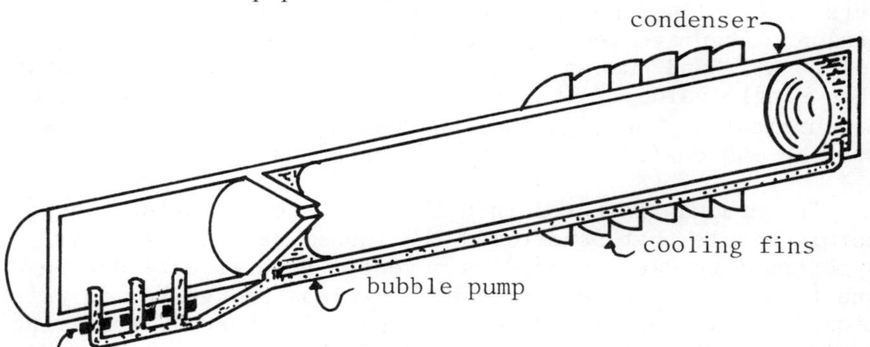

b) Three-dimensional view of the variable-
   conductance heat pipe

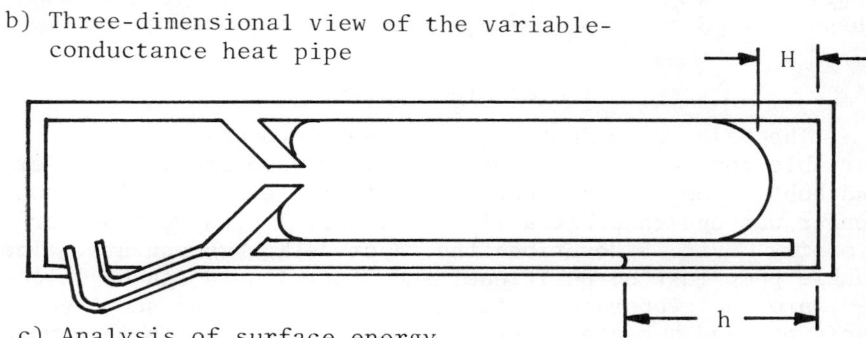

c) Analysis of surface energy

Fig. 5   Detailed design of a 0-g bubble-pump-injected
         variable-conductance heat pipe.

2) The Bond numbers (ratio of gravitational energy to surface energy) must be of equal magnitude.

3) The container and amount of liquid in the simulator must be geometrically similar to that of the prototype.

In most heat-pipe designs, contact angles near zero are necessary to maximize surface tension forces. Methanol was used as a test fluid, since it wets the test fixture readily, i.e., near zero contact angle. The Bond number [Eq. (1)] was varied by controlling the angle ϕ of the apparatus:

$$B = \rho g\, D_c^2\, \sin\phi / \sigma \qquad (1)$$

The characteristic length is chosen as heat-pipe condenser diameter (D_c). The shape of the rubber gasket is similar to that of the proposed heat pipe.

Apparatus plate spacing is approximately 0.14 cm and is sufficiently within the maximum allowable for valid simulators[1]. Leveling the apparatus requires the utmost in precision. Accurate levels were used which indicated angle of tilt. All Bond number readings are reported as ranges rather than single values. This reflects the uncertainty in these measurements. The high and low readings of the Bond number are the 99% confidence limits on the actual Bond number.

Liquid was injected into the test section using a hypodermic needle that penetrated the rubber gasket. Another hypodermic needle was used as a vent. Edge effects due to the triangular groove formed by the glass and gasket do exist. Liquid can wick back and forth along this channel, resulting in long-term equilibrium shapes different from reality. Generally this edge effect flow is not significant if the liquid shape is analyzed immediately after system perturbation.

Heat-Pipe Design

The major components of a 0-g bubble-pump-injected variable-conductance heat pipe are the evaporator, condenser, and bubble pump. A cylindrical heat-pipe evaporator and condenser was chosen arbitrarily for simplicity as well as for symmetry. Fig. 5 describes the major design components. The bubble pump must be positioned such that liquid can be injected into the evaporator. Adequate liquid must be supplied to the pump, and a heater must be mounted near the bubble pump exhaust. The operation of this design is affected significantly by wetting liquid shapes in the 0-g environment. When a surface wetting liquid is placed in a cylindrical tube, its shape

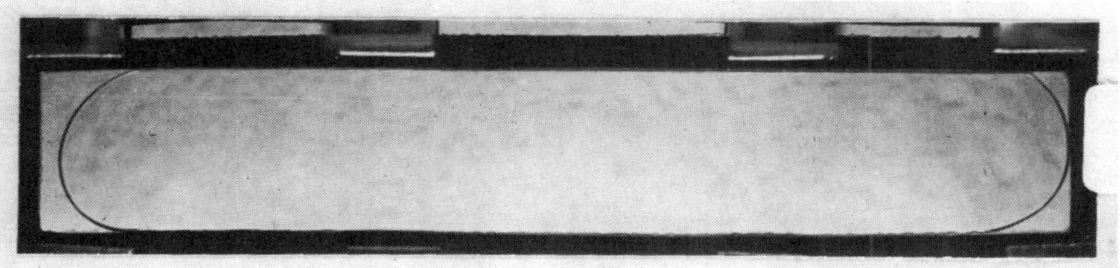

Fig. 6d   Photograph of a simulation of a cylindrical tube,
           10% full of liquid, Bond number range is 0 to 1.3
           (tube diameter 2.28 cm, tube length 13.8 cm).

Fig. 6e   Photograph of a simulation of a cylindrical tube,
           g force is in the negative y direction, Bond num-
           ber range is 21 to 23 (tube diameter 2.28 cm, tube
           length 13.8 cm).

Fig. 6f   Photograph of a simulation of a cylindrical tube,
           near circular meniscus curvature, Bond number
           range is 0 to 1.3 (tube diameter 2.28 cm, tube
           length 13.8 cm).

imposed on the system, the liquid should remain there. If
the simulator is disturbed, liquid will flow randomly about
by attaching to one wall (see Fig. 6c). When equilibrium is
reached at near zero Bond number, the configuration of Fig. 6b
results, where liquid resides near each corner. This configuration should be stable at B near zero. Because of an edge
effect that allows communication of liquid from one liquid
pool to the other on the simulator, some change in liquid pool

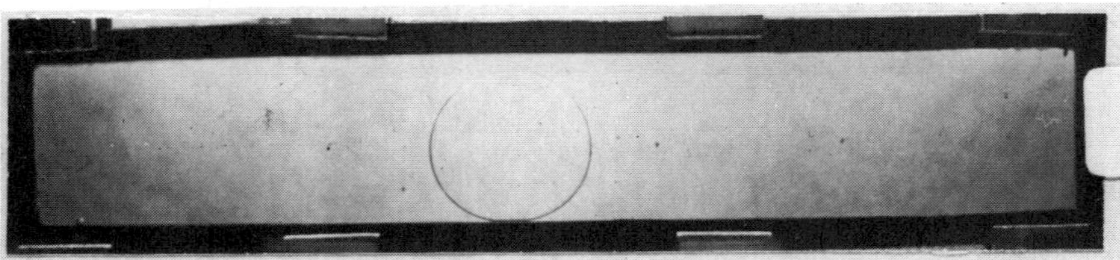

Fig. 6a   Photograph of a simulation of a
          cylindrical tube, 80% full of liquid,
          Bond number range 0 to 1.3 (tube diame-
          ter 2.28 cm, tube length 13.8 cm).

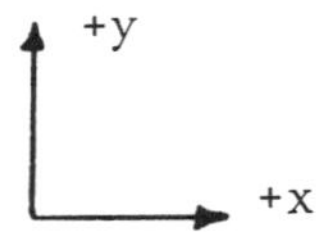

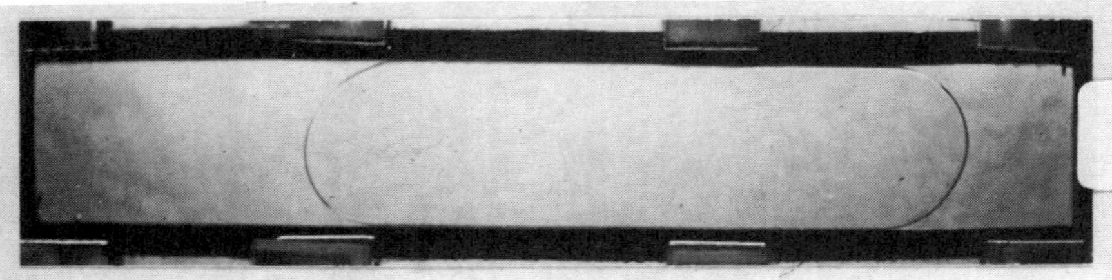

Fig. 6b   Photograph of a simulation of a cylindrical tube,
          50% full of liquid, Bond number range 0 to 1.3
          (tube diameter 2.28 cm, tube length 13.8 cm).

Fig. 6c   Photograph of a simulation of a cylindrical tube,
          liquid motion under varying g force from positive
          to negative x direction (tube diameter 2.28 cm,
          tube length 13.8 cm).

at small Bond number is shown in Fig. 6a. This has been verified experimentally in Ref. 5 from drop tower experiments. In
this case, a large amount of liquid, near 80% of the volume of
the tube, is shown. In a heat pipe, less liquid is required,
so that room remains for vapor transport. Figure 6b shows
a smaller amount of liquid (approximately 50% by volume) for
a Bond number range from 0 to 1.3. The simulator shows liquid
residing at each end of the tube. If no other forces are

Fig. 6g Photograph of a simulation of a cylindrical tube, g force is in the negative x direction, Bond number range is 5.4 to 6.4 (tube diameter 2.28 cm, tube length 13.8 cm).

Fig. 6h Photograph of a simulation of a cylindrical tube, g force is in the negative x direction, Bond number range is 21 to 23 (tube diameter 2.28 cm, tube length 13.8 cm).

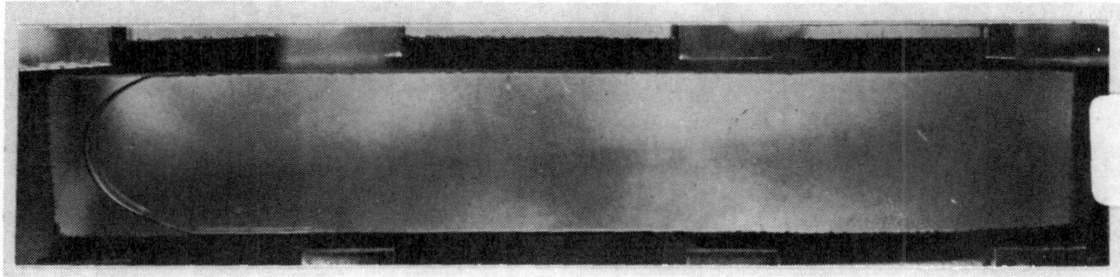

Fig. 6i Photograph of a simulation of a cylindrical tube, g force is in the positive x direction, Bond number range is 2.7 to 3.9 (tube diameter 2.28 cm, tube length 13.8 cm).

shape was observed which should not occur on the prototype. It should be noted from these experiments that liquid transport from one end of the tube to the other almost always takes place via the walls and not by large drops that break from the liquid pools. Fig. 6d shows a cylindrical tube simulator with approximately 10% liquid, as would be expected in a heat pipe. Again liquid pools collected at the ends of the tube. The

direction of g forces as well as the magnitude of the Bond
number significantly affect the liquid shape in the tube.
Fig. 6e shows a tube with a Bond number range of 21-23 in
the negative y direction. Liquid tends to lie against the
tube walls, since there is insufficient capillary force to
keep the liquid at the ends. Figures 6f-6h show liquid shapes
at varying Bond numbers. As the Bond number increases, meniscus curvature decreases, as is expected. The 1-g shape
would be approximately flat. Figure 6i shows liquid at the
end of the tube but with the g force in the plus x direction.
There appears to be some capillary pressure capability in the
liquid reservoir which results in a resistance to movement in
the positive x direction.

These observations of fluid behavior in a cylindrical
tube yield information on fluid behavior in a heat pipe of
similar shape. An important fact to notice from these experiments is that liquid movement generally occurs along the walls
and in one continuous shape. During all simulator experiments, no separate spheroids of liquid were observed drifting
in the simulator. With a wetting fluid, all liquid tends to
coalesce into one or two liquid pools that attach themselves
to the wall of the tube.

An important design requirement is to assure that liquid
flow to the evaporator is controlled solely by the bubble pump.
Consequently, a liquid flow blockage device (baffle) is placed
near the evaporator to prevent liquid from bypassing the
bubble pump. The baffle diameter should be small to limit
liquid flow to the evaporator when g forces are in the negative x direction. The bubble pump intakes are positioned in
areas where excess liquid accumulates to insure adequate supply to the bubble pump. The bubble pump is primed at all
times to insure that the flow is maintained. The following
design conditions must be met:

$$D_b < D_c \qquad (2)$$

To demonstrate this requirement, suppose that liquid is
in the condenser and the bubble pump is not primed (see Fig.
5c). Assuming that the system will attempt to seek a minimum
energy configuration, a change in liquid height (ΔH) in the
condenser is hypothesized. The change in energy on the bubble
pump surface (Δs) will equal the gain in energy due to covering the area (Δa) with liquid minus the loss in energy due to
the loss in vapor wetted area:

$$\Delta s = \Delta a \; \sigma_{ls} - \Delta a \; \sigma_{vs} \qquad (3)$$

From the Young - Dupre equation,

$$\sigma_{vs} - \sigma_{ls} = \sigma_{lv} \cos\Theta \tag{4}$$

Substituting into Eq. (3)

$$\Delta s = - \Delta a\, \sigma_{lv} \cos\Theta \tag{5}$$

Likewise, for the condenser,

$$\Delta S = - \Delta A\, \sigma_{lv} \cos\Theta \tag{6}$$

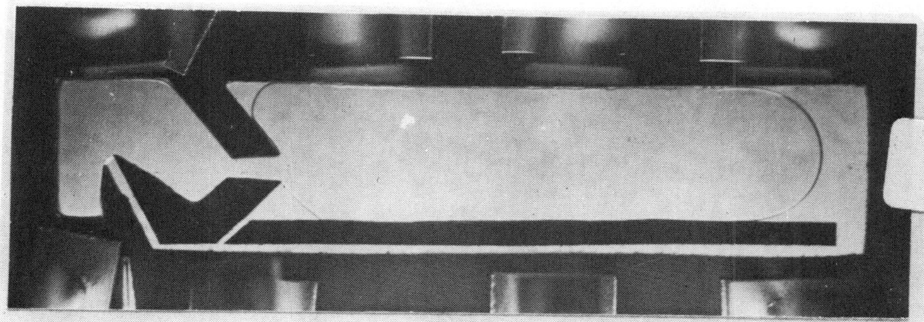

Fig. 7a Simulated hydrodynamic operation of a
0-g bubble-pump-injected variable-
conductance heat pipe (L=15.1 cm,
D_c=2.28 cm, D_i=0.15 cm, D_r=0.076 cm,
D_{fb}=0.38 cm), g force near 0, Bond
number range is 0 to 1.2.

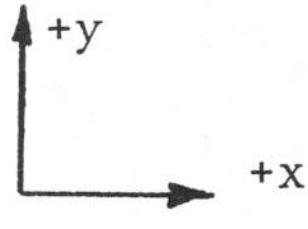

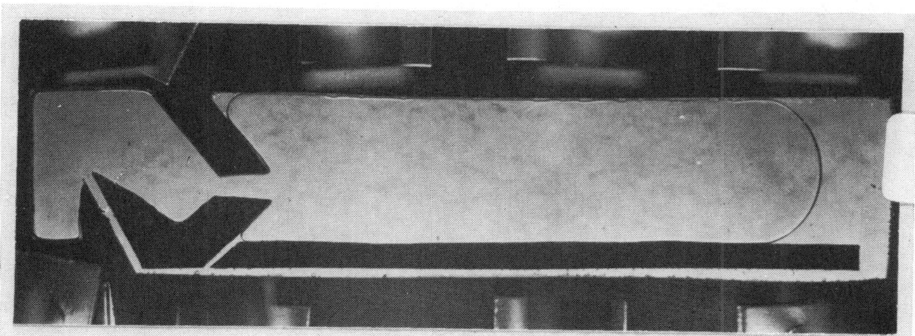

Fig. 7b Simulated hydrodynamic operation of a 0-g bubble-
pump-injected variable-conductance heat pipe
(L=15.1 cm, D_c= 2.28 cm, D_i=0.15 cm, D_r=0.076 cm,
D_{fb}=0.38 cm), g force in the positive x direction,
Bond number range is 3.1 to 5.4.

and, since

$$\Delta H = -\Delta h \; D_b^2/D_c^2 \tag{7}$$

then

$$\Delta s + \Delta S = \pi \Delta h \; \cos\Theta \; [D_b(D_b/D_c - 1)] \tag{8}$$

If $\cos\Theta$ and Δh are positive, then a negative value of $(\Delta s + \Delta S)$ will occur when Eq. (2) is satisfied. This indicates that liquid should be drawn toward the evaporator via the

Fig. 7c   Simulated hydrodynamic operation of a 0-g bubble-pump-injected variable-conductance heat pipe (L=15.1 cm, D_c=2.28 cm, D_i=0.15 cm, D_r=0.076 cm, D_{fb}=0.38 cm), g force in the negative y direction, Bond number range is 14.4 to 16.7.

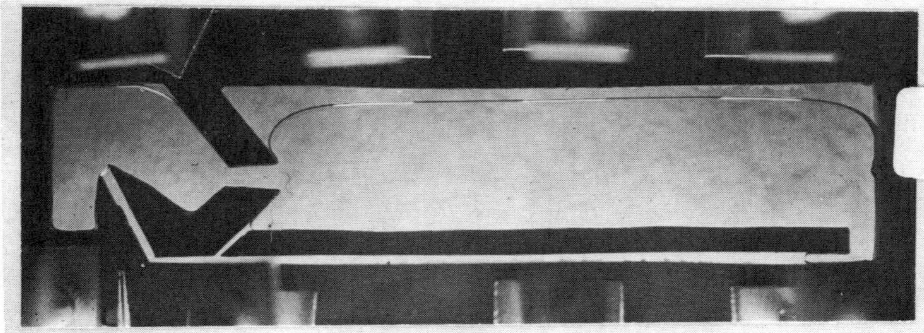

Fig. 7d   Simulated hydrodynamic operation of a 0-g bubble-pump-injected variable-conductance heat pipe (L=15.1 cm, D_c=2.28 cm, D_i=0.15 cm, D_r=0.076 cm, D_{fb}=0.38 cm), g force in the positive y direction, Bond number range is 14.4 to 16.7.

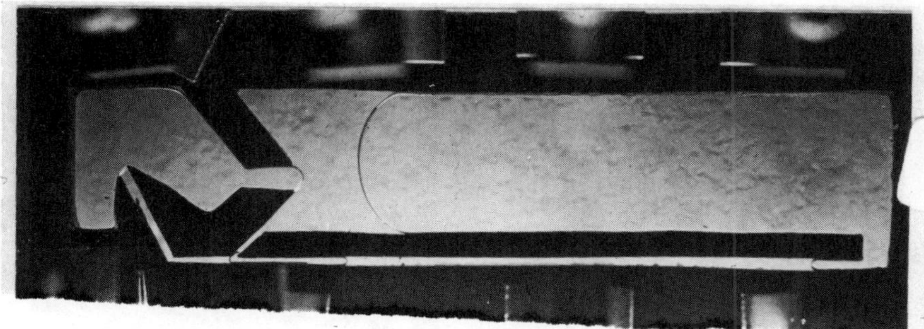

Fig. 7e   Simulated hydrodynamic operation of a 0-g bubble-
          pump-injected variable-conductance heat pipe
          (L=15.1 cm, D_c=2.28 cm, D_i=0.15 cm, D_r=0.076 cm,
          D_{fb}=0.38 cm), g force in the negative x direction,
          Bond number range is 14.4 to 16.7.

bubble pump from the condenser reservoir. This was observed
experimentally, as shown in Fig. 7a. Fig. 7a shows the proposed bubble pump heat-pipe simulator at Bond number near
zero. Liquid resides at the condenser end and at the baffle
end of the heat pipe. The bubble pump injector can draw from
either of these reservoirs. Because of the nonsymmetry of the
design, liquid was placed artificially at the top baffle. In
the prototype, liquid communication would exist between liquid
in the baffle and that in the injector tube. In this configuration, when liquid was injected from the bubble pump into the
evaporator, liquid was restored successfully to the injector,
and heat-pipe operation was maintained. The injected liquid
will evaporate and be driven through the flow baffle opening
into the condenser. The resultant condensate will supply
both reservoirs if communication exists. For B near zero,
operation of this design appears feasible. There is sufficient
back-pressure in the bubble pump tube to allow liquid to
inject through the evaporator end. As discussed earlier,
liquid shapes for Bond numbers different from zero and at
various directions may have significant effects on heat-pipe
operation.

     Fig. 7b shows the simulator at a Bond number range of
3.1 to 5.4 with the g forces in the plus x direction. This
results in the recession of the liquid surface deep into the
bubble pump injector throat. If the liquid surface is below
the injector heater, the pump will not operate, and variable
conductance will fail. Therefore, there is a wicking limit
that is observed when g forces are in the plus x direction.
The limit is

$$B_{max} = 4\ D_c^2\ \cos\Theta/\ D_B\ (L\text{-}H) \tag{9}$$

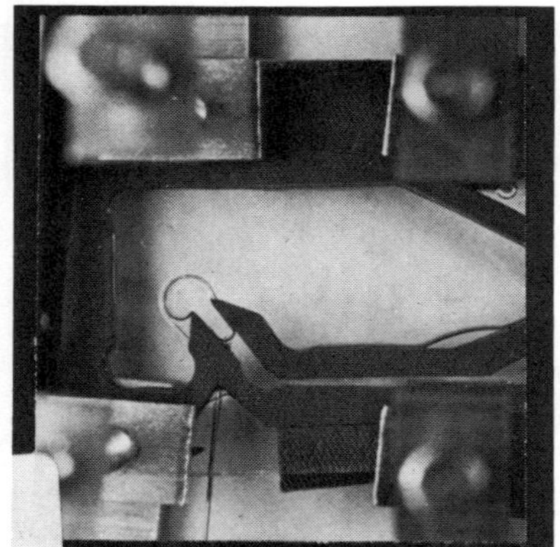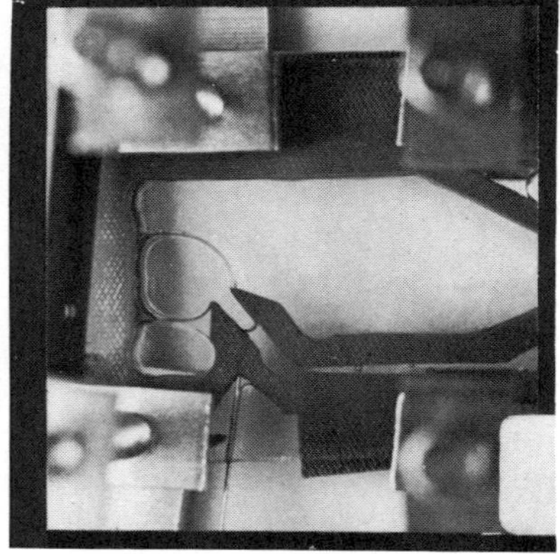

Fig. 8   Bubble pump injection sequence.

Notice the recession of the liquid in the reservoir at the flow baffle (Fig. 7b). This is why D_r must be less than D_i to prevent desaturation or depriming of the bubble pump.

When g forces are in the minus y direction, the condition of Fig. 7c results. Here liquid floods the bubble pump but does not pose a problem except when liquid is injected inadvertently. However, with proper liquid fill, this should not occur. A capillary limit does exist for this configuration, however, and is shown below:

$$B_{max} = 4\ D_c^2\ \cos\Theta/\ D_B\ (Y_d)   \tag{10}$$

For positive g forces in the y direction, we see the liquid configuration of Fig. 7d. Because of the unsymmetrical heat-pipe simulator, we unfortunately do not observe priming of the injector tube because of lack of communication from the top to the bottom of the tube. This communication would exist in the heat pipe in the apex groove of the baffles. Here the liquid wicking limit would depend on the flow baffle groove angle to the heat-pipe wall. A liquid flow pressure drop limit also would be observed, as liquid did flow along this communication channel to the injector.

The liquid configuration when g forces are in the negative x direction appears to be the most difficult to control because of liquid entering through the flow baffle. If g forces are applied suddenly, the liquid configuration of Fig. 7e results. Here the liquid reservoir in the condenser has blocked the baffle channel and is not allowing liquid to inject into the heat pipe inadvertently. However, when injection pressure is supplied, the vapor will be driven through

this restriction and form a channel. This may penetrate the
liquid reservoir and result in liquid being injected inadvertently through the bubble pump into the evaporator. Depending on the diameter of the heat-pipe injector D_i, the
amount of uncontrolled liquid being injected may be small.
However, this situation should be considered when negative-
x-direction g forces result. In the case when the negative-x
g force is added very slowly, significant overinjection of
liquid occurs.

It should be noted that the bubble pump operation should
not be affected significantly if the g forces are of small
duration. However, it appears that long-duration g forces in
the negative x direction are most troublesome. G forces in
the other directions can be tolerated up to certain magnitudes,
as indicated in some of the proposed limit equations.

Another important design requirement is that the bubble
pump should inject liquid and restore during idle mode. Figure 8
shows a bubble pump injection sequence from the simulator. As
the bubble exhausts from the injector, it expands to many
times its initial size and tends to distribute liquid uniformly before contacting the wick. After the bubble was injected,
liquid was restored to the injector, which now is ready for
the next signal to inject liquid.

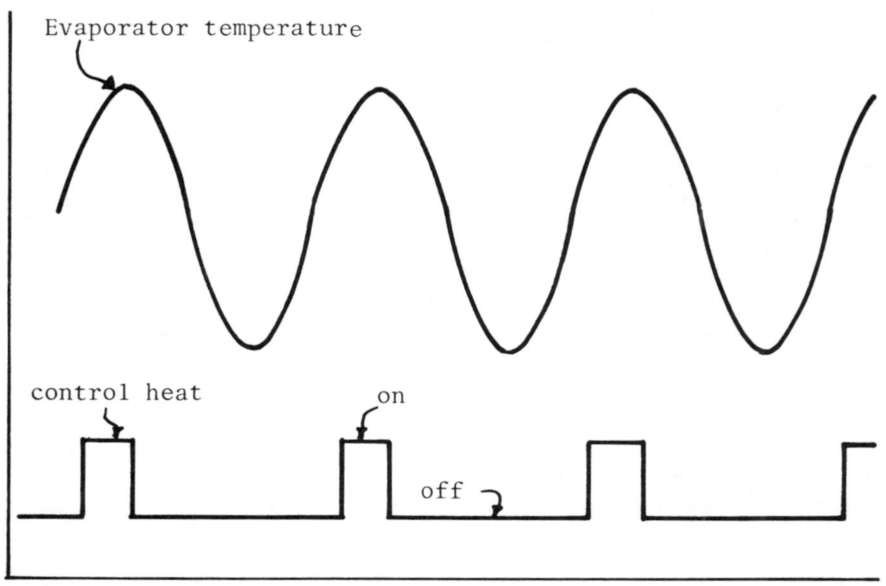

Fig. 9 Evaporator temperature and control heat vs time.

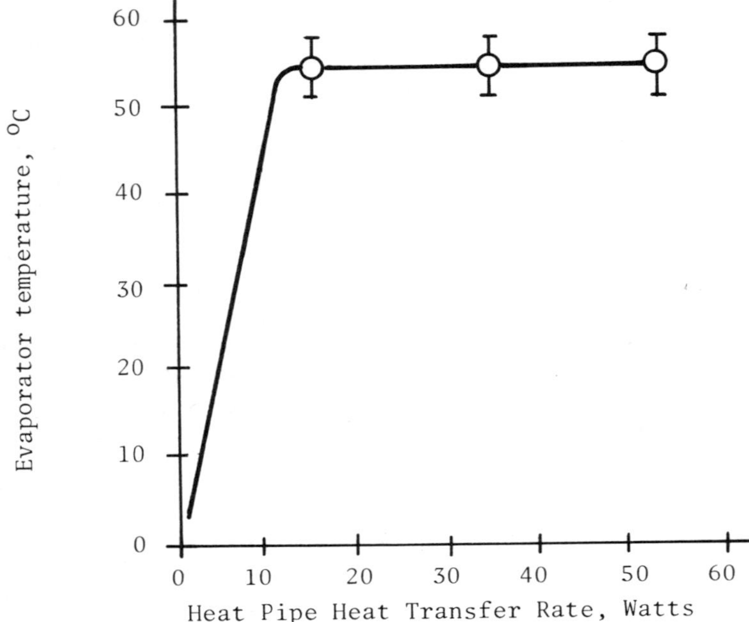

Fig. 10   Typical thermal control performance of the
1-g bubble-pump-injected variable-conductance
heat pipe.

Thermal Control

Thermal control is attained by injecting liquid into the
heat pipe at the proper time. Figure 9 is a graph of evaporator temperature and control heat vs time. As the evaporator
temperature increases beyond a set point, a signal is sent to
the control heater to activate. Control heat is applied until
the bubble has injected liquid, which is apparent when the
first derivative of evaporator temperature with respect to
time becomes negative. At that instant, control heat is removed, and the system coasts until the injected liquid is
exhausted. When the evaporator temperature begins to rise
above the set point again, the cycle continues.

Figure 10 shows thermal performance of the 1-g version
of the bubble-pump-injected variable-conductance heat pipe.
Evaporator temperature is controlled by the amount of liquid
injected, as well as the frequency. At low heat-transfer
rates, small injectors and low injection frequencies are used.
At high heat-transfer rates, larger injectors and higher injection frequencies are used. Precise temperature control is
attained with small injectors pulsing at higher frequencies.

Activation time of the system is on the order of 1 to 2 sec, depending on the heat capacity of the injector. Generally injector heat requirements do not exceed 1% of the total heat transferred in the system. More detailed analysis of this system is available in Ref. 1.

The thermal control range for this system is much greater than other variable conductance heat pipes. Varying liquid flow is often more effective than utilizing condenser blockage since parasitic heat loss may be uncontrollable. When the bubble pump heat pipe is shut off, the only mechanism of heat transfer is axial conduction and thermal radiation from the heat source. In other types of heat pipes, liquid and vapor remain in circulation resulting in additional parasitic heat transfer.

The bubble pump enhances heat pipe heat transfer. The pumping rate of bubble pump injectors is often greater than many capillary wicks. The result is a greater thermal control range.

One characteristic of the bubble pump heat pipe is the discrete liquid injections that occur during normal operation. This tends to introduce a sinusodial-type variation in evaporator temperature as shown in Fig. 9. This variation can be controlled by varying liquid injection frequency, liquid volume per injection, and by increasing the thermal mass of the evaporator.

Conclusions

From hydrodynamic experiments on a two-dimensional 0-g simulator, it appears that the bubble-pump-injected variable-conductance heat pipe should operate at Bond numbers near zero. For higher Bond numbers, greater than 4 to 5, g-force direction may be a consideration in the design of such a heat pipe. Wide ranges of thermal control can be achieved with single or multiple injectors. Noncondensable gas accumulation in wicks or arteries is not a problem with this system. The bubble pump heat pipe appears to be insensitive to both source and sink temperature variations, as verified in Ref. 1.

References

[1] Roberts, C. C., "A Variable Conductance Heat Pipe Using Bubble Pump Injection," Second International Heat Pipe Conference, Bologna, Italy, 1976.

[2] Chisholm, D., "The Anti-Gravity Thermosyphon," Multi-Phase

Flow Systems, Inst. of Chemical Engineers Symposium Ser. 38, 1974.

[3] Basiulis, A., "VBP Heat Pipes for Energy Storage," 11th Intersociety Energy Conversion Conference, 1976.

[4] Hartnett, J. P. and Irvin, T. F., Jr., Advances in Heat Transfer, Vol. 9, Academic Press, New York, p. 89.

[5] Olson, W. A., "Simulator for Static Liquid Configuration in Propellant Tanks Subject to Reduce Gravity," NASA TND 3249, May 1966.

RE-ENTRANT GROOVE HEAT PIPE

W. Harwell*
Grumman Aerospace Corporation, Bethpage, N. Y.

and

W. B. Kaufman† and L. K. Tower†
NASA Lewis Research Center, Cleveland, Ohio

Abstract

This paper describes theoretical and experimentally verified heat pipe characteristics of an axially grooved aluminum
extrusion with a re-entrant groove profile. The extrusion is
13 mm diam with 20 axial grooves, each groove consisting of a
nominal 0.8-mm-diam channel with a 0.2-mm-wide passageway
connecting the channel to the hollow core. A computer program
was written to compute the zero-gravity heat-transport capability of the extrusion. A heat pipe was fabricated and its
performance characteristics measured. The characteristics of
the pipe with ammonia at 20°C are zero-gravity pumping limit
143 W-m; static wicking height 21.5 mm; evaporator and condenser coefficients 7300 and 20,500 W/m^2 -°C, respectively.

I. Introduction

Although accounts of the heat-pipe principle were published as early as 1945 by Gaugler[1] and later by Thompson[2] in
1960, active development of the principle is not considered to
have started until Grover et al.[3] independently realized the
potential of heat pipes. Grover's paper was followed quickly by
a number of papers deriving the basic theory.[4-6] The initial
approach equated the capillary pumping to the system viscous
losses with simplified gravitational effects. Frank[6] developed
an optimization procedure for maximum transport capacity in

Presented as Paper 77-773 at the AIAA 12th Thermophysics
Conference, Albuquerque, N. Mex., June 27-29, 1977.

* Senior Engineer.
† Aerospace Technologist.

zero g based on grooves of specified nondimensional shape. Subsequent advances include consideration of real groove geometry (fillets, etc.) with numberical integration along the groove to determine fill requirements for zero-g applications,[7] visualization of liquid distribution in grooves,[8] groove-by-groove analysis backed up by neutron photography to determine puddle characteristics,[9] inclusion of shear at the liquid/vapor interface,[10] and zero-g flight data.[11]

In the early days of the heat pipe, axial grooves were produced by machining or broaching the groove into the inside surface of thick-wall tubes. It was not until 1967 that the first practical, low-cost grooved heat pipe tubing was produced[8] through the efforts of Grumman Aerospace Corporation working with the French Tube Corporation. This tubing was produced by a swaging process. Under contract to NASA Goddard Space Flight Center (GSFC), Grumman built three isothermalizers for the Orbiting Astronomical Observatory (OAO-B) with Freon-21 as the working fluid.[9] Further efforts to upgrade heat-pipe performance resulted in building a heat pipe that used ammonia as the working fluid for the OAO-C spacecraft. This pipe still is functioning after 4 yr in orbit.[12]

The successful performance of the heat pipe aboard the OAO-C led NASA to baseline similar pipes as the prime thermal control system for the ATS-F spacecraft, and 55 swaged axial groove ammonia heat pipes were built into the north, south, and transverse panels of the spacecraft.[13] Swaged tubing had a number of disadvantages, including the following:

1) Poor ground testability. The swaging process results in poor uniformity (axial and circumferential) in groove dimensions, as well as relatively wide grooves.

2) Poor surface condition. The swaging process invariably results in crevices and can result in inclusions that may not be removed in the cleaning process.

These disadvantages were recognized early by NASA and as part of the ATS-F program a development of extruded grooved heat pipes was begun. As a result, Universal Alloy Corporation, La Habra, Calif., produced an extruded axially grooved tube.[10] At approximately the same time, Battelle Institute also produced an axially grooved extrusion[14] with lower transport capacity than the NASA extrusion.

The NASA GSFC extrusion (referred to in the open literature as the Micro Extrusion or the ATS extrusion) was a considerable improvement over the original swaged tube, being

cleaner and considerably more uniform. However, it still had
poor ground testability; i.e., it was very tilt-sensitive. At
this point, NASA Lewis performed some tradeoff studies on im-
proving the performance of axially grooved heat pipes, particularly to improve static wicking height without significantly
reducing transport capability. The result of these tradeoff
studies was the NASA Lewis re-entrant groove extrusion, produced by Universal Alloy Corporation with the enthusiastic
support of the late Earl Bell. Unfortunately, only a few
lengths of this extrusion were produced for test purposes.
This paper details the analysis and testing of this extrusion.

II. Analytical Model

This phase of the program included taking measurements to
determine the characteristic shape of the axial grooves in the
delivered NASA Lewis re-entrant groove extrusion, writing a
computer code to estimate the transport capability of the extrusion, and finally exercising the computer code para-
metrically to predict the transport capability of the extrusion
with ammonia working fluid at room temperature.

Characteristic Dimensions

Three lengths of extrusion were examined in this investigation, each length being approximately 0.9 m long and having
20 internal, longitudinal grooves. Each groove consisted of
an entrance passage plus a circular trough. Visual examination of the pipe lengths did not show any rifling or axial
twisting of the grooves. 2-cm-long samples were cut from each
end of each extrusion length, mounted in an epoxy, and
polished to expose the sharp groove profile. Photographic
enlargements of one of the samples are shown in Fig. 1.

Fig. 1 Photographic enlargement of re-entrant grooved extrusion.

Table 1   Re-entrant groove characteristic dimensions

PARAMETER		PIPE 2			PIPE 3		
		MIN	AV	MAX	MIN	AV	MAX
OUTSIDE DIAMETER	D_O	12.62	12.67	12.72	12.65	12.68	12.70
ROOT DIAMETER	D_R	11.29	11.37	11.45	11.31	11.38	11.47
CORE DIAMETER	D_C	9.22	9.28	9.34	9.25	9.28	9.29
ENTRANCE WIDTH	W_T	0.239	0.267	0.309	0.221	0.269	0.256
GROOVE OPENING	W_B	0.185	0.218	0.241	0.191	0.226	0.256
GROOVE DIAMETER	D_{G_1}	0.792	0.808	0.826	0.808	0.826	0.846
	D_{G_2}	0.782	0.798	0.810	0.787	0.803	0.815
	D_{G_3}	0.790	0.810	0.640	0.800	0.813	0.841
WALL THICKNESS	W_{ALL}	0.622	0.640	0.693	0.589	0.643	0.704

DIMENSIONS IN MILLIMETERS.

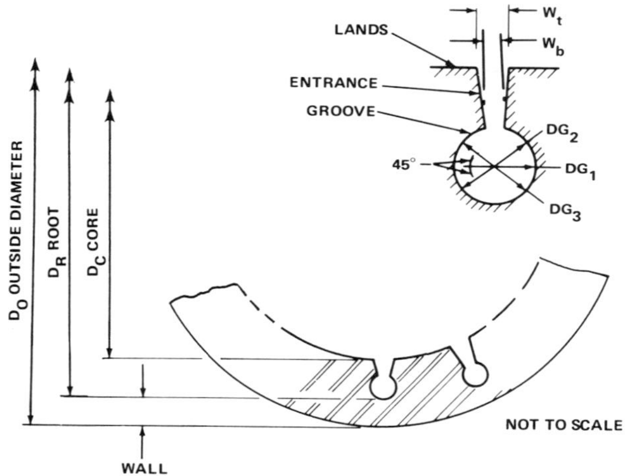

The characteristic dimensions of two of the pipes and grooves measured from photographic enlargements (20x) under a 10-power microscope are reported in Table 1. The average values reported are the simple arithmetic average, whereas the "min" and "max" are the minimum and maximum values measured. The characteristic dimensions are explained in the figure beneath Table 1:

1) The cores are eccentric by approximately 0.05 mm. This in itself is not important other than that it increases the local hoop stress.

2) Individual grooves are not circular, and some are offset (skewed) about their respective entrance passageways. This offset should have no effect on the groove pumping capability.

3) Most significantly, the groove entrance passageways
are convergent. The average taper of the entrance passageway
was calculated as 5 deg 46 min.

The internal cross-sectional area of two of the pipes
was computed from measured data in three ways: outside
diameter, length, and weight, 0.8110 cm^2, volume of water
required to fill each pipe, 0.8129 cm^3, and average dimensions,
0.7910 cm^2.

Computer Program

A simple computer program was written based upon hydrodynamic phenomen, all thermodynamic processes being ignored.
As is usual with heat pipe analysis, the starting point was to
assume a pressure balance:

$$\Delta P_{cap} = \Delta P_\ell + \Delta P_v + \Delta P_{v/\ell} \text{ shear}$$

where

ΔP_{cap} = capillary pressure difference

ΔP_ℓ = liquid-phase pressure drop

ΔP_v = vapor-phase pressure drop

$\Delta P_{v/\ell}$ shear = additional pressure drop in the liquid
                      phase due to the countercurrent vapor flow

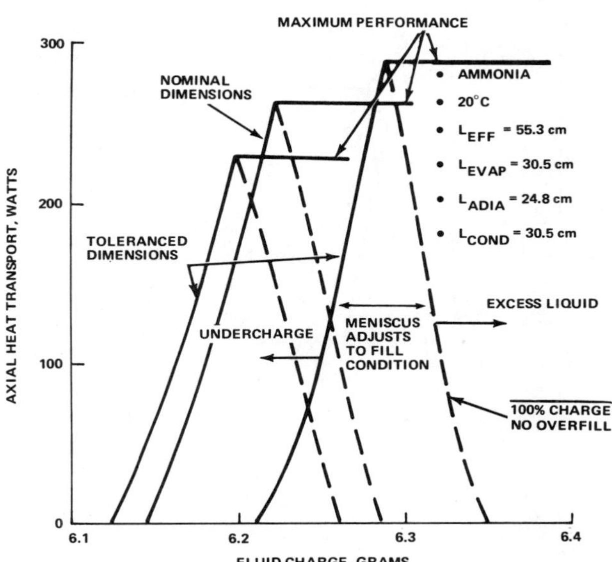

Fig. 2 Effect of measured-groove dimensional tolerances.

    The assumptions used in deriving the governing equations
were: 1) all grooves had the average groove geometry;
2) uniform circumferential and axial heat input and rejection;
3) one-dimensional laminar flow in the liquid phase (grooves);
4) one-dimensional laminar or turbulent flow in the vapor phase
(core); and 5) local two-dimensional curvature of the meniscus.
A numerical integration technique was used to compute the zero-
gravity pumping capability with a given charge condition.
Additional parameters are evaluated in the program, including
static wicking height, sonic heat flux, and entrainment limit.

Analytical Results

    The performance of the re-entrant groove with ammonia at
20°C was computed for both nominal and toleranced groove
dimensions; see Fig. 2. The curve for each set of dimensions
shows a maximum transport capacity at a peak charge value
corresponding to a flat meniscus at the condenser end and a
fully depressed meniscus at the evaporator end. The drop in
axial heat transport for charge values below the peak is a
real loss in capacity, caused by meniscus recession in the
condenser to accommodate the reduced volume of liquid present.
The drop in axial heat transport shown for charge values above

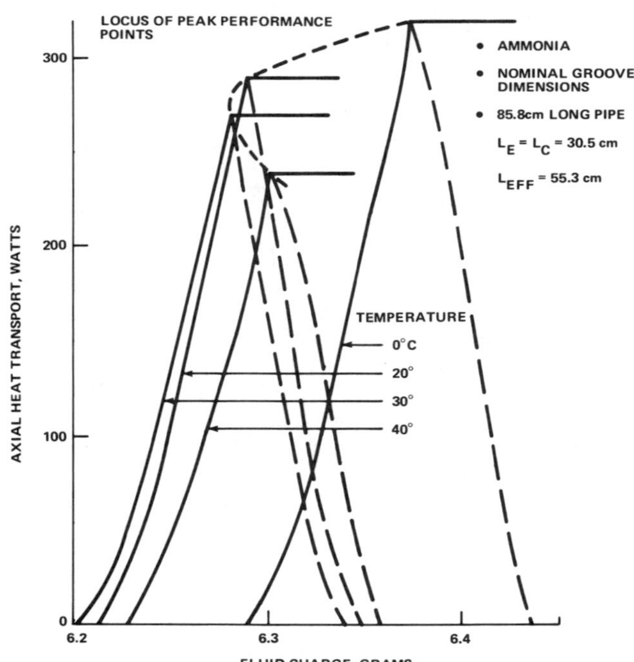

Fig. 3 Performance map for ammonia working fluid.

the peak is associated with a rise in liquid level (less depressed meniscus) in the evaporator to accommodate the increased liquid volume, with a corresponding reduction in capillary force and liquid flow rate. With an overcharged pipe, the axial heat-transport limit would equal the maximum values shown, but the excess liquid could not be contained in the grooves and would show up as free liquid in the condenser, blocking some of the vapor space. An additional point to note from this figure is the sensitivity of this heat pipe to dimensional tolerances.

Figure 3 shows that, as the temperature increases for a nominally dimensioned re-entrant groove charged with ammonia, the fluid charge for maximum performance initially decreases and then starts to increase again. This is due to the 6:1 vapor-to-liquid volume ratio of the re-entrant groove.

III.   Experimental

Detailed performance characteristics for a heat pipe fabricated from the NASA Lewis re-entrant axially grooved extrusion then were determined experimentally with ammonia as the working fluid. These performance characteristics included

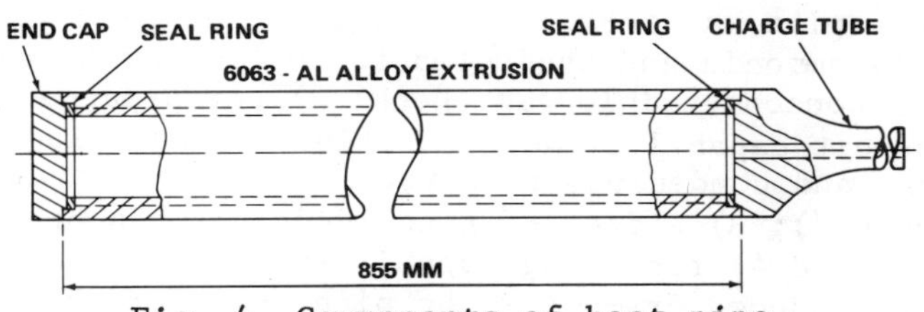

Fig. 4   Components of heat pipe.

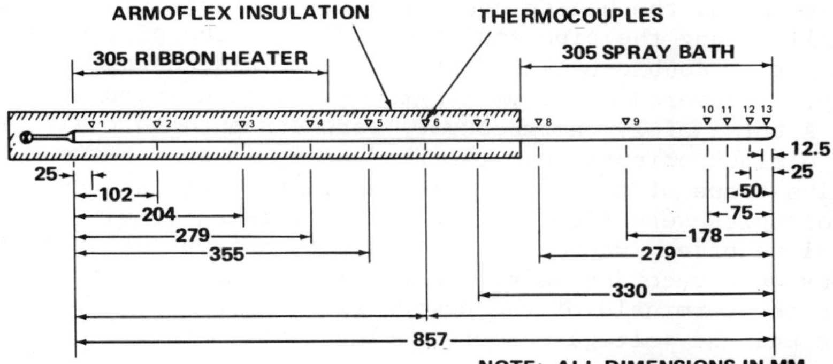

Fig. 5   NASA Lewis re-entrant groove heat-pipe test instrumentation.

transport capability as a function of adverse tilt and fluid
charge and limited repriming tests.

Heat-Pipe Design

The burst pressure of the extrusion was determined using
the pipe length designated No. 1. This pipe length was
thinner-walled than the length used to fabricate the heat pipe
and was estimated to give pessimistic burst pressure data
(approximately 5 to 10% lower). Two 230-mm long burst samples
were cut from each end of this pipe length. One end of each
burst sample was capped using flared fittings, whereas the
other end was coupled to a hydraulic test facility again using
flared fittings. Both samples burst at 190±7 atm. This is in
good agreement with the calculated burst pressure for the
extrusion in the T6 temper. Each sample failed along a
longitudinal rip at the base of a groove, and, coincidentally,
both failures initiated approximately 50 mm from their capped
ends. Assuming a factor of safety of 4 on design pressure,
this would limit operation (or storage temperature) of the
pipe with ammonia to 85°C in the T6 condition and 36°C in the
T6 as welded condition. To insure reasonable working margins,
it was decided to heat-treat the pipe to the T6 condition.

Test Assembly

A heat pipe was fabricated (Fig. 4), charged with Ultra
High Purity ammonia and subjected to the following series of
tests: 1) noncondensible gas check; 2) performance (maximum
heat-transport capability as a function of heat pipe tilt and
evaporator, and condenser, temperature drops as function of
heat transport); 3) effect of noncondensible gas on pumping
capability; and 4) repriming capability from both mechanical
and thermal-induced dry-out. The pipe used for these tests
was initially deliberately over-charged with 9.2 g of ammonia
and then valved off. It was instrumented with a heater and
thermocouples, the heater consisting of Nichrome ribbon wrapped
helically along the pipe and insulated from the pipe by a 0.5-
mil layer of double-backed Kapton tape. This heater, as shown
on Fig. 5, covered a 305-mm length of the pipe at the valve
end. A total of 13 thermocouples were strapped to the pipe at
the 3 o'clock orientation, the bead being located between con-
secutive turns of the heater element in the heated zone. The
thermocouples were recorded continuously using a multipoint
Bristol recorder, whereas the steady-state temperature distributions were recorded using a digital voltmeter. The voltage
across the terminals of the heater ribbon was varied by a
Variac and the voltage measured using a voltmeter.

The heat sink for the noncondensible gas check was a
stirred alcohol bath with liquid nitrogen bubbled through it.
The heat sink for all other tests was a cold-water-spray bath.

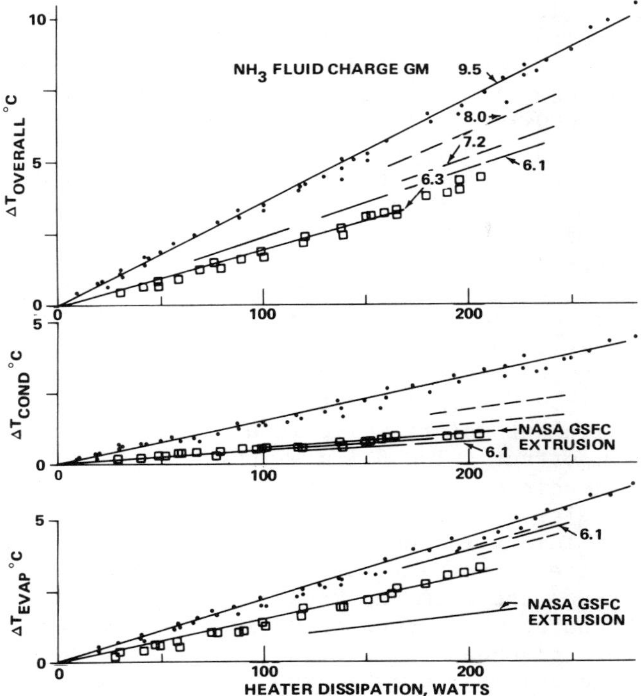

Fig. 6 Heat-pipe temperature difference as a function of heat transport and fluid charge.

Experimental Results

The test data for the performance tests were examined for incipient dry-out by plotting the evaporator temperature difference against heater dissipation, the evaporator temperature difference being defined as the difference in temperature between the end evaporator thermocouple (T/C 1) and the average of the three thermocouples in the adiabatic section (5, 6, and 7). The data indicated a linear change in temperature difference with heat input with a very sharply defined incipient dry-out point. All data recorded below the incipient dry-out condition at each heat-pipe tilt have been presented in Fig. 6, the evaporator, condenser, and overall temperature differences being based upon the average temperatures in each region of the pipe:

$$\Delta T_{evap} = \bar{T}_{evap} - \bar{T}_{adia}; \quad \Delta T_{cond} = \bar{T}_{adia} - \bar{T}_{cond}$$

$$\Delta T_{overall} = \bar{T}_{evap} - \bar{T}_{cond} = \Delta T_{evap} + \Delta T_{cond}$$

There is an obvious effect of charge condition on both the
evaporator and condenser temperature drops. The dependence
of the condenser temperature drop on charge was not unexpected;
the puddle reduces the area available for condensation; hence
the temperature drop should increase with increasing charge at
constant heater dissipation. However, increasing tilt at constant charge also reduces the surface area covered by the
puddle, and one would expect the temperature drop to decrease
with increasing tilt at constant heater dissipation and fluid
charge. No trend with tilt was detected. The dependency of
the evaporator temperature drop on charge is more difficult
to explain but can be explained qualitatively by the local
change in meniscus profile due to the puddle extending along
the pipe. As heat transfer in the evaporator occurs through
the meniscus liquid film local to the solid boundary, then the
meniscus attachment point, which is a function of charge, becomes very important. The 6.1-g ammonia charge represents an
undercharge condition. With this charge, the meniscus retreats
into the circular section of the groove, and hence the evaporator film coefficient is less than the coefficient with 6.3-g
ammonia. In the condenser, the condensate immediately runs
into the groove; the film thickness is slightly thinner than
in the 100% charge condition; there is slightly greater area
available for condensation to occur, and hence, the condenser
film coefficient is higher than in the 100% charge. The heat-
transfer coefficients for zero-g application based upon the
test data obtained with 100% charge, i.e., 6.3-g ammonia, and
referenced to the area corresponding to the pipe core diameter,
are 1) evaporator film coefficient, 7300 W/m^2-°C; and 2) condenser film coefficient, 20,500 W/m^2-°C. For comparison, the
corresponding film coefficients for the NASA GSFC extrusion,
computed from unreported Grumman test data, are 11,600 W/m^2-°C
for the evaporator and 16,800 W/m^2-°C for the condenser. If
one adopts the view that heat transfer to the liquid in the
evaporator occurs primarily in the meniscus attachment regions
and, for a given wall and fluid combination, is similar from
pipe to pipe, then for n grooves

$$Q = 2\ nl_{ev}Cg\Delta T$$

where C_g is some characteristic dimension of the meniscus and
depends on wall and liquid properties. Since

$$Q = h_{ev}\ \pi\ D_v\ l_{ev}\ \Delta T$$

substituting gives

$$h_{ev} = (2C_g/\pi)\ (n/D_v)$$

from which one might expect h_{ev} to vary directly with n/D_v.

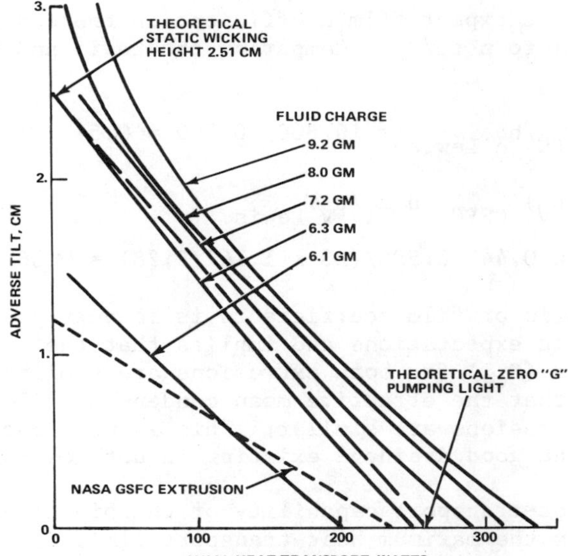

Fig. 7   Performance limit as a function of adverse tilt.

Comparing the Lewis and GSFC extrusions,

$$h_{ev,GSFC}/h_{ev,Lewis} = 11,600/7,500 = 1.59$$

$$(n/D_v)_{GSFC}/(n/D_v)_{Lewis} = (27/8.585)/(20/9.28) = 1.46$$

The ratio of film coefficients is thus within 10% of what
might have been expected, which is good agreement, considering
experimental accuracy. It also implies that the C_g values
for both extrusions are very similar and supports the generally
held view that in the evaporator heat transfer occurs local to
the meniscus attachment point.

Condensation generally is held to occur on the fin lands,
with subsequent drainage into the grooves. If one adopts the
view that condensation occurs on the fin tips and is propor-
tions to fin tip area,

$$Q = C_{cg} \, n \, t_f \, l_c \, \Delta T_c$$

where C_{cg} is an effective film coefficient.   Since

$$Q = h_c \, \pi \, D_v \, l_c \, \Delta T_c$$

one can write

$$h_c = (C_{cg}/\pi) \, (n \, t_f/D_v)$$

One then might expect film coefficient in the condenser to be
proportional to n t_f/D_v. Comparing the Lewis and GSFC extrusions,

$$h_{c, \text{GSFC}}/h_{c,\text{Lewis}} = 16,800/20,500 = 0.57$$

$$(n\ t_f/D_v)_{\text{GSFC}}/(n\ t_f/D_v)_{\text{Lewis}}$$

$$= (27 \times 0.447/8.585/(20 \times 1.162/9.28) = 0.56$$

The ratio of film coefficients is in very good agreement
with analytic expectations and implies that the effective film
coefficients (C_{cg}) for both extrusions are similar, which in
turn means that the effective mean condensate film thicknesses
for both extrusions are similar. This is what one would expect with the good drainage existing in both extrusions.

The heat-transport capability of the pipe is summarized
in Fig. 7 as the maximum heat-transport limit as a function of
heat pipe tilt. As one would expect, the maximum transport
capability at constant tilt decreased with decreasing fluid
inventory. The concave shape of the performance curves with
overcharged pipes is typical of grooved heat pipe data and
is due to the puddle decreasing the effective length of the
condensate return path. On reducing the fluid charge from
6.3 to 6.15 g, the performance fell off dramatically, indicating a transition from fully charged to an undercharged
condition. From Fig. 3, it will be seen that a decrease in
charge of 0.1 g is sufficient to reduce the heat transport
from maximum to zero. The test data are in good agreement with
the theory (100% charge is approximately 6.3 gm at 20°C).

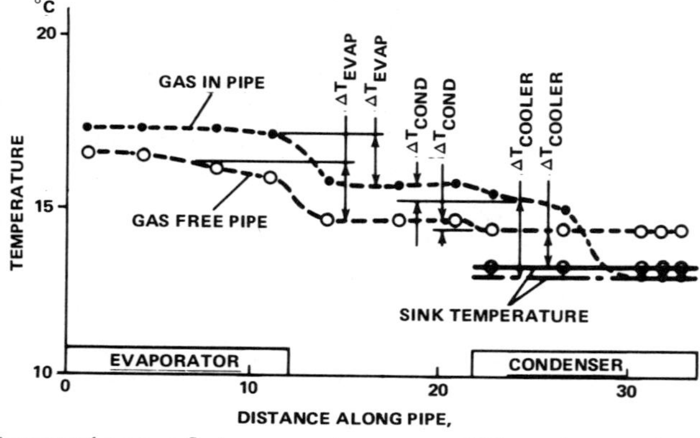

Fig. 8 Comparison of temperature profile along pipe with and
without inert gas.

Also shown in Fig. 7 is the line joining the zero-g predicted performance point to the static wicking height. These
two points can be considered as 1) the pumping capability of
the pipe in 1 g at zero tilt if puddle flow can be eliminated
(that is, no overfill or no groove drainage), and 2) the tilt
at which the pumping capability decreases to zero, again
assuming that the puddle can be eliminated. The predicted
performance limit is in excellent agreement with the test data.
Figure 7 also includes a summary of the test data for a heat
pipe manufactured from the NASA GSFC extrusion. The NASA
Lewis extrusion has a slightly higher zero-tilt pumping
capability and approximately double the static wicking height
when compared to the NASA GSFC extrusion. In these respects,
the Lewis extrusion has superior ground-test characteristics.
In addition, it must be noted that the Lewis extrusion requires only approximately 67% of the ammonia charge required
by the GSFC extrusion. This means that, although the Lewis
extrusion is very sensitive to undercharge, it is less sensitive to overcharge, a 6% overcharge being equivalent (in zero
g) to a liquid slug filling 1% of the pipe in the Lewis extrusion and 2.1% in the GSFC extrusion.

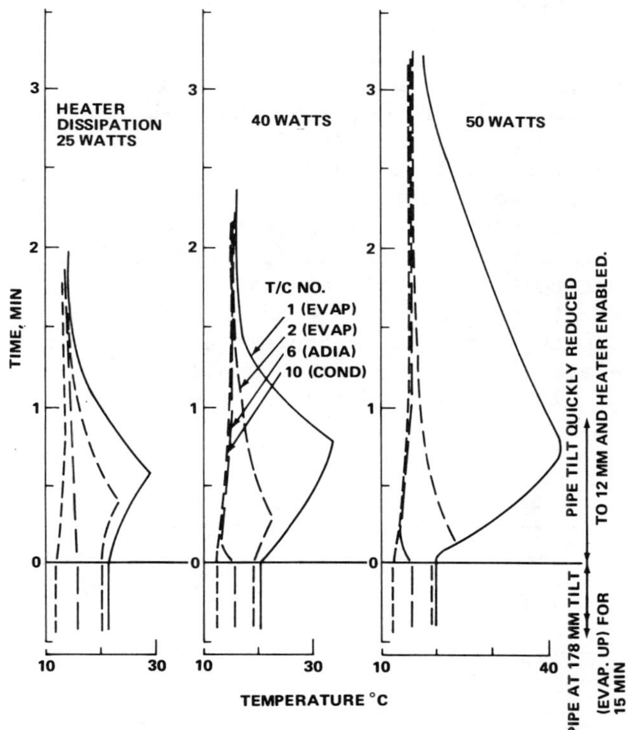

Fig. 9 Repriming from a mechanically induced dry-out.

The effect of deliberately injecting some dry nitrogen into the pipe is shown in Fig. 8. The gas converted the heat pipe into a reservoirless Variable Conductance Heat Pipe (VCHP). AT 15°C, the gas occupied part of the condenser, decreasing the effective pipe length by approximately 15% at dry-out. As noted with other grooved VCHPs, the presence of the gas decreased the maximum watt-meter transport capability of the heat pipe but maintained the same maximum axial heat transport. Although the reason for the reduction in $\int$ Qdx is not well understood at the present time, the fact that Qmax is not reduced makes this more of an academic than a practical problem. The NASA Lewis extrusion should be perfectly acceptable for gas-loaded VCHPs, in contrast with arterial heat pipes, where degraded performance has been encountered due to gas blockage of the arteries.

Repriming tests were performed to investigate the rate at which the pipe would recover due to a mechanical or a mechanical-thermal depriming sequence. The mechanical depriming, for example, would be equivalent to despinning a spacecraft soon after launch before any temperature gradients could be set up. The mechanical-thermal deprime would correspond to despinning a spacecraft after temperature gradients had been established. The test data are summarized in Fig. 9 and 10. In Fig. 9, the repriming of the pipes (under load) from a purely mechanical deprime are compared for 25-, 40-, and 50-W heater powers. In all cases, the pipe quickly primed and operated as an isothermalizer. One further test was carried out in this series with 60 W of heater power. The last thermocouple on the evaporator (T/C 1) did not recover fully and was about 15°C above the rest of the pipe 15 min after decreasing the pipe tilt to 12 mm. The heat-transport limit for this tilt from Fig. 7 is 140 W. The inability to reprime at 60 W is therefore not a simple transport limit. Possible mechanisms are poor surface wetting (perhaps related to boiling heat transfer) or upper grooves not priming. In Fig. 10, repriming after a mechanical-thermal

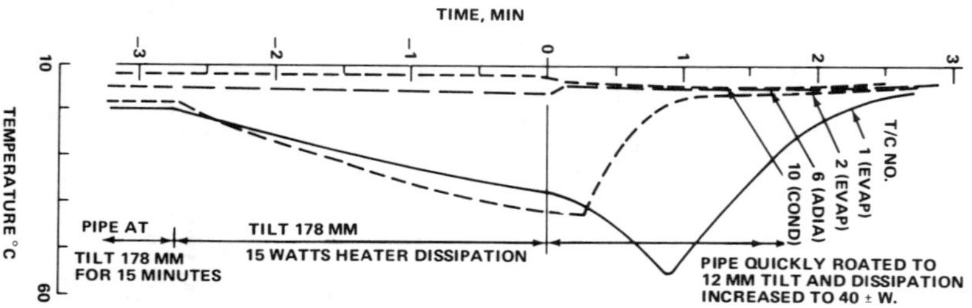

Fig. 10 Comparison of repriming from a mechanical and a mechanical/ thermal dry-out.

dry-out is presented. At 40 W dissipation, the pipe quickly primed, at 50 W, the pipe took 25 min to recover, at 60 W, the pipe did not recover. These data are in good agreement with what one would predict from the self priming index for the re-entrant groove extrusion. The reader is directed to Ref. 15 for the derivation and explanation of self-priming index.

IV. Conclusions

Examination of the Lewis extrusion both visually and by measurement showed that the extrusion had no discernible rifling and little variation in groove geometry both circumferentially (groove to groove) and axially (along a groove). The groove dimensions were very close to the original tradeoff design values, an excellent recommendation for the accuracy of the extrusion process. The main disadvantage with this particular extrusion is the thin wall; the pipe o.d. was fixed by the available condenser and heater blocks of the Lewis test fixtures; the pipe i.d. was limited by fabrication of the extrusion mandrel. The fabricators of the extrusion (Universal Alloy) are confident, however, that the extrusion can be produced with a thicker wall (larger o.d.). A thicker wall combined with heat treatment after welding would allow this extrusion to be used in any application envisaged for an extruded aluminum heat pipe.

The analytical predictions of transport capability are in excellent agreement with the experimental data and confirm the original tradeoff studies, which predicted the superior capability of the re-entrant groove extrusion. The measured evaporator and condenser film coefficients with ammonia are also in good agreement with what one would have predicted from data recorded from the NASA GSFC extrusion, the measured values being evaporator coefficient 7300 W/m^2-°C, and condenser coefficient 20,500 W/m^2°-C. Other relevant data are as follows: static wicking height, 2.15 cm at 20°C; zero-gravity pumping limit, 143 W-m at 20°C; and fluid charge, 7.3 g/m length of pipe at -20°C. These test data and theoretical predictions indicate that the NASA Lewis extrusion would be an excellent candidate for any axially grooved heat-pipe application in which an aluminum pipe could be used. Potential fluids are ammonia, butane, ethane, methane, and hydrogen.

References

[1] Gaugler, R.S., "Capillary Heat Transfer Devices for Refrigerating Apparatus," U.S. Patent 2,448,261, disclosure date April 30, 1945.

[2] Thompson, D.C., Patent disclosure 48148, Feb. 10, 1960.

[3] Grover, G.M., Cotter, T.P., and Erickson, G.F., "Structures of Very High Thermal Conductance," Journal of Applied Physics, Vol. 35, 1964, p. 1990.

[4] Cotter, T.P., "Theory of Heat Pipes," Los Alamos Scientific Lab., LA-3246-MS, 1965.

[5] Grover, G.M., Bohdansky, J., and Busse, C.A., "The Use of a New Heat Removal System in Space Thermionic Power Supplies," Joint Nuclear Research Center, Ispra, Italy, EUR2229.E, 1965.

[6] Frank, S., "Optimization of Grooved Heat Pipe," Intersociety Energy Conversion Conference, 1967.

[7] Green, M., "An Improved Method of Calculating Frictional Pressure Losses Along Grooved Heat Pipes," Grumman Aerospace Corp., Rept. ADN04-02-69.1, 1969.

[8] Kosowski, N. and Kosson, R., "Experimental Performance of Grooved Heat Pipes at Moderate Temperatures," Progress in Astronautics and Aeronautics: Fundamentals of Spacecraft and Thermal Design, Vol. 29, edited by J.W. Lucas, AIAA, New York, 1972.

[9] Bilenas, J. and Harwell, W., "OAO Heat Pipes - Design Analysis and Testing," American Society of Mechanical Engineers, Paper 70-HT/SpT-9, 1970.

[10] Schlitt, K.R., Kirkpatrick, J.P., and Brennan, P.J., "Parametric Performance of Extruded Axial Grooved Heat Pipes from 100 to 300°K," Progress in Astronautics and Aeronautics: Heat Transfer with Thermal Control Applications, Vol. 39, edited by M.M. Yovanovich, AIAA, New York, 1975, pp. 215-234.

[11] McIntosh, R., Ollendorf, S., and Harwell, W., "International Heat Pipe Experiment," AIAA Paper 75-726, 1975.

[12] Harwell, W., Edelstein, F., McIntosh, R., and Ollendorf, S., "Orbiting Astronomical Observatory Heat Pipe Flight Performance Data," Progress in Astronautics and Aeronautics: Thermophysics and Spacecraft Thermal Control, Vol. 35, edited by R.G. Hering, AIAA, New York, 1974, pp. 445-465.

[13] Berger, M.D. and Kelly, W.H., "Application of Heat Pipes to the ATS-F Spacecraft," American Society of Mechanical Engineers, Paper 75-ENAS-46, 1975.

[14] Eggers, P.E., private communication.

[15] Brennan, P. J., Kroliczek, E. J., Jen, H., and McIntosh, R. "Axially Grooved Heat Pipes — 1976," AIAA Paper 77-747; also published elsewhere in this volume.

Chapter II—Heat Transfer

INFLUENCE OF REFRACTIVE INDEX ON EMITTANCE:
FINITE MEDIUM

B. F. Armaly*
University of Missouri, Rolla, Rolla, Mo.

Abstract

The directional and hemispherical emittances of an isothermal absorbing-emitting and isotropically scattering finite medium, which has a refractive index larger than unity, have been determined for a case in which the upper interface is smooth (specular) and the lower interface (substrate) is diffuse. An exponential kernel approximation is used to develop a closed-form approximate solution to the governing radiative transfer equation. Algebraic relations for the intensity, flux distribution through the medium, and emittance are presented as functions of the medium's refractive index, scattering albedo, optical thickness, and properties of the substrate. An increase in the medium's refractive index or optical thickness can either increase or decrease the apparent emittance, depending on the magnitudes of the scattering albedo and the substrate properties. Nongray behavior can be evaluated readily by utilizing the proposed approximate solution.

Introduction

The magnitude and the directional behavior of the radiant energy emitted from a semitransparent finite medium are two important parameters that are used in remote sensing and radiative transfer studies. They are affected by the medium's refractive index, the scattering and absorption coefficients, the optical thickness, the reflective and emissive

---

Presented as Paper 77-743 at the AIAA 12th Thermophysics Conference, Albuquerque, N. Mex., June 27-29, 1977. Copyright © American Institute of Aeronautics and Astronautics, Inc., 1977. All rights reserved.

*Professor of Mechanical Engineering, Member of Thermal Radiative Transfer Group, Department of Mechanical and Aerospace Engineering.

characteristics of the substrate, and the temperature distribution through the medium. Several studies have been conducted to examine the influence of scattering when the medium's refractive index is unity.[1-4] Others have been pursued for the purpose of examining the influences of refractive index and temperature distribution through a treatment of a nonscattering isothermal[5,6] and nonisothermal[7,8] medium. In many applications where the scattering particles are embedded in a medium whose refractive index is greater than air, such as colloids in suspensions and pigments in glass, plastics, and surface coatings, both the scattering and the medium's refractive index play important roles.

The emittance from an isothermal isotropically scattering finite medium with a refractive index greater than unity and two specular boundaries was examined by Turner and Love[9] and Merriam and Viskanta.[10] The nonisothermal case has been investigated by Roux and Smith.[11] The emittance of the substrate has not been accounted for in these studies. The investigators employed a numerical solution, which, when applied to a nongray radiative transfer problem, becomes lengthy and complicated. The objectives of the present study are to develop an approximate, closed-form solution for the intensity and flux distribution and for the emittance so that it can be seen how the various optical parameters affect the results

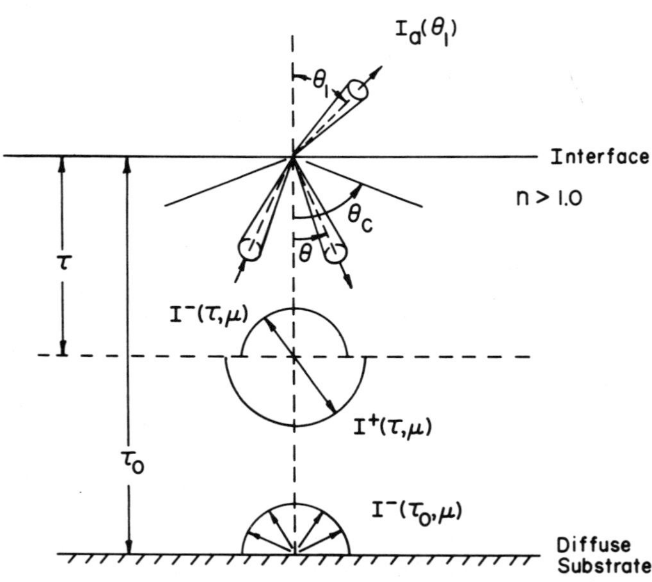

Fig. 1 Schematic of coordinate system.

and to reduce the required calculation time for a nongray problem significantly.

Formulation

The system considered in the present study is shown schematically in Fig. 1. It is equivalent to an isothermal absorbing-emitting and isotropically scattering layer of a semitransparent coating on a diffuse substrate. It is bounded from above by either air or a vacuum (a medium with a unity refractive index) and is not subjected to any external radiation. This upper interface is considered to be smooth, and its reflection and refraction characteristics are governed by Snell's law and the Fresnel relations. The substrate is at the same temperature as the coating, and it emits and reflects radiation diffusely. The following development is applicable on a monochromatic basis and can be used directly for a gray medium.

The radiation intensity distribution within the finite medium is governed by[12]

$$I^+(\tau,\mu) = I^+(0,\mu) \exp \frac{-\tau}{\mu}$$
$$+ \int_0^\tau S(t,\tau_0) \exp\left[-\frac{(\tau-t)}{\mu}\right] \frac{dt}{\mu} \tag{1}$$

and

$$I^-(\tau,\mu) = I^-(\tau_0,\mu) \exp\left[\frac{-(\tau_0-\tau)}{\mu}\right]$$
$$+ \int_\tau^{\tau_0} S(t,\tau_0) \exp\left[\frac{-(t-\tau)}{\mu}\right] \frac{dt}{\mu} \tag{2}$$

in which $S(t,\tau_0)$ is the isotropic source function given by

$$S(\tau,\tau_0) = (1-\omega)n^2 I_b(T) + \frac{\omega}{2}\left[\int_0^1 I^+(\tau,\mu)\, d\mu\right.$$
$$\left.+ \int_0^1 I^-(\tau,\mu)\, d\mu\right] \tag{3}$$

The terms $I^+(\tau,\mu)$ and $I^-(\tau,\mu)$ represent the intensities in the positive and negative directions, respectively. The symbol τ represents the optical depth: $\tau = \int_0^x \beta dx$; β is the extinction coefficient. The symbol τ_0 signifies the optical thickness: $\tau_0 = \int_0^L \beta dx$; L is the physical thickness of the medium, and θ is the polar direction measured from the vertical axis with $\mu = \cos\theta$. The symbol ω is the scattering albedo, n the medium's refractive index, and $I_b(T)$ the blackbody intensity at temperature T. For a smooth upper interface, the intensity at the boundary is given by

$$I^+(0,\mu) = \rho(\mu)\, I^-(0,\mu) \tag{4}$$

and the intensity leaving the diffuse substrate is given by

$$I^-(\tau_0,\mu) = J(\tau_0)/\pi \tag{5}$$

The symbol $\rho(\mu)$ is the interface reflectance, which results from radiation incident from the medium in the direction of μ governed by the Fresnel relations, and $J(\tau_0)$ is the radiosity of the diffuse substrate, which is given by

$$J(\tau_0) = n^2\,\varepsilon_b\pi I_b(T) + 2\pi\rho_b\!\int_0^1 I^+(\tau_0,\mu)\mu d\mu \tag{6}$$

The symbols ε_b and ρ_b represent the hemispherical emittance and reflectance of the substrate, respectively. For an opaque emitting and reflecting substrate (OERS), these properties are related by

$$\varepsilon_b = 1 - \rho_b \tag{7}$$

For a transparent nonemitting but reflecting substrate (TNERS), the reflectance is related to the transmittance τ_b by

$$\rho_b = 1 - \tau_b \tag{8}$$

By making appropriate substitutions in Eq. (3), the source function can be expressed by

$$[S(\tau,\tau_0)/n^2 I_b(T)] = (1-\omega)\Phi(\tau,\tau_0)$$

$$+ \omega\psi(\tau,\tau_0)[J(\tau_0)/n^2\pi I_b(T)] \tag{9}$$

in which

$$\Phi(\tau,\tau_0) = 1 + \frac{\omega}{2}\int_0^{\tau_0}\Phi(t,\tau_0)[R_1(\tau+t)$$

$$+ E_1(|\tau-t|)]dt \tag{10}$$

and

$$\psi(\tau,\tau_0) = \frac{1}{2}\,[R_2(\tau_0+\tau) + E_2(\tau_0-\tau)]$$

$$+ \frac{\omega}{2}\int_0^{\tau_0}\psi(t,\tau_0)[R_1(\tau+t) + E_1(|\tau-t|)]dt \tag{11}$$

The functions $E_1(t)$ and $E_2(t)$ are the exponential integrals, and $R_1(t)$ and $R_2(t)$ are functions associated with the interface reflectance defined by

$$R_i(t) = \int_0^1 \rho(\mu) \exp\left(\frac{-t}{\mu}\right)\mu^{i-2}d\mu; \quad i=1,2,3\ldots \tag{12}$$

If the refractive index of the medium is unity, the interface function $R_i(t)$ diminishes, and the equations governing the source function reduce to the simpler and more familiar form.[13] The dimensionless radiosity, $\bar{J}(\tau_0) = J(\tau_0)/n^2\pi I_b(T)$, can be expressed in terms of the preceding functions by

$$\bar{J}(\tau_0) = \frac{[\varepsilon_b + 2(1-\omega)\rho_b K_1(\tau_0)]}{[1 - 2\rho_b R_3(2\tau_0) - 2\omega\rho_b K_2(\tau_0)]} \tag{13}$$

in which

$$K_1(\tau_0) = \int_0^{\tau_0} \phi(t,\tau_0)[R_2(\tau_0 + t) + E_2(\tau_0 - t)]dt \tag{14}$$

and

$$K_2(\tau_0) = \int_0^{\tau_0} \psi(t,\tau_0)[R_2(\tau_0 + t) + E_2(\tau_0 - t)]dt \tag{15}$$

The intensity distribution within the medium and the emittance from the medium can be calculated once the solution to the source function, Eq. (9), is found. The directional emittance $\varepsilon(\mu_1)$ of the medium is given by

$$\varepsilon(\mu_1) = [1-\rho(\mu)]I^-(0,\mu)/n^2 I_b(T) \tag{16}$$

The symbol μ_1 is equal to $\cos\theta_1$. In this expression, θ_1 is the direction as measured from the normal on the upper side of the interface. The directions on both sides of the interface are related by Snell's law:

$$\mu^2 = 1-(1-\mu_1^2)/n^2 \tag{17}$$

The hemispherical emittance ε_H is given by

$$\varepsilon_H = 2\int_0^1 \varepsilon(\mu_1)\mu_1 d\mu_1 \tag{18}$$

Approximate Solution

The exact numerical solutions for Eqs. (10, 11, and 13) can be obtained; however, they are not presently available in the published literature. The applications of these solutions to determine the intensity and flux distribution in a nongray problem could become quite lengthy and complicated, because they require the repeated solution of integral equations at each wavelength and then integration over the entire spectrum. For this reason, an approximate but closed-form solution has been developed. This was done by using an exponential approximation for the interface functions and the exponential integral. This approximate solution provides algebraic expressions for the intensity distribution through the medium. From this, the emittance and the flux distribution can be evaluated readily. This method, which was used previously by Armaly et al.[14] on a semi-infinite medium, provides reasonable approximations whose errors are smaller than 10% in the hemispherical emittance. This study is, in effect, an extension of the previous one,[14] which is applicable to a finite medium.

A common exponential approximation for the exponential integral is given by

$$E_2(t) \approx \exp(-qt) \tag{19}$$

in which q is a constant equal to 2 or to $\sqrt{3}$. The value of q = 2 is used in the present study, because it provides a

Table 1 Coefficients for approximating the interface functions $R_2(t)$, $B_2(t)$

n	a_1	b_1	a_2	b_2	a_3	b_3
1.33	0.365	6.443	0.322	1.780	0.316	1.18
1.4	0.378	6.241	0.347	1.700	0.271	1.16
1.5	0.400	5.898	0.368	1.610	0.228	1.13
1.6	0.413	5.735	0.391	1.550	0.196	1.11
1.7	0.428	5.514	0.407	1.495	0.168	1.10
1.8	0.435	5.435	0.419	1.465	0.143	1.08
1.9	0.442	5.361	0.431	1.440	0.128	1.07
2.0	0.453	5.210	0.435	1.412	0.114	1.07

better approximation for the case of unity refractive index.[3] Similarly, the function $R_2(t)$, which appears in Eq. (11), is approximated by two exponential terms:

$$R_2(t) \simeq a_1 \exp(-b_1 t) + a_2 \exp(-b_2 t) \qquad (20)$$

One exponential term produces a poor approximation for $R_2(t)$. The constants a_1, b_1, a_2, and b_2 are functions of the medium's refractive index and were evaluated by equating the function $R_2(t)$ and its three moments at zero to the approximate function and its three moments, producing four equations and four unknowns. This approach is similar to but produces a better approximation than the one utilized earlier by Armaly et al.[14] The coefficients are tabulated in Table 1 as a function of the refractive index and produce an error of less than 20% in the range of optical thicknesses $0<\tau<15$. This magnitude of error is much smaller than the errors from the exponential approximation for the exponential integral, which has been used frequently in similar radiative transfer problems.

An additional function, which will appear later in the development, is defined by

$$B_i(t) = \int_0^1 [1-\rho(\mu)] \, \exp\!\left(\frac{-t}{\mu}\right)\mu^{i-2} d\mu; \quad i=1,2,3... \qquad (21)$$

This set of functions accounts for the interface transmittance of radiant energy from the upper interface of the medium and is approximated by one exponential term. The approximation for $B_2(t)$ is given by

$$B_2(t) \simeq a_3 \exp(-b_3 t) \qquad (22)$$

This approximation provides errors of smaller than 5% in the optical thickness range between $0<\tau<15$. The constants a_3 and b_3 are tabulated in Table 1 as a function of the medium's refractive index and were evaluated by a method similar to the one described above for $R_2(t)$. The three approximations for $E_2(t)$, $R_2(t)$, and $B_2(t)$ are needed in order to provide a closed-form solution to the present problem. Approximate expressions for either $E_1(t)$, $R_1(t)$, $B_1(t)$ or $E_3(t)$, $R_3(t)$, and $B_3(t)$ that appear later in the development are obtained by either integrating or differentiating the approximate relation for $E_2(t)$, $R_2(t)$, and $B_2(t)$.

The proposed approximation for the exponential integral and the interface functions is used to transform the kernel of the integral equations, Eqs. (10) and (11), to a separable kernel. A standard solution procedure then is used to obtain a closed-form approximate solution to these integral equations. These solutions are

$$\Phi(\tau,\tau_0) = C_1\exp(-b\tau) + C_2\exp(b\tau)$$
$$+ C_3\exp(-b_1\tau) + C_4\exp(-b_2\tau) + 1/(1-\omega) \qquad (23)$$

and

$$\psi(\tau,\tau_0) = C_5\exp(-b\tau) + C_6\exp(b\tau)$$
$$+ C_7\exp(-b_1\tau) + C_8\exp(-b_2\tau) \qquad (24)$$

in which

$$b^2 = q^2(1-\omega) \qquad (25)$$

The coefficients C_1 through C_8 are governed by relations described in the Appendix.

The solutions for the source functions now can be used with the proposed approximations for the interface functions and the exponential integral to develop closed-form solutions for the intensity and flux distribution through the medium and for the emittance from the medium. These solutions are

$$I^+(\tau,\mu)/n^2 I_b(T) = \overline{J}(\tau_0)\{\rho(\mu)\ \exp[-(\tau_0+\tau)/\mu]$$
$$+ (\omega/\mu)\ \exp(-\tau/\mu)[\rho(\mu)Z_2(0,\tau_0,1/\mu) +P_2(\tau,1/\mu)]\}$$
$$+ [(1-\omega)/\mu]\ \exp(-\tau/\mu)\ [\rho(\mu)Z_1(0,\tau_0,1/\mu)$$
$$+ P_1(\tau,1/\mu)] \qquad (26)$$

$$I^-(\tau,\mu)/n^2 I_b(T) = \overline{J}(\tau_0)\ \exp(\tau/\mu)\{\exp(-\tau_0/\mu)$$
$$+ (\omega/\mu)Z_2(\tau,\tau_0,1/\mu)\}$$
$$+ [(1-\omega)/\mu]\ \exp(\tau/\mu)Z_1(\tau,\tau_0,1/\mu) \qquad (27)$$

$$F^+(\tau,\tau_0)/n^2\pi I_b(T) = 2\ \overline{J}(\tau_0)\{R_3(\tau_0+\tau)$$
$$+ \omega[a_1\ \exp(-b_1\tau)Z_2(0,\tau_0,b_1)$$
$$+ a_2\ \exp(-b_2\tau)Z_2(0,\tau_0,b_2) + \exp(-q\tau)P_2(\tau,q)]\}$$

$$+ 2(1-\omega)[a_1 \exp(-b_1\tau)Z_1(0,\tau_0,b_1)$$

$$+ a_2 \exp(-b_2\tau)Z_1(0,\tau_0,b_2) + \exp(-q\tau)P_1(\tau,q)] \tag{28}$$

$$F^-(\tau,\tau_0)/n^2 \pi I_b(T) = 2 \bar{J}(\tau_0)\{E_3(\tau_0-\tau)$$

$$+ \omega \exp(q\tau)Z_2(\tau,\tau_0,q)\}$$

$$+ 2(1-\omega) \exp(q\tau)Z_1(\tau,\tau_0,q) \tag{29}$$

$$\varepsilon(\mu_1) = [1 -\rho(\mu)]\{\bar{J}(\tau_0)[\exp(-\tau_0/\mu)$$

$$+ (\omega/\mu)Z_2(0,\tau_0,1/\mu)]$$

$$+ [(1-\omega)/\mu]Z_1(0,\tau_0,1/\mu)\} \tag{30}$$

$$\varepsilon_H = 2\bar{J}(\tau_0)n^2[B_3(\tau_0) + a_3\omega Z_2(0,\tau_0,b_3)]$$

$$+ 2(1-\omega)n^2 a_3 Z_1(0,\tau_0 b_3) \tag{31}$$

The functions $K_1(\tau_0)$ and $K_2(\tau_0)$ that appear in Eqs. (13) and (14), respectively, become equivalent to

$$K_1(\tau_0) = a_1 \exp(-b_1\tau_0)Z_1(0,\tau_0,b_1)$$

$$+ a_2 \exp(-b_2\tau_0)Z_1(0,\tau_0,b_2) + \exp(-q\tau_0)P_1(\tau_0,q) \tag{32}$$

and

$$K_2(\tau_0) = a_1 \exp(-b_1\tau_0)Z_2(0,\tau_0,b_1)$$

$$+ a_2 \exp(-b_2\tau_0)Z_2(0,\tau_0,b_2) + \exp(-q\tau_0)P_2(\tau_0,q) \tag{33}$$

The functions Z and P appearing in the preceding expressions are given by

$$Z_1(\tau,\tau_0,x) = \int_\tau^{\tau_0} \Phi(t) \exp(-tx)\, dt \tag{34}$$

$$Z_2(\tau,\tau_0,x) = \int_\tau^{\tau_0} \psi(t) \exp(-tx)\, dt \tag{35}$$

$$P_1(\tau,x) = \int_0^\tau \Phi(t) \exp(tx)\, dt \tag{36}$$

$$P_2(\tau,x) = \int_0^\tau \psi(t) \exp(tx)\, dt \tag{37}$$

The algebraic equivalents for these functions are given in
the Appendix.

Results and Discussion

The proposed approximate solutions are used to examine
the influences of both the refractive index and the properties
of the substrate on the directional and hemispherical emittance of the medium. In the research, two types of diffuse
substrates were examined: an opaque emitting and reflecting
substrate (OERS) and a transparent nonemitting but reflecting
substrate (TNERS). The directional emittance is presented
in Figs. 2 and 3 for a medium with an optical thickness of
unity. This property is directly proportional to the interface transmittance which decreases with angle (rapidly beyond
60 deg) and with medium's refractive index. The scattering
contributions to the magnitude of the directional emittance,
however, increases as the angle increases. This is due to
the increase in the optical depth in that direction which,
for the case of the (TNERS) substrate, causes the emittance to
increase. The two opposing trends of the interface transmittance and the scattering contributions can in some cases

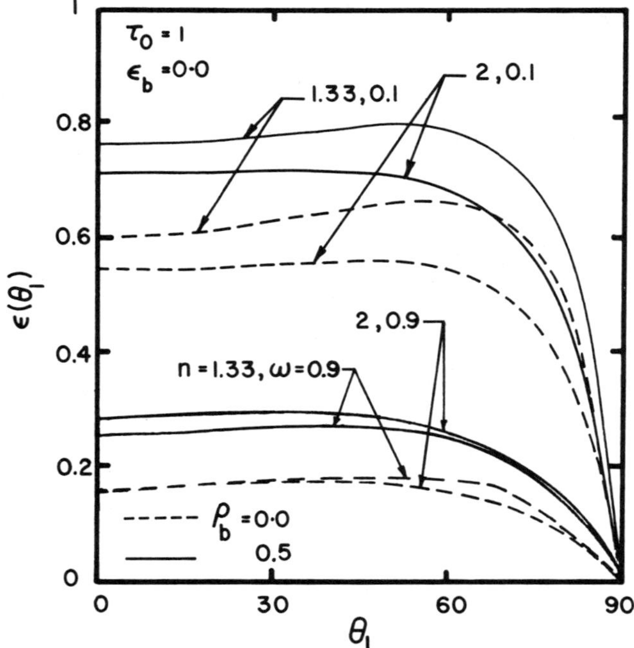

Fig. 2 Directional emittance (TNERS).

cause the directional emittance to increase with angle as can be seen in Fig. 2. The behavior of the interface transmittance dominates at angles larger than 60 deg. For a given refractive index and substrate reflectance, the directional emittance decreases as the scattering albedo increases, because the backscattering increases, and, as expected, its magnitude for an OERS substrate is always higher than the corresponding TNERS substrate case.

The ratio of the directional emittance at any angle to the one at an angle of zero (normal emittance ε_N), $\varepsilon(\theta_1)/\varepsilon_N$, appears to be almost (errors less than 10%) independent of the medium's refractive index, scattering albedo, optical thickness, and properties of the substrate. This fact can be used with Figs. 4 and 5, in which the normal emittance is presented as a function of optical thickness, to approximate the directional emittance from a medium having an optical thickness other than unity.

When the optical thickness is equal to zero, which is equivalent to a nonscattering and nonabsorbing medium, the

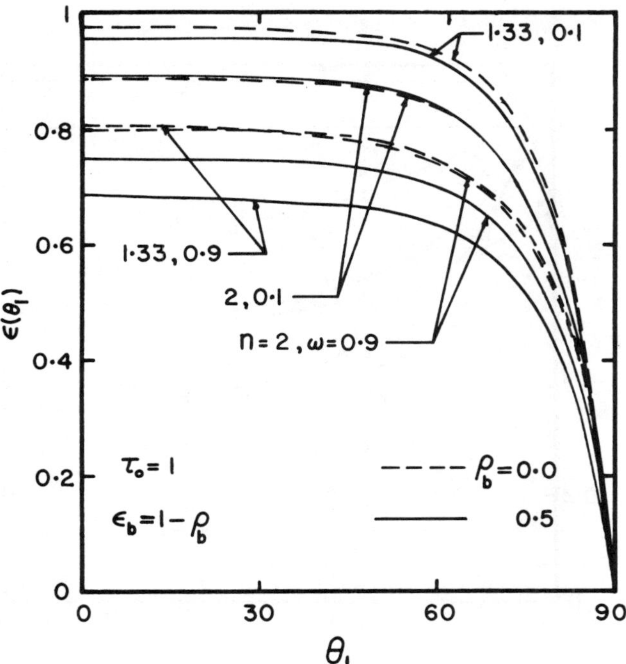

Fig. 3 Directional emittance (OERS).

B. F. ARMALY

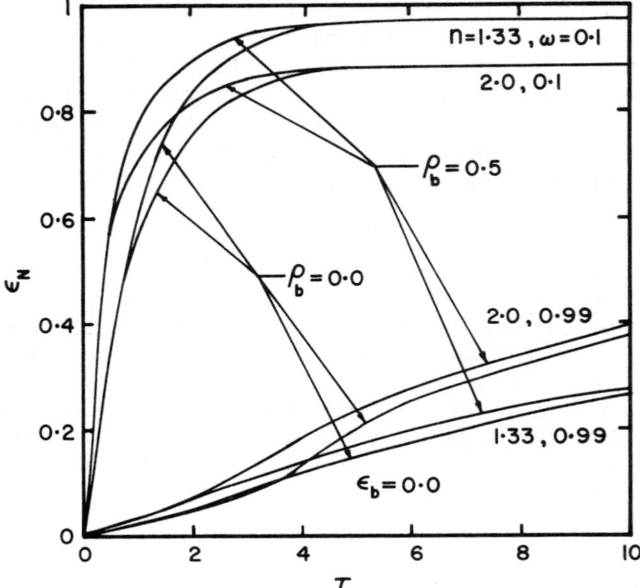

Fig. 4   Normal emittance (TNERS).

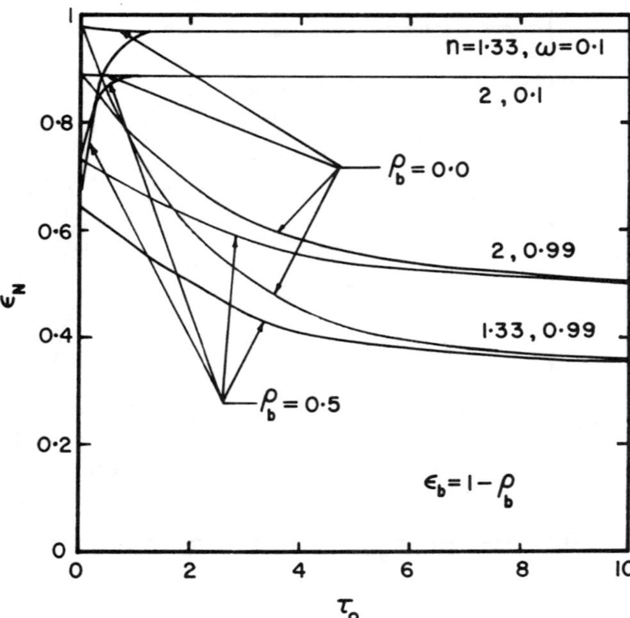

Fig. 5   Normal emittance (OERS).

interface causes the normal emittance to become equivalent
to

$$\varepsilon_N\big|_{\tau_o=0} = [1-\rho(0)]\varepsilon_b/[1-\rho\rho_b] \tag{38}$$

where $\rho(0)$ is the normal reflectance and ρ is the hemispherical reflectance of the interface back to the medium.

The hemispherical emittance is presented in Figs. 6 and
7. When the optical thickness is equal to zero, it becomes
equivalent to

$$\varepsilon_H\big|_{\tau_o=0} = n^2\varepsilon_b(1-\rho)/(1-\rho\rho_b) \tag{39}$$

This magnitude could be interpreted as being due to only the
refractive index and the substrate emittance. As the role of
both scattering and attenuation increases (larger optical
thickness and scattering albedo), the hemispherical emittance
can either increase or decrease to its limiting value of the
semi-infinite medium.[14] The magnitude of the optical thickness at which the predicted emittance is equivalent to that
of a semi-infinite medium increases as the refractive index,
and the scattering albedo increases because these parameters

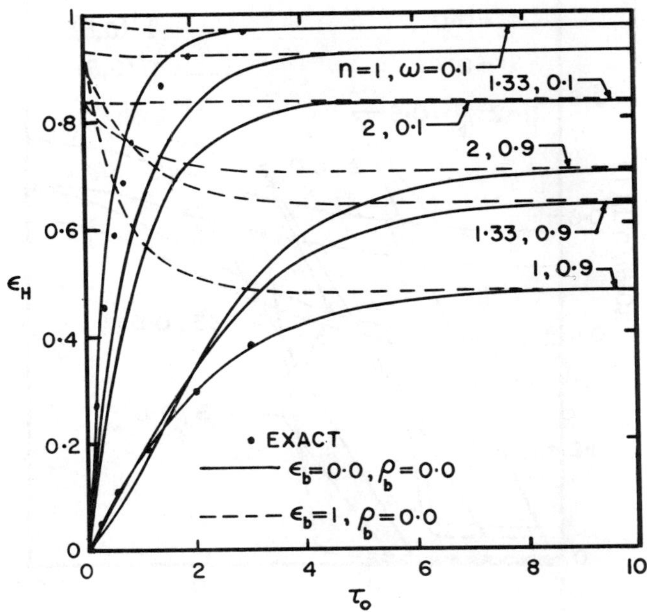

Fig. 6   Hemispherical emittance, ρ_b=0.0.

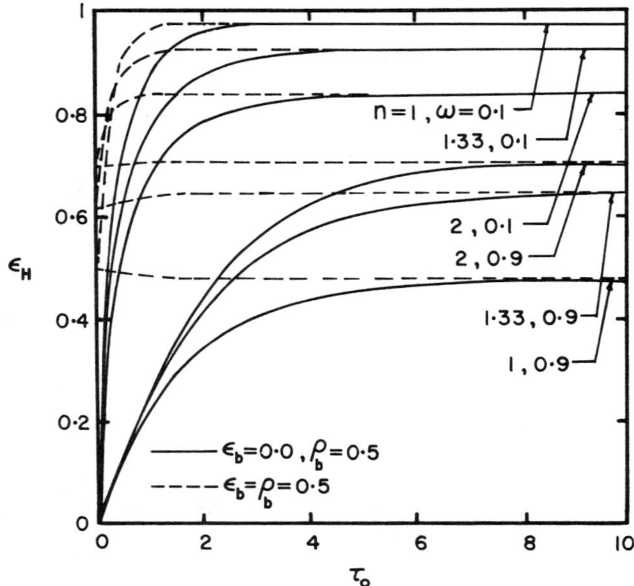

Fig. 7   Hemispherical emittance, $\rho_b=0.5$.

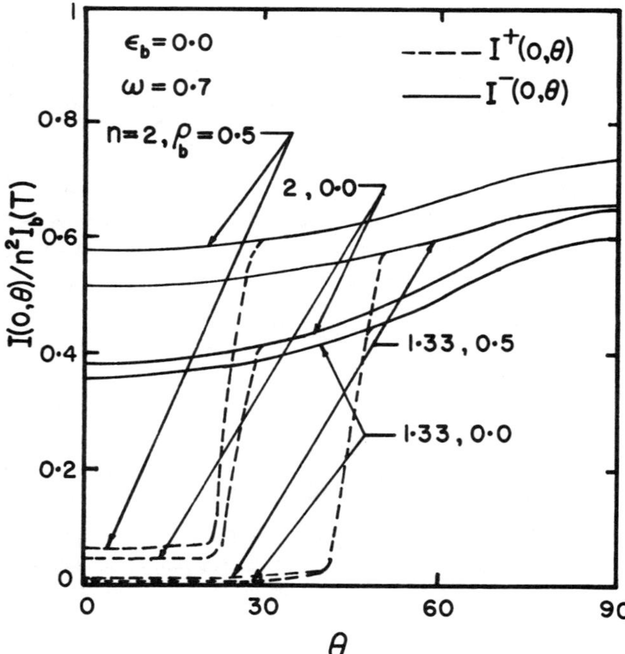

Fig. 8   Intensity at the interface.

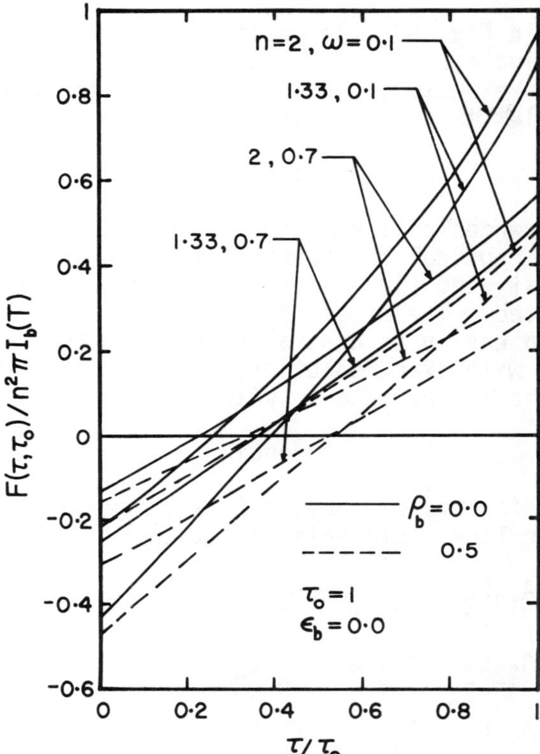

Fig. 9 Flux distribution.

influence the effective optical thickness of the medium. An increase in the refractive index causes a decrease in the magnitude of the hemispherical emittance when the scattering albedo is small, and the reverse is true when the scattering albedo is large. Results for the case when the medium's refractive index is unity are included for comparison with the available exact solution.[2] Errors between the exact and the approximate values are smaller than 15% over a wide range of optical parameters.

The intensity at the interface, $\tau=0.0$, within the medium is presented in Fig. 8 as a function of directions. It is clear from this figure that a significant portion of the energy within the critical angle is transmitted across the interface. An increase in the refractive index decreases the transmitted fraction. The energy outside the critical angle is totally reflected back into the medium. The flux distribution given by

$$F(\tau,\tau_o)/n^2\pi I_b(T) = [F^+(\tau,\tau_o) - F^-(\tau,\tau_o)]/n^2\pi I_b(T) \qquad (40)$$

is presented in Fig. 9. The positive flux $F^+(\tau,\tau_0)$ increases, and the negative flux $F^-(\tau,\tau_0)$ decreases as the optical depth increases. As expected, because energy is leaving the medium at $\tau=0$, the flux is negative at the interface.

Conclusions

The influences of the medium's refractive index, scattering albedo, optical thickness, and substrate properties on the intensity and flux distribution and on the emittance are demonstrated. The closed-form solution is simple and can be applied easily to nongray problems. From comparisons when possible with exact solutions, the approximate results are believed to be accurate to within 20%.

Appendix

The coefficients C_1 through C_4 are governed by the following four linear algebraic relations:

$$A_{11}C_1 + A_{12}C_2 + A_{13}C_3 + A_{14}C_4 = G_1$$

$$A_{21}C_1 + A_{22}C_2 + A_{23}C_3 + A_{24}C_4 = G_2$$

$$A_{31}C_1 + A_{32}C_2 + A_{33}C_3 + A_{34}C_4 = G_3$$

$$A_{41}C_1 + A_{42}C_2 + A_{43}C_3 + A_{44}C_4 = G_4$$

The coefficients C_5 through C_8 are governed by the following four linear algebraic relations:

$$A_{11}C_5 + A_{12}C_6 + A_{13}C_7 + A_{14}C_8 = H_1$$

$$A_{21}C_5 + A_{22}C_6 + A_{23}C_7 + A_{24}C_8 = H_2$$

$$A_{31}C_5 + A_{32}C_6 + A_{33}C_7 + A_{34}C_8 = H_3$$

$$A_{41}C_5 + A_{42}C_6 + A_{43}C_7 + A_{44}C_8 = H_4$$

The only difference between the first four and second four algebraic equations is in the nonhomogeneous terms. The coefficients in the preceding equations are given below:

$$A_{11} = \{(1-\exp[-(b+b_1)\tau_0])/(b+b_1)\}$$

$$A_{12} = \{(\exp[(b-b_1)\tau_0]-1)/(b-b_1)\}$$

$$A_{13} = \{[[1-\exp(-2b_1\tau_0)]/2b_1] - [2q^2/a_1b_1(b_1^2-q^2)]$$
$$- [2/\omega a_1 b_1]\}$$

$$A_{14} = \{[1-\exp[-(b_1+b_2)\tau_0]]/(b_1+b_2)\}$$

$$A_{21} = \{[1-\exp[-(b+b_2)\tau_0]]/(b+b_2)\}$$

$$A_{22} = \{[\exp[(b-b_2)\tau_0]-1]/(b-b_2)\}$$

$$A_{23} = \{[1-\exp[-(b_1+b_2)\tau_0]]/(b_1+b_2)\}$$

$$A_{24} = \{[[1-\exp(-2b_2\tau_0)]/(2b_2)] - [2q^2/a_2b_2(b_2^2-q^2)]$$
$$- [2/\omega a_2 b_2]\}$$

$$A_{31} = \{\exp(-2b\tau_0)/(b+q)\}$$

$$A_{32} = -\{1/(b-q)\}$$

$$A_{33} = \{\exp[-(b+b_1)\tau_0]/(b_1+q)\}$$

$$A_{34} = \{\exp[-(b+b_2)\tau_0]/(b_2+q)\}$$

$$A_{41} = 1/(b-q)$$

$$A_{42} = -[1/(b+q)]$$

$$A_{43} = 1/(b_1-q)$$

$$A_{44} = 1/(b_2-q)$$

$$H_1 = -\exp(-b_1\tau_0)/\omega b_1$$

$$H_2 = -\exp(-b_2\tau_0)/\omega b_2$$

$$H_3 = \exp(-b\tau_0)/\omega q$$

$$H_4 = 0$$

$$G_1 = \{[\exp(-b_1\tau_0)-1]/b_1(1-\omega)\}$$

$$G_2 = \{[\exp(-b_2\tau_0)-1]/b_2(1-\omega)\}$$

$$G_3 = -\{\exp(-b\tau_0)/q(1-\omega)\}$$

$$G_4 = 1/(1-\omega)q$$

The functions Z_1 and Z_2 are given by

$$Z_1(\tau,\tau_0,x) = C_1 W_1(\tau,\tau_0,x) + C_2 W_2(\tau,\tau_0,x)$$
$$+ C_3 W_3(\tau,\tau_0,x) + C_4 W_4(\tau,\tau_0,x) + W_5(\tau,\tau_0,x)$$

$$Z_2(\tau,\tau_0,x) = C_5 W_1(\tau,\tau_0,x) + C_6 W_2(\tau,\tau_0,x)$$
$$+ C_7 W_3(\tau,\tau_0,x) + C_8 W_4(\tau,\tau_0,x)$$

in which

$$W_1(\tau,\tau_0,x) = \{\exp[-(b+x)\tau] - \exp[-(b+x)\tau_0]\}/(b+x)$$

$$W_2(\tau,\tau_0,x) = \{\exp[(b-x)\tau_0] - \exp[(b-x)\tau]\}/(b-x)$$

$$W_3(\tau,\tau_0,x) = \{\exp[-(b_1+x)\tau] - \exp[-(b_1+x)\tau_0]\}/(b_1+x)$$

$$W_4(\tau,\tau_0,x) = \{\exp[-(b_2+x)\tau]$$
$$- \exp[-(b_2+x)\tau_0]\}/(b_2+x)$$

$$W_5(\tau,\tau_0,x) = \{\exp(-\tau x) - \exp(-\tau_0 x)\}/(1-\omega)x$$

The functions P_1 and P_2 are given by

$$P_1(\tau,x) = C_1 Y_1(\tau,x) + C_2 Y_2(\tau,x) + C_3 Y_3(\tau,x)$$
$$+ C_4 Y_4(\tau,x) + Y_5(\tau,x)$$

$$P_2(\tau,x) = C_5 Y_1(\tau,x) + C_6 Y_2(\tau,x) + C_7 Y_3(\tau,x)$$
$$+ C_8 Y_4(\tau,x)$$

in which

$$Y_1(\tau,x) = \{1 - \exp[-(b-x)\tau]\}/(b-x)$$

$$Y_2(\tau,x) = \{\exp[(b+x)\tau] - 1\}/(b+x)$$

$$Y_3(\tau,x) = \{1 - \exp[-(b_1-x)\tau]\}/(b_1-x)$$

$$Y_4(\tau,x) = \{1 - \exp[-(b_2-x)\tau]\}/(b_2-x)$$

$$Y_5(\tau,x) = \{\exp(\tau x) - 1\}/[x(1-\omega)].$$

Acknowledgment

This work was supported in part by the National Science Foundation, ENG 75-06237.

References

[1] Edward, R. H. and Bobco, R. P., "Radiant Heat Transfer from Isothermal Dispersion with Isotropic Scattering," *Journal of Heat Transfer*, Ser. C, Vol. 87, Nov. 1967, p. 300.

[2] Crosbie, A. L., "Emittance of an Isothermal, Isotropic Scattering Medium," *AIAA Journal*, Vol. 11, Aug. 1973, p. 1203.

[3] Armaly, B. F. and El-Baz, H. S., "Influence of Substrate Properties on the Apparent Emittance of an Isothermal, Isotropically Scattering Medium," *Journal of Heat Transfer*, Vol. 18, Oct. 1977, pp. 419-424.

[4] Dayan, A. and Tien, C. L., "Radiative Transfer with Anisotropic Scattering in an Isothermal Slab," *Journal of Quantitative Spectroscopy and Radiative Transfer*, Vol. 16, Feb. 1976, p. 113.

[5] Francis, J. E. and Love, T. J., "Effect of Optical Thickness on the Directional Transmittance and Emittance of a Dielectric," *Journal of the Optical Society of America*, Vol. 56, June 1966, p. 779.

[6] Viskanta, R. and Johnson, R. O., "Directional and Hemispherical Radiation Characteristics of Isothermal, Semi-Transparent Plates," *Proceedings of the Second National Indian Heat and Mass Transfer Conference*, Kanpur, India, 1973, p. C-1.

[7] Chupp, R. E. and Viskanta, R., "Thermal Emission Characteristics of a Nonisothermal Dielectric Coating on a Conductor Surface," *AIAA Journal*, Vol. 8, March 1970, pp. 551-557.

[8] Anderson, E. E., "Estimating the Effective Emissivity of Nonisothermal, Diathermanous Coating," *Journal of Heat Transfer*, Ser. C, Vol. 97, Aug. 1975, p. 480.

[9] Turner, W. D. and Love, T. J., "Directional Emittance of a One-Dimensional Absorbing-Scattering Slab with Reflecting Boundaries," *AIAA Progress in Astronautics and Aeronautics: Thermal Control and Radiation*, Vol. 31, edited by C. L. Tien, 1973, pp. 389-395.

[10] Merriam, R. L. and Viskanta, R., "Radiative Characteristics of Absorbing, Emitting and Scattering Media on Opaque Substrate," *Journal of Spacecraft and Rockets*, Vol. 5, Oct. 1968, pp. 1210-1215.

[11] Roux, J. A. and Smith, A. M., "Radiative Transport Analysis for Plane Geometry with Isotropic Scattering and Arbitrary Temperature," *AIAA Journal*, Vol. 12, Sept. 1974, pp. 1273-1277.

[12] Viskanta, R., "Radiation Transfer and Interaction of Convections with Radiation Heat Transfer," *Advances in Heat Transfer*, Vol. 3, edited by T. F. Irvine, Jr. and J. P. Hartnett, 1966, p. 175.

[13] Armaly, B. F. and El-Baz, H. S., "Radiative Source Functions for a Slab: Approximate Solution," *Journal of Quantitative Spectroscopy and Radiative Transfer*, Vol. 18, July 1977, pp. 65-68.

[14] Armaly, B. F., Lam, T. T., and Crosbie, A. L., "Emittance of Semi-Infinite Absorbing and Isotropically Scattering Medium with Refractive Index Greater than Unity," *AIAA Journal*, Vol. 11, Nov. 1973, pp. 1498-1502.

FINITE-ELEMENT METHODOLOGY FOR THERMAL ANALYSIS OF CONVECTIVELY COOLED STRUCTURES

Earl A. Thornton[*]
Old Dominion University, Norfolk, Va.

and

Allan R. Wieting[†]
NASA Langley Research Center, Hampton, Va.

Abstract

A finite-element method for steady-state thermal analysis of convectively cooled structures is presented. The method is based on representing the coolant passages by finite elements with fluid bulk temperature nodes and fluid/structure interface nodes. Four finite elements are described: two basic elements (mass transport and surface convection) and two integrated elements for applications to discrete tube and plate-fin convectively cooled structural configurations. Comparative finite-element and lumped-parameter thermal analyses demonstrate the practicality of utilizing finite-element methodology for thermal analysis of realistic convectively cooled structures.

Nomenclature

A	= flow cross-sectional area
A_s	= surface convection area
$[B]$	= finite-element interpolation matrix for temperature gradients
C_p	= coolant specific heat
h	= convective heat-transfer coefficient

Presented as Paper 77-187 at the AIAA 15th Aerospace Sciences Meeting, Los Angeles, Calif. Jan. 24-26, 1977.

[*]Associate Professor.

[†]Research Engineer, Structures and Dynamics Division.

$[H]$ = convective conductance matrix between a surface and a fluid with known temperature, defined in Eq. (B2)

k = thermal conductivity

$\underline{\underline{K}}, \underline{\underline{K}}_S, \underline{\underline{K}}_F$ = thermal conductivity tensor; subscripts S and F refer to the solid and fluid, respectively

$[K_h]$ = surface convection conductance matrix, defined in Eq. (6)

$[K_v]$ = mass transport conductance matrix, defined in Eq. (4)

ℓ = fin height; see Fig. 7

L = length

m = fin parameter, defined in Eq. (8)

$\dot{m}$ = fluid mass flow rate $\dot{m} = \rho A V$

n = normal direction to coolant passage surface

$[N]$ = finite-element interpolation matrix for temperature

q = heat flux

Q = finite-element nodal heat load (positive into node)

t = fin thickness

T = temperature

T_b = coolant bulk temperature

T_w = wall temperature of coolant passage

T_o = coolant entrance temperature, Fig. 7

T_1, T_2 = specified wall temperatures, Fig. 7

u, v, w = components of coolant velocity vector $\vec{V}$

x, y = Cartesian coordinates

$\vec{V}$ = coolant velocity vector

w = width of heat exchanger wall, Fig. 7

α = parameter defined in Fig. 7

ρ = fluid density

η = fin efficiency factor, defined in Eq. (7)

<u>Subscripts</u>

I, J, K, L, M, N = finite-element node numbers

Introduction

The NASA Langley Research Center is conducting research programs focused on the development of convectively cooled structures for high-speed aircraft[1] and scramjet propulsion systems.[2] Typical convectively cooled structural configurations are shown schematically in Fig. 1. Convectively cooled structures, which operate in a severely hostile thermal environment compared to conventional structures for low-speed

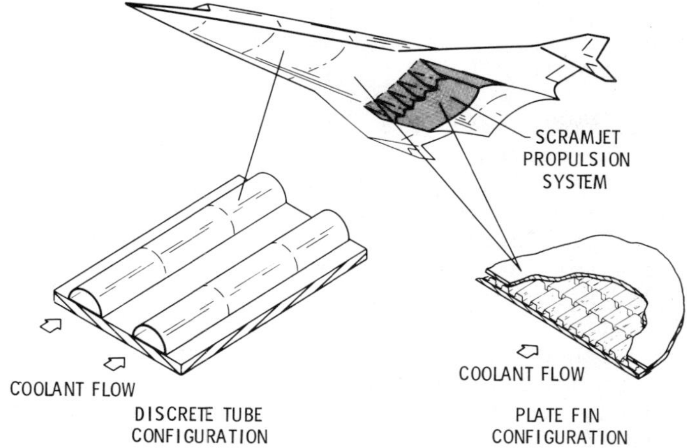

Fig. 1   Schematic of typical convectively cooled structural
configurations.

aircraft, require combined thermal and structural analyses to
obtain viable, optimal thermal-structural configurations.

Practical structures may be analyzed thermally using two
alternative methods: 1) the finite-difference lumped-
parameter technique such as used in the Martin Interactive
Thermal Analyses System (MITAS),[3] or 2) the finite-element
procedure such as used in NASTRAN.[4] The lumped-parameter
approach is well established and has been used successfully
on convectively cooled structures.[2] The finite-element procedure, a more recent development, is an attractive alternate
analysis technique because of its capability to perform both
thermal and structural analyses of general structures.

Frequently, combined thermal-structural analyses join a
lumped-parameter thermal analysis and a finite-element
structural analysis. However, because the analysis methods
differ, an efficient interface between the two solutions is
difficult to achieve. Since most structural analyses are
performed with the finite-element method, a unified integrated
analysis procedure based on this method is a desirable goal.

Most finite-element thermal computer programs do not have
the capability to analyze convectively cooled structures. In
past applications of the finite-element method to convectively
cooled structures,[5-7] the coolant fluid has been modeled as a
convective boundary condition with specified coolant temperatures and convective film coefficients. This procedure is not
a practical alternative, in general, because the coolant

temperature is usually unknown and the convective film coefficient may be strongly dependent on the coolant temperature and
other flow parameters.

A literature search revealed that the finite-element
methodology for convection heat transfer had not been developed.
Recently, it has come to the authors' attention that a pro-
prietary finite-element thermal analyzer, ANSYS (Swanson
Analysis Systems, Inc., Elizabeth, Pa. 15037), has a thermal-
flow pipe element that may be used to analyze some convectively cooled structures. The development of the ANSYS
thermal-flow pipe element has not been documented in the open
literature. Consequently, a program was undertaken to develop
finite-element methodology for steady-state thermal analyses
of convectively cooled structures. The analytical requirements for the previously mentioned programs[1,2] served as the
focal point for this study. However, the results are applicable to a broad class of convection systems.

This paper presents 1) concepts for the finite-element
representation of the convective heat-transfer modes, 2) four
new convective finite elements, and 3) comparative finite-
element and lumped-parameter analyses to demonstrate and
verify the new analysis capability.

Methodology Development

Finite-Element Formulation

Thermal analysis of convectively cooled structures includes coupled conduction and convective heat transfer in a
region consisting of a solid structure and a moving fluid.
The problem may be formulated mathematically in terms of the
energy equations of the solid and fluid assuming incom-
pressible flow.[8] For steady-state heat transfer and by
neglecting viscous energy dissipation in the fluid, the
temperature $T(x,y,z)$ satisfies

$$\vec{\nabla} \cdot (\underline{\underline{K}}_S \cdot \vec{\nabla}T) = 0 \tag{1}$$

for the solid region and

$$\rho c_p \, \vec{\nabla} \cdot \vec{\nabla}T - \vec{\nabla} \cdot (\underline{\underline{K}}_F \cdot \vec{\nabla}T) = 0 \tag{2}$$

for the fluid region. The thermal properties of the solid
($\underline{\underline{K}}_S$) and fluid ($\underline{\underline{K}}_F$ and c_p) are temperature dependent. The
velocity vector $\vec{V}$ [Eq. (2)] specifies the fluid motion as
a function of the spatial coordinates and, in general, is un-

known. Equations (1) and (2) must be solved subject to specified boundary conditions on the external surfaces of the region and appropriate continuity conditions at the solid/fluid interface.

In the most general approach to thermal analysis of convectively cooled structures, the steady-state velocity distribution of the fluid first is determined by solving the continuity and momentum equations of fluid flow. With the fluid velocity distribution known, Eqs. (1) and (2) then are solved simultaneously for the temperature distribution in the solid/fluid region. Combined flow-thermal finite-element analyses[8] have been performed for some highly idealized cooled structures. In these analyses, the fluid region is subdivided across its width and downstream directions into a number of elements connecting nodes where the fluid velocity components are computed in the flow analysis. In realistic convectively cooled structures, however, a complete flow analysis is impractical because of the great complexity of the fluid flow in the coolant passages.

In the present approach, a simplified finite-element solution procedure is developed by employing a number of assumptions customarily used in practical heat-transfer analysis.[9] The thrust of the assumptions is to eliminate the computation of the fluid velocity distribution. The assumptions are as follows: 1) The thermal energy state of the fluid is characterized by the fluid bulk temperature, which varies only in the flow direction; 2) The fluid velocity is represented by a mean velocity V, which varies only in the flow direction. The mass flow rate in a coolant passage then is given by $\dot{m} = \rho A V$, where A is the flow cross-sectional area; 3) Fluid conduction heat transfer in the flow direction is negligible compared to heat transfer by mass transport; 4) A convection heat-transfer coefficient h is defined such that the heat transfer between the coolant passage surface and the coolant is expressed by

$$q = h \left(T_w - T_b \right) = - k \left. \frac{\partial T}{\partial n} \right|_{\text{wall}}$$

The convection heat-transfer coefficient is input as a function of the fluid bulk temperature.

These assumptions permit a single finite element to represent the total cross-sectional area of the flow passage. A typical fluid element thus is characterized by fluid bulk temperature nodes and fluid/solid interface nodes that connect to elements representing the solid structure.

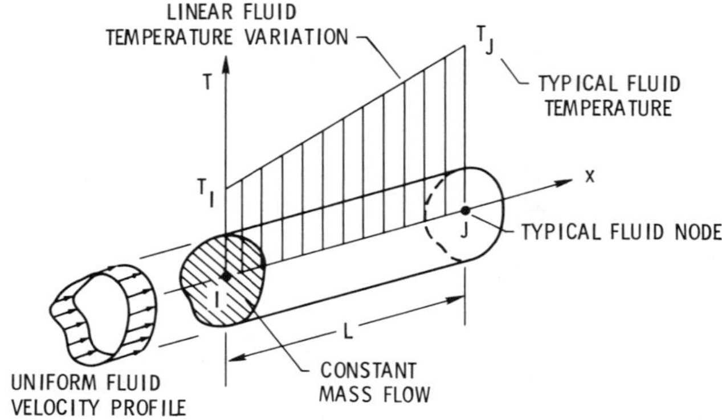

Fig. 2 Mass transport element.

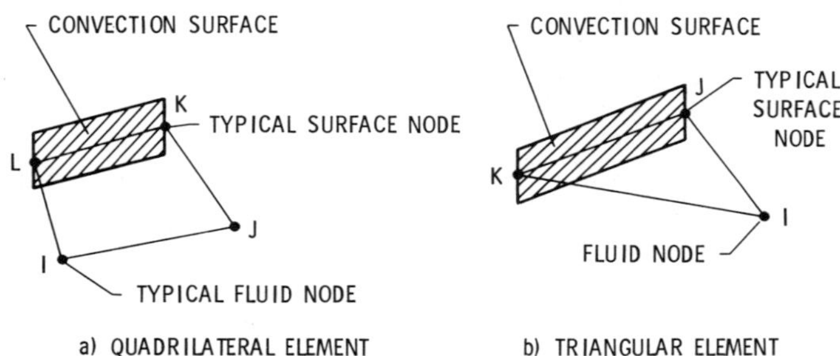

Fig. 3 Surface convection elements with unknown fluid
temperatures.

Basic Elements

Convective finite elements are developed for the fluid
to represent both of the terms in Eq. (2). The first term is
represented by the mass transport convection element shown in
Fig. 2, and the second term is represented by the surface convection elements with unknown fluid temperatures shown in
Fig. 3.

Mass Transport Element. The mass transport element
(Fig. 2) represents the energy transported downstream due to
the fluid flow. Utilizing assumptions 1 and 2, the finite-
element matrix equation describing the element energy balance
is derived. The details of the derivation are in Appendix A.
Denoting the nodal fluid bulk temperatures as T_I and T_J
and the nodal heat loads as Q_I and Q_J, then the finite-
element matrix equation relating the temperatures and heat

loads is

$$[K_v] \begin{Bmatrix} T_I \\ T_J \end{Bmatrix} = \begin{Bmatrix} Q_I \\ Q_J \end{Bmatrix} \tag{3}$$

where the element mass transport conductance matrix $[K_v]$ is

$$[K_v] = \frac{\dot{m}C_p}{2} \begin{bmatrix} -1 & 1 \\ -1 & 1 \end{bmatrix} \tag{4}$$

The mass transport conductance matrix of Eq. (4) is indefinite and asymmetric. This situation is an important dis-. tinction from the finite-element analysis of elastic structures that produce positive semidefinite symmetric matrices. The indefinite character of the mass transport conductance matrix means that it is possible to obtain zero terms on the diagonal in the global conductance matrix for some assemblies (e.g., in series) of mass transport elements. The asymmetry of the conductance matrix means that advantage cannot be taken of symmetry as in finite-element structural analysis. Approximate methods are described in Ref. 8 for treating the asymmetrical conductance matrix as an initial flux vector to permit use of symmetric matrix solution techniques.

<u>Surface Convection Elements</u>. To represent the heat transfer between fluid/solid interface and the fluid with unknown temperature, surface convection elements are developed using assumption 4. In previous finite-element heat-transfer analyses (such as Ref. 4), convection heat transfer between a surface and a fluid customarily is represented as a boundary condition since the fluid temperatures are assumed to be known. The quadrilateral and triangular surface convection elements shown in Fig. 3 have unknown temperatures at both fluid and wall surface nodes. The triangular element is a special case of the quadrilateral element when the fluid nodal temperatures at nodes I and J are equal. For the quadrilateral element, the finite-element matrix equation relating the nodal heat loads is

$$[K_h] \begin{Bmatrix} T_I \\ T_J \\ T_K \\ T_L \end{Bmatrix} = \begin{Bmatrix} Q_I \\ Q_J \\ Q_K \\ Q_L \end{Bmatrix} \tag{5}$$

where the surface convection conductance matrix $[K_h]$ is

$$[K_h] = \frac{hA_s}{6} \begin{bmatrix} 2 & 1 & -1 & -2 \\ 1 & 2 & -2 & -1 \\ -1 & -2 & 2 & 1 \\ -2 & -1 & 1 & 2 \end{bmatrix} \tag{6}$$

and A_s is the convection area of the coolant passage surface. Details of the derivation of this equation and a similar equation for the triangular element are given in Appendix B.

Integrated Elements

The mass transport element and the surface convection
elements shown in Figs. 2 and 3 are used in a general way for
a variety of convective thermal applications. Also, the
elements are used to develop two integrated convective finite

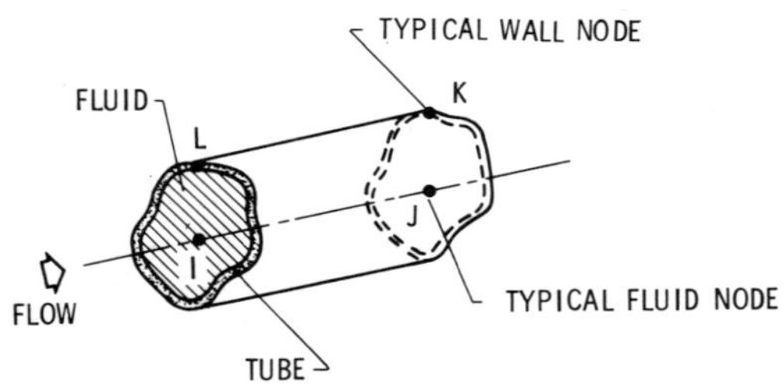

Fig. 4   Integrated tube/fluid element.

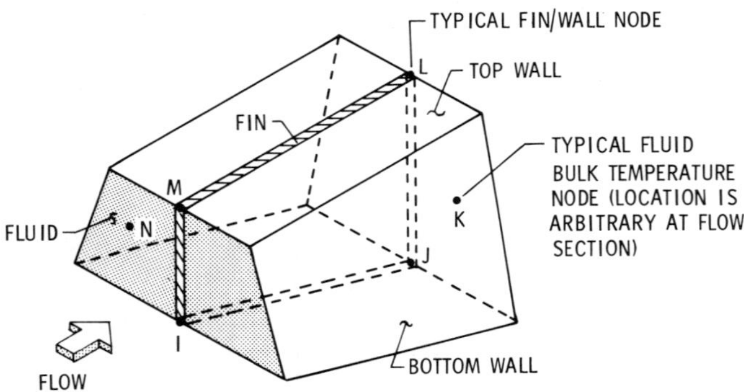

Fig. 5   Integrated plate-fin/fluid element.

elements that are of greater practical value in many applications. The first integrated element is the tube/fluid element shown in Fig. 4; the second integrated element is the plate-fin/fluid element shown in Fig. 5.

<u>Tube/Fluid Element</u>. The tube/fluid element (Fig. 4) consists of fluid within a thin tube of constant thickness and constant, arbitrary cross section. The element has two fluid nodes I and J and two tube nodes L and K. The following heat-transfer modes are represented in the element: 1) axial conduction in the tube between nodes L and K; 2) convection between the internal tube surface and the enclosed fluid (nodes L, K, and nodes I, J); 3) mass transport convection due to fluid flow from I to J; and 4) heat transfer between the external tube surface and a surrounding medium, which may be represented by specifying a heat flux or the medium temperature and convective film coefficient. Input data needed to describe the element include tube geometry and thermal conductivity, fluid thermal properties, and mass flow rate.

The element formulation includes options to 1) compute fluid pressure drops including flow friction and acceleration effects, and 2) modify the convective heat-transfer coefficient for large fluid-property variations across the flow passage.[9]

<u>Plate-Fin/Fluid Element</u>. The plate-fin/fluid element (Fig. 5) consists of two walls (plates) connected by internal fins. For convenience, a single plain fin is shown; in practice, other fin configurations (e.g., offset or pin fins) may be represented by using an equivalent thickness and surface area for the single fin. The fluid flows along both sides of the fins through an arbitrary flow cross section (shown in the figure as a trapezoidal cross section for convenience), which may vary linearly along the element. The element has six nodes: two nodes to represent the fluid bulk temperatures (nodes N and K) and four fin/wall nodes (I, J, L, and M). The fluid node locations are arbitrary at a given flow section, since fluid conditions are assumed to vary only in the flow direction. The following heat-transfer modes are represented in the element: 1) two-dimensional conduction in the fin between the nodes I, J, L, and M; 2) convection between the wall surfaces (top nodes M and L; bottom nodes I and J) and the fluid (nodes N and K); 3) convection between the fin surfaces (nodes I, J, L, and M) and the fluid (nodes N and K); and 4) mass transport convection due to fluid flow from N to K. Input data include plate-fin geometry, thermal properties of the fin and fluid, and mass flow rate.

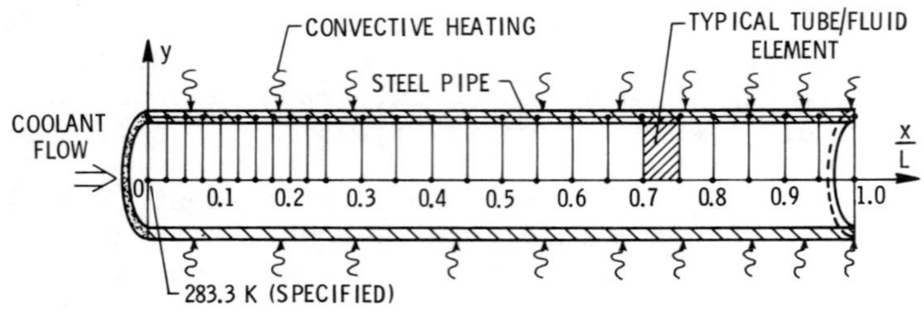

TEMPERATURE (K)

| | PIPE | | | WATER COOLANT | |
x/L	MITAS	FINITE ELEMENT		MITAS	FINITE ELEMENT
0.0	337.1	337.1		283.3*	283.3*
0.1	338.6	339.4		291.1	291.1
0.2	343.2	343.1		298.3	298.5
0.3	347.2	346.9		306.1	305.4
0.4	351.1	350.8		312.6	312.0
0.5	354.8	354.5		318.7	318.3
0.6	358.4	358.2		324.6	324.2
0.7	361.8	361.7		330.1	329.9
0.8	365.1	365.1		335.3	335.2
0.9	368.0	368.1		340.2	340.2
1.0	369.6	369.9		343.6	344.8

*SPECIFIED

Fig. 6 Convectively heated water-cooled pipe.

Element formulation includes options to 1) compute fluid pressure drops, 2) modify the convective heat-transfer coefficient for large fluid-property variations across the flow passage, and 3) modify the fin convective heat transfer by an efficiency factor η, which accounts for deviations in the fin temperature profile from the assumed linear profile. Here, the fin efficiency factor is defined to be

$$\eta = \frac{2}{m\ell} \; \frac{\cosh m\ell - 1}{\sinh m\ell} \tag{7}$$

where

$$m = \sqrt{2h/tk} \tag{8}$$

Thermal Analysis Program

A finite-element thermal analysis computer program[10] has been written to apply and verify the four convective finite

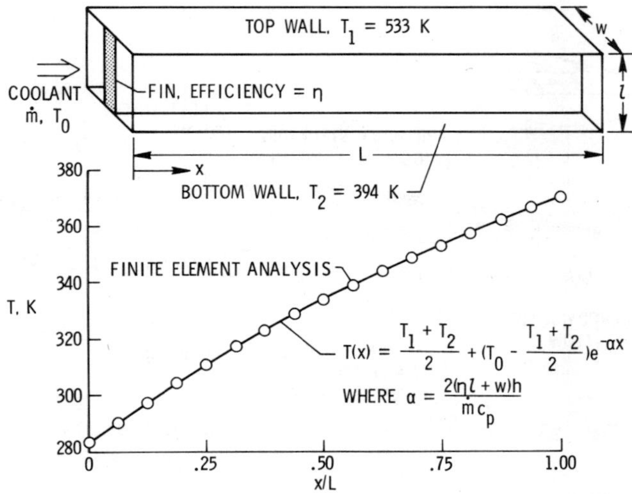

$$T(x) = \frac{T_1 + T_2}{2} + (T_0 - \frac{T_1 + T_2}{2})e^{-\alpha x}$$

$$\text{WHERE } \alpha = \frac{2(\eta l + w)h}{\dot{m} c_p}$$

Fig. 7 Calculated coolant temperatures in a simplified heat
exchanger passage (constant thermal properties).

elements previously described. In addition to these elements,
the program element library includes conventional one-dimensional (rod) and two-dimensional (quadrilateral) conduction/
surface-convection elements. A general banded matrix solution
technique is used to handle the asymmetric mass transport conductance matrix, Eq. (4). Temperature-dependent thermal properties are permitted in tabular form.

Applications

In this section, the convective finite elements are
applied to several convectively cooled structures of increasing complexity: 1) a convectively heated/cooled pipe,
2) a simplified heat exchanger, and 3) a scramjet fuel-injection strut. These applications demonstrate the use of the
elements independently and in combination with conventional
conduction elements for the thermal analysis of idealized and
realistic structures. Finite-element and lumped-parameter
calculated temperatures are compared to verify the new finite-
element methodology.

Convectively Heated/Cooled Pipe

Results of an application of the tube/fluid thermal element to a water-cooled pipe convectively heated by a surrounding medium are presented in Fig. 6. The entrance temperature of the coolant is specified; the downstream coolant
temperatures and pipe wall temperatures are to be determined.

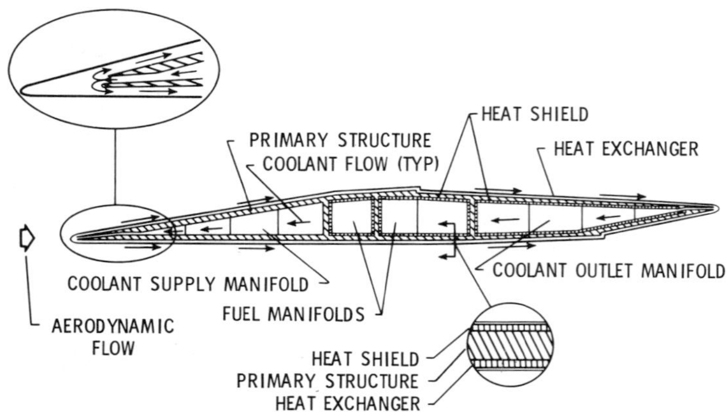

Fig. 8   Scramjet fuel-injection strut cross section (chordwise).

Calculated temperatures are in excellent agreement with results from an equivalent lumped-parameter analysis (MITAS)[3]
as shown in the table in Fig. 6. The finite-element thermal
analysis input data and results are presented in Ref. 10.

<u>Simplified Heat Exchanger</u>

     A thermal analysis of a simplified heat exchanger with
the plate-fin/fluid element is presented in Fig. 7. The top
and bottom walls have specified temperatures, and the fluid
bulk temperature is to be determined.

     Fluid temperatures calculated by the finite-element
method for constant thermal parameters are compared with a
closed-form analytical solution in Fig. 7. Agreement is excellent, with the maximum error less than 0.06%. Although it
cannot be detected from Fig. 7, the finite-element coolant
temperatures oscillated about the analytical solution. This
behavior also was observed in coolant temperatures predicted by
the tube/fluid finite element. As finite-element meshes have
been refined, the amplitudes of the oscillation errors have
decreased. The oscillatory behavior of the finite-element
coolant temperatures thus appears to be a characteristic of the
convergence behavior of the elements. The finite-element
thermal analysis input data and results are presented in
Ref. 10.

<u>Scramjet Fuel-Injection Strut</u>

     A hydrogen-cooled scramjet fuel-injection strut (Fig. 8)
is analyzed to demonstrate the capability of the new elements
for thermal analysis of realistic structures. The strut is

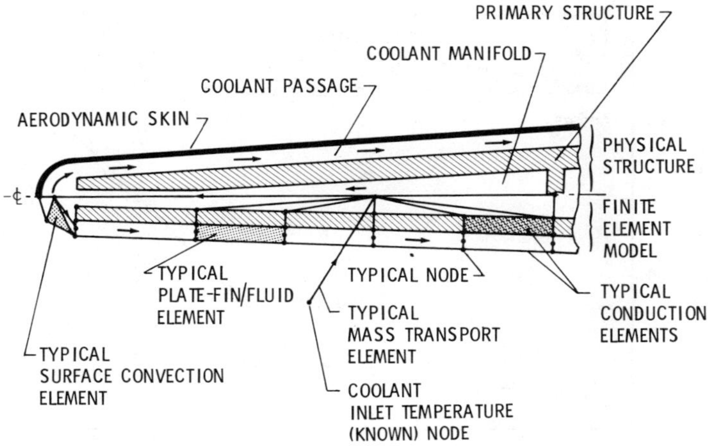

Fig. 9 Model of forward portion of strut cross section.

considered because its unsymmetrical configuration and heating
leads to complex thermal behavior and because a lumped-parameter MITAS analysis was available[6] for comparison. The strut
is subjected to unsymmetric, nonuniform aerodynamic heating on
both sides and is cooled internally by hydrogen flowing from
internal manifolds through heat exchangers bonded to the external surface of the primary structure. Further details of
the fuel-injection strut are given in Ref. 2.

The strut was analyzed using the mass transport, surface
convection, and plate-fin/fluid elements developed in this
study. No attempt was made to develop a finite-element model
equivalent to the very detailed MITAS model presented in Ref.
6. Rather, a coarse discretization was made to demonstrate
the capability of the finite elements to represent the overall thermal behavior of a realistic structure. The forward
portion of the finite-element model of the strut cross section
is shown in Fig. 9. This portion is typical of the modeling
procedure used for the entire strut. The hydrogen flow
entering the coolant manifold from the hydrogen tank is represented by a mass transport element. Heat transfer between
the walls of the coolant supply manifold and the hydrogen is
represented by triangular surface convection elements. The
hydrogen flow from the manifold to the leading edge and from
there to the port and starboard wall coolant passages also is
modeled using mass transport elements. The coolant passages
are represented with plate-fin/fluid elements. Conduction
heat transfer is represented by quadrilateral elements in the
walls of the primary structure and by rod elements in the interior bulkheads and the aerodynamic skin. The asymmetrical

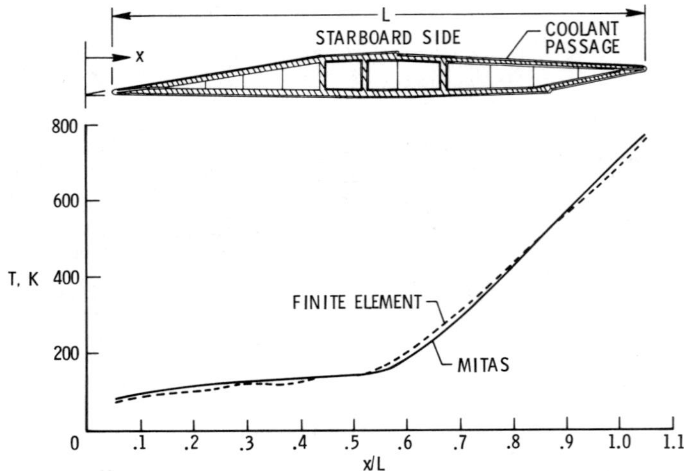

Fig. 10 Comparison of finite-element and MITAS starboard coolant temperature for strut cross section.

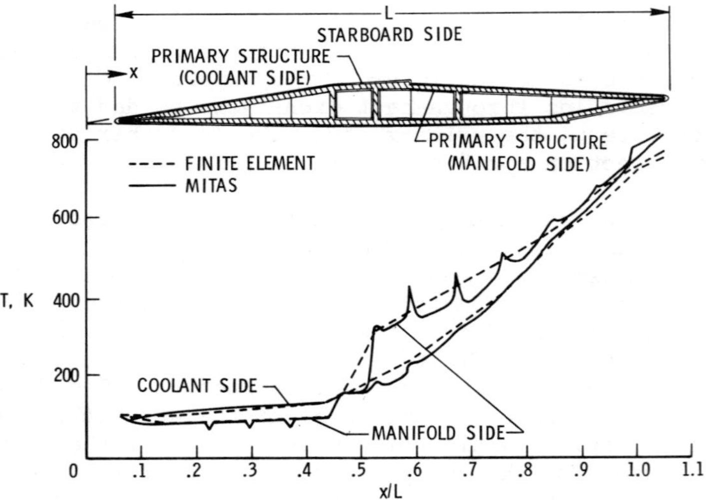

Fig. 11 Comparison of finite-element and MITAS starboard primary structure temperatures for strut cross section.

nonuniform aerodynamic heating is represented as a convective boundary condition.

Finite-element and MITAS-calculated temperatures for the coolant (starboard passage) and primary structure (starboard wall) are presented in Figs. 10 and 11, respectively. Agreement between coolant temperatures and agreement for trends of structural temperatures are good, but details of the structural temperatures are missing from the finite-element results because of the coarse discretization. For example, on the coolant side of the primary structure (Fig. 11) the structural temperatures are predicted quite well. However, on the manifold side, the agreement is not as good because of the highly localized heating at the bulkheads. The good agreement with the MITAS predictions for the salient features of the temperature distributions in this complex thermal problem shows the practicality and versatility of finite-element methodology for thermal analysis of realistic convectively cooled structures.

Concluding Remarks

A finite-element method for steady-state thermal analysis of convectively cooled structures has been developed. Finite elements characterized by fluid bulk temperature nodes and fluid/solid interface nodes were developed using assumptions employed in practical heat-transfer analysis. Two basic finite elements were developed to represent the principal heat-transfer modes in the coolant passages: 1) a mass transport element for energy transport downstream due to fluid flow, and 2) a surface convection element for heat transfer between the fluid passage surfaces and the fluid with unknown temperature. By using these basic elements, integrated finite elements were developed for applications to discrete tube and plate-fin convectively cooled structural configurations.

The finite-element method was applied to several convectively cooled structures of increasing complexity. Calculated temperatures were found to be in excellent agreement with lumped-parameter calculations made with equivalent mathematical models. A scramjet fuel-injection strut was analyzed using mass transport elements, surface convection elements, and plate-fin/fluid elements in combination with standard conduction elements. The analysis clearly demonstrates the capability of the new elements to model complex heat-transfer modes in a realistic structure. The applications show for steady-state heat transfer that the new finite elements may be used with the same confidence as the well-established lumped-parameter approach and, therefore, demonstrates the potential for development of a unified, integrated thermal/structural analysis capability for convectively cooled structures.

The separation of the convective heat-transfer mechanism into mass transport and surface convection elements provides a general approach to the analysis of a variety of convectively cooled structures. The integrated element concept, which incorporates conduction and convection heat-transfer modes into a single element, simplifies the mathematical model for practical configurations. The applications show that it is advantageous to have both basic and integrated thermal/fluid elements available in finite-element thermal analyzers for convectively cooled structures.

Appendix A: Mass Transport Conductance Matrix

The mass transport conductance matrix [Eq. (4)] is derived from the general finite-element equation representing the mass transport term in Eq. (2). From Ref. 8, the mass transport conductance matrix is given in general terms by

$$[K_v] = \int_{vol} \rho C_p [N]^T \begin{Bmatrix} u \\ v \\ w \end{Bmatrix}^T [B] \, d(vol) \qquad (A1)$$

where $[N]$ is the interpolation matrix relating the temperature at a general point within the element to the element nodal temperatures, and $[B]$ is the interpolation matrix relating the thermal gradient vector within the element to the nodal temperatures. The components of the velocity vector $\vec{V}$ in Eq. (2) are expressed in Eq. (A1) by $[u,v,w]$. The integration is performed over the volume of the element.

Based on assumption 1 in the text, the coolant bulk temperature varies only in the downstream direction, and it may be expressed in terms of the nodal bulk temperatures. From assumption 2, the only nonzero velocity component is $u = V$, where V denotes the mean fluid velocity. With these substitutions, the integral definition of the conductance matrix [Eq. (A1)] can be evaluated to yield Eq. (4).

Appendix B: Surface Convection Conductance Matrices

The derivation of conductance matrices for the quadrilateral and triangular surface convection elements are based upon assumptions 3 and 4 in the text. These assumptions permit expressing the heat transfer between the coolant passage surface in terms of a convective coefficient h.

Quadrilateral Element

The nodal heat loads entering the element at the wall

surface nodes (Fig. 3a) may be expressed by

$$\left\{ \begin{matrix} Q_K \\ \\ Q_L \end{matrix} \right\} = [H] \left\{ \begin{matrix} T_K - T_J \\ \\ T_L - T_I \end{matrix} \right\} \tag{B1}$$

where $[H]$ denotes the surface convection conductance matrix derived in finite-element theory for convective heat transfer between a rectangular flat surface and a fluid with known temperatures. From Ref. 4,

$$[H] = \frac{hA_s}{6} \begin{bmatrix} 2 & 1 \\ & \\ 1 & 2 \end{bmatrix} \tag{B2}$$

By conservation of energy, the nodal heat loads at the fluid nodes may be written as

$$\left\{ \begin{matrix} Q_J \\ \\ Q_I \end{matrix} \right\} = - \left\{ \begin{matrix} Q_K \\ \\ Q_L \end{matrix} \right\} = - [H] \left\{ \begin{matrix} T_K - T_J \\ \\ T_L - T_I \end{matrix} \right\} \tag{B3}$$

Substituting Eq. (B2) into Eqs. (B1) and (B3) and combining the results into a single matrix equation gives

$$[K_h]_{quad} \left\{ \begin{matrix} T_I \\ T_J \\ \\ T_K \\ T_L \end{matrix} \right\} = \left\{ \begin{matrix} Q_I \\ Q_J \\ \\ Q_K \\ Q_L \end{matrix} \right\} \tag{B4}$$

where the surface convection conductance matrix is given in Eq. (6).

Triangular Element

The conductance matrix for the triangular surface convection element shown in Fig. 3b may be derived in a similar way. The result is

$$[K_h]_{triangle} = \frac{hA_s}{6} \begin{bmatrix} 6 & -3 & -3 \\ -3 & 2 & 1 \\ -3 & 1 & 2 \end{bmatrix} \qquad (B5)$$

References

[1] Nowak, R. J. and Kelly, H. N., "Actively Cooled Airframe Structures for High-Speed Flight," _Proceedings of the AIAA/ASME/SAE 17th Structures, Structural Dynamics, and Materials Conference_, King of Prussia, Pa., May 5-7, 1976, pp. 209-217.

[2] Wieting, A. R. and Guy, R. W., "Thermal-Structural Design/Analysis of an Airframe-Integrated Hydrogen-Cooled Scramjet," _Journal of Aircraft_, Vol. 13, March 1976, pp. 192-197.

[3] "Martin Interactive Thermal Analyses System, Version 1.0," Martin-Marietta Corp., Denver, Colo., MDS-SPLPD-71-FD238 (REV3), March 1972.

[4] Lee, H. P., "NASTRAN Thermal Analyzer, Vol. I: The NASTRAN Thermal Analyzer Manual," NASA Goddard Space Flight Center, X-322-76-16, Dec. 1975.

[5] Zelazny, S. W., Manhardt, P. D., and Baker, A. J., "Fluid-Thermal-Structural Analysis of Chemical Laser Nozzles," AIAA Paper 74-1140, San Diego, Calif., Oct. 1974.

[6] Thornton, E. A. and Wieting, A. R., "Comparison of NASTRAN and MITAS Nonlinear Thermal Analyses of a Convectively Cooled Structure," _NASTRAN: Users' Experiences, Fourth Colloquium_, NASA TM X-327 8, Hampton, Va., Sept. 1975.

[7] Hamed, A., Baskharone, E., and Tabakoff, W., "Temperature Distribution Study in a Cooled Radial Inflow Turbine Rotor," AIAA Paper 76-44, Washington, D.C., Jan. 1976.

[8] Hsu, M. B. and Nickell, R. E., "Coupled Convective and Conductive Heat Transfer by Finite Element Methods," _Finite Element Methods in Flow Problems_, edited by J. T. Oden, O. C. Zienkiewicz, R. H. Gallagher, C. Taylor, _International_

Symposium on Finite Element Methods in Flow Problems UAH Press, University of Alabama, Huntsville, Ala., 1974, pp. 427-449.

[9]Kays, W. M. and London, A. L., *Compact Heat Exchangers*, 2nd ed., McGraw-Hill, New York, 1964.

[10]Thornton, E. A., "TAP1:A Finite Element Program for Steady-State Thermal Analysis of Convectively Cooled Structures," NASA CR-145069, 1976.

EFFECT OF A CONDUCTING WALL ON A
STRATIFIED FLUID IN A CYLINDER

Constance W. Miller*
University of California, Berkeley, Calif.

Abstract

The effect of thermally conducting side walls of a finite thickness ℓ on a stable stratified fluid in a cylinder of radius R and height h was investigated. Experimentally, the centerline temperature profile was measured as a function of time for a thermocline that was established in two different enclosures, one aluminum ($\ell/R=0.013$), and one glass ($\ell/R=0.026$). The degradation of the thermocline was ten times faster in the aluminum tank. The temperature field and flow velocity then were computed using a finite difference method with a variable grid. In addition, it was necessary to decouple the energy equation in the fluid from that in the wall. The computations were done for Rayleigh numbers up to 10^9 (based on h/2); for h/R=0.25, 0.5, 1.0; and for $\ell/R=0.015$ to 0.06. The natural convection generated at the fluid/wall interface increased for larger Rayleigh numbers, decreased for larger values of H/R, and increased proportional to $(\ell/R)^{1/2}$.

Nomenclature

A = aspect ratio, H/R
a = wall thickness to
    radius ratio, ℓ/R
g = gravity
h = height of cylinder
H = half height of
    cylinder, h/2
k = thermal conductivity

Presented as Paper 77-792 at the AIAA 12th Thermophysics Conference, Albuquerque, New Mexico, June 27-29, 1977.

*Assistant Professor, Dept. of Mechanical Engineering; currently at Earth Sciences Div., Lawrence Berkeley Lab, Univ. of California, Berkeley.

ℓ = enclosure wall thickness
N = transformed coordinate
ΔN = grid spacing
p = pressure
Pr = Prandtl number, ν/α
R = inner radius of cylinder
R_2 = outer radius of cylinder, $R+\ell$
r = radial coordinate
Ra = Rayleigh number, $g\beta\Delta T H^3/\nu_f\alpha_f$
T = temperature
T_H = temperature of hot fluid
T_L = temperature of cold fluid
$\Delta T = T_H - T_L$
$T_M = (T_H + T_L)/2$
t = time
v = velocity
z = axial coordinate

α = thermal diffusivity
β = coefficient of expansion
δ = boundary layer thickness
ρ = density
μ = viscosity
ν = μ/ρ
ψ = stream function
ω = vorticity

Subscripts

f = fluid
s = wall
r = radial component
z = axial component

Nondimensional Variables

$r^* = r/R$, $z^* = z/H$
$v_r^* = v_r H/\alpha_f$, $v_z^* = v_z R/\alpha_f$
$t^* = t\alpha_f/RH$, $\omega^* = \omega RH/\alpha_f$
$\psi^* = \psi/\alpha_f H$
$\theta = (T-T_M)/(T_H-T_L)$

Introduction

The problem that has been studied is the effect of thermally conducting side walls of a tank on the rate of heat

transfer in the enclosed stratified fluid. The relevance of
the problem is to the storage of thermal energy as sensible
heat in liquids. Water commonly is used at moderate temperatures (20°-150°C), because of its availability and high heat
capacity, and liquids with low vapor pressures can be used at
higher temperatures. A liquid is desirable because it can be
used directly in a cycle, avoiding the degradation of thermal
energy which exists with a heat exchanger and avoiding the
cost of the heat exchanger.

To reduce storage costs further, the hot and cold fluids
might be stored in the same tank. The heated fluid is stored
on top of the colder, recycled fluid. A natural separation
will exist in the liquid due to the stable thermocline that
is established. If the conductivity of the fluid is low (for
water k = 0.6 W/m-°C), the degradation of the thermocline due
to the heat transfer between the hot and cold fluids is low.
However, if the fluid is stored in a container made from a material of a thermal conductivity vastly different from the
fluid, convection currents will be generated at the fluid/
wall interface inside the container, causing degradation of
the thermocline at a faster rate than anticipated.

Estimates of the thermocline decay based on conduction
alone have been done[1], and the calculated rate of decay is
slow for a liquid such as water. However, convection effects,
even though small, can have a significant effect. The problem
of the transient response of a stratified fluid to a change in
the wall temperature has been important in the storage of
cryogenic fluids. In 1964, Clark[2] reviewed the research on
the temperature response of the enclosed fluid to a constant
heat influx at the wall. Here the stratification of the fluid
played an important role in the design of partially liquid
filled pressure vessels. The heated liquid at the surface
vaporized, increasing the pressure in the tank. To reduce
the pressure, more mixing of the enclosed fluid is needed.
However, for the problem considered here of storing a fluid
of two different temperatures in the same tank, less mixing
is desired.

To study this effect of the wall on the heat transfer in
the enclosed fluid, a numerical model of the fluid flow was
determined. Barakat and Clark[3] did present a numerical solution of the two dimensional, laminar, transient free convection flow in a partially filled liquid container with an arbitrary wall temperature. However, the solution did not take
into account the heat transfer within the wall, and the solution used a uniform grid which caused problems at higher Rayleigh numbers. As the boundary layer became thin, the uniform
grid introduced errors in the stream function calculations.

The numerical solution presented here takes into account both the boundary layer regime and the recirculation zone. The heat transfer within the side walls was also included. In addition, experimental temperature measurements and velocity magnitudes were made in the laboratory to establish the importance of this problem. For a storage system, the rate of change in the temperature stratification of the fluid can be compared then to the desired storage time.

Experimental Measurements

Experimental work was done to establish that this problem did exist and could be quite important. A measurement of the centerline temperature profile as a function of time was made for thermoclines in two different tanks, one made from aluminum ($k=160$ W/m-^{o}C) and one from glass ($k=0.8$ W/m-^{o}C). Both tanks were cylindrical.

The dimensions of the aluminum tank were: $h=0.36$m, $R=0.13$ m, and $\ell=0.64$cm. For the glass tank, the dimensions were: $h=0.29$m, $R=0.16$m, and $\ell=1.6$ cm. The bottom of the tank was filled with cold water and the top half with hot. To minimize the mixing between the two fluids, a removable membrane was used in the filling process. The initial thermocline obtained was as close to a step function as possible. The entire tank was insulated to prevent heat loss.

Once the steepest thermocline was obtained, temperature measurements as a function of time and position were made. Figures 1 and 2 show the centerline temperature profile as a function of time for the two cases. The axisymmetric numerical solution that neglects convection effects but includes the conduction in the wall is also included in the plots. For the aluminum tank, the conduction only solution does not match the experimental measurements. In 45 min ($t^{*}=0.03$), the fluid temperature in the aluminum tank is almost uniform. Thermal energy is transported from the top of the tank to the bottom through the highly conducting walls. The fluid near the wall at the bottom of the enclosure is heated and starts to rise. The convection currents increase the heat transfer accounting for the faster degradation of the thermocline. The fluid at the bottom of the tank is increasing in temperature at the same rate that the fluid on the top is cooling off, indicating the main heat transfer is within the fluid itself. There is little loss in energy from the tank.

In the glass tank (Fig. 2), there is little heat transfer between the hot and cold water. The temperature of the hotter fluid cools off slowly due to the heat loss from the tank over

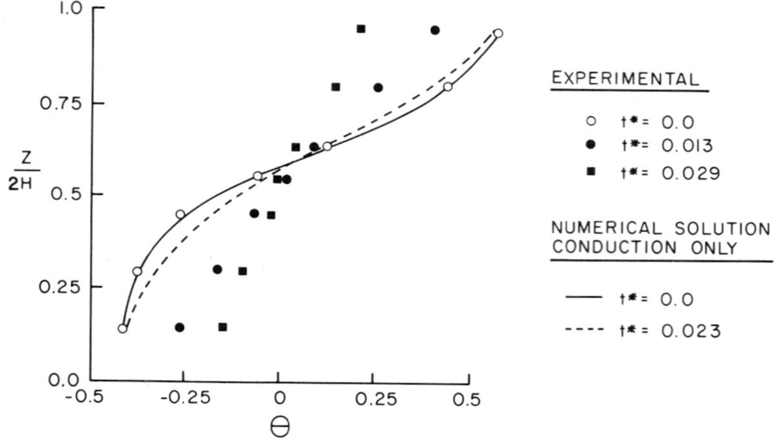

Fig. 1   Experimental and numerical results (assuming conduction only) of the centerline temperature profile as a
function of time for an insulated aluminum tank.

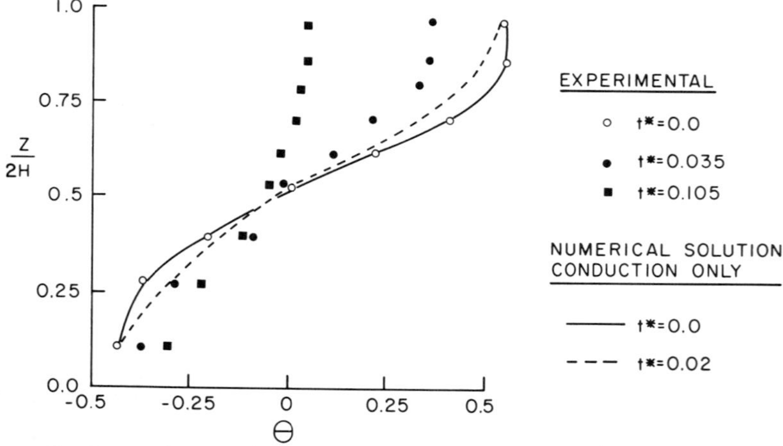

Fig. 2   Experimental and numerical results (assuming conduction only) of the centerline temperature profile as a
function of time for an insulated glass tank.

the six hours that the temperature measurements were made.
The fluid at the bottom heats up very little and in fact the
temperature measurements there match the conduction—only solution quite well. Because the conductivity of the glass wall
is approximately equal to water, there is no temperature
gradient in the radial direction, and convection currents are
negligible.

To estimate the rate of decay of the thermocline, the
fluid velocity and the boundary—layer width at the wall must

be known. To determine a rough estimate of the fluid velocities, the Baker discharge technique was used. For the aluminum tank, an average axial velocity of 1mm/sec was measured
over a region of 2-4 mm near the wall. These values may seem
quite small, but one can estimate the proportion of fluid that
decays to the mean temperature using these values. The volume
of fluid that moves to the center region in the bottom of the
tank is $(2\pi R)\,\delta v_z$. Dividing by the total volume of the colder
fluid in the bottom of the tank gives the percent being mixed
as $(2\pi R)\pi v_z/\delta R^2 H = 2v_z\delta/RH$. For the aluminum tank considered,
this is just 2%/minute (using $v_z = 1$ mm/sec and $\delta = 4$ mm).
However, as the fluid mixes, the driving force, because of
the temperature difference, decreases and the fluid velocity
will decrease also. Exact measurements of the fluid velocity
and boundary layer width are being done now and will be presented at a later time.

Of course, these experimental measurements were for a
small laboratory-scale tank, and the original engineering
application that started this work was a large thermal storage
tank. However, the intent was to understand the flow conditions and the controlling parameters in the small enclosure
first before looking at the more difficult problem of a large
tank. On this smaller tank, it has been possible to solve the
flow numerically and to make transient fluid velocity measurements. The numerical solution will be presented here.

Numerical Formulation

The main thrust of this initial work was to obtain a
numerical solution to the bouyancy driven flow induced by the
thermally conducting walls. A vertical cylinder, as considered in the experimental measurements, was used. The flow is
axisymmetric. The flow could also be considered to consist
of two regimes, a thin boundary layer at the wall and a large
recirculating zone in the rest of the tank. To be able to
take both of these effects into account, a variable grid system was used. Near the wall, a grid spacing small enough to
resolve the boundary layer (on the order of 1% of the tank
radius) was needed, whereas a much larger grid could be used
at the center of the cylinder where velocity gradients are
smaller. The grid variation was achieved through a transformation instead of an arbitrary grid change. For an arbitrary
grid change, the accuracy in the finite difference approximation is of lower order than for a uniform grid. In addition,
the conduction in the wall had to be solved separately from
the conduction in the fluid. A matching of the heat transfer
at the fluid/wall interface was necessary. This allowed for
using different time steps and different grid spacings in the

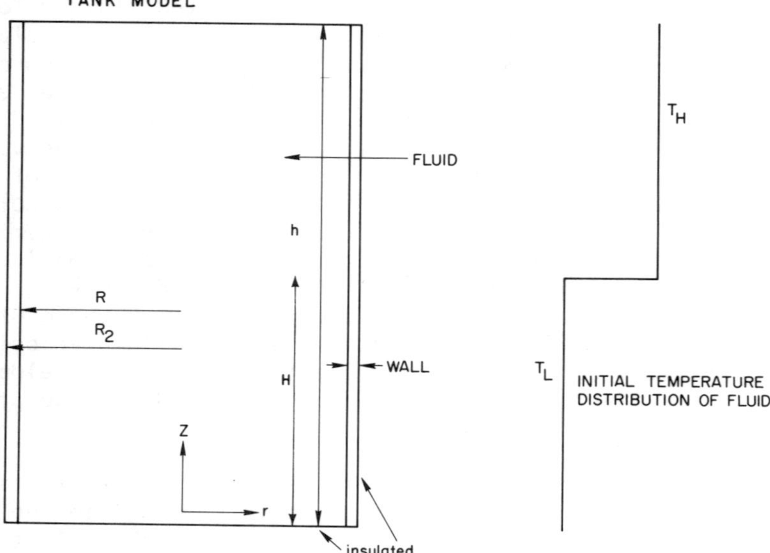

Fig. 3   Tank model used for numerical calculations and the
assumed initial thermocline.

fluid and the wall. By taking into account these points, a
numerical model was developed that solved the flow up to
$Ra=10^9$. An outline of the numerical model and important considerations are given below. Results of the calculations are
also reported.

The problem that was solved corresponds to the enclosure
with the specified boundary conditions shown in Fig. 3. The
wall thickness of the enclosure can be varied in relation to
the size of the enclosure. Any initial thermocline can be
assumed, but a step function was considered. For a perfectly
insulated tank, it will be necessary to only solve half of the
flow because of the symmetry. Then to determine the destabi-
lizing effect of the thermally conducting walls, the Navier
Stokes equations of mass, momentum, and energy were solved in
the fluid, and the energy equation was solved in the wall.
Using the Boussinesq approximation to relate the density
change to the temperature $(\rho=\rho_o[1-\beta(T-T_o)])$, the equations
solved in the fluid were

$$\vec{\nabla} \cdot \vec{v} = 0$$

$$\rho_f[\frac{\partial \vec{v}}{\partial t} + \vec{\nabla} \cdot (\vec{v}\,\vec{v})] = \rho_f\vec{g} - \vec{\nabla}p + \mu_f\nabla^2\vec{v}$$

$$\rho_f C_f[\frac{\partial T}{\partial t} + \vec{\nabla} \cdot (\vec{\nabla}T)] = k_f\nabla^2 T$$

In the wall, $\rho_s C_s (\partial T / \partial t) = k_s \nabla^2 T$.

The equations were nondimensionalized using the half height and the radius of the cylinder, the thermal diffusivity of the water evaluated at the median temperature, and the temperature difference of the stored fluid $(T_H - T_L)$. By taking the curl of the momentum equation and by introducing the stream function concept, the nondimensionalized equations are

$$\frac{\partial v_r}{\partial r} + \frac{v_r}{r} + \frac{\partial v_r}{\partial z} = 0 \tag{1}$$

$$\frac{\partial \omega}{\partial t} + \frac{\partial}{\partial r}(v_r \omega) + \frac{\partial}{\partial z}(v_z \omega) = \frac{-RaPr}{A} \frac{\partial \Theta}{\partial r} +$$
$$Pr\{A^{-1} \frac{\partial^2 \omega}{\partial z^2} + A \frac{\partial}{\partial r} \frac{1}{r} \frac{\partial}{\partial r} r\omega\} \tag{2}$$

$$\frac{\partial \Theta}{\partial t} + \frac{1}{r} \frac{\partial}{\partial r}(rv_r \Theta) + \frac{\partial}{\partial z}(v_z \Theta) = A \frac{\partial^2 \Theta}{\partial z^2} + \frac{1}{A}\left\{\frac{1}{r} \frac{\partial}{\partial r} r \frac{\partial \Theta}{\partial r}\right\} \tag{3}$$

The *'s have been left out for ease. The stream function and vorticity are determined from

$$v_r = \frac{-1}{r} \frac{\partial \psi}{\partial z} \quad ; \quad v_z = \frac{1}{r} \frac{\partial \psi}{\partial r} \tag{4}$$

$$\omega = \frac{1}{A} \frac{\partial v_r}{\partial z} - A \frac{\partial v_z}{\partial r} = \frac{-1}{A} \frac{1}{r} \frac{\partial^2 \psi}{\partial z^2} - A \frac{\partial}{\partial r} \frac{1}{r} \frac{\partial \psi}{\partial r} \tag{5}$$

The nondimensionalized energy equation in the wall is

$$\frac{\partial \Theta}{\partial t} = \frac{\alpha_s}{\alpha_f} \left\{\frac{1}{A} \frac{\partial^2 \Theta}{\partial z^2} + \frac{A}{a} \frac{1}{r} \frac{\partial}{\partial r} r \frac{\partial \Theta}{\partial r}\right\} \tag{6}$$

The r in this equation has been nondimensionalized by the wall thickness.

The Rayleigh number based on the height of the enclosure and the maximum temperature difference of the fluid in the enclosure is quite large, on the order of 10^8 to 10^9 for any laboratory experiment (Ra=10^9 for a ΔT=50°C in the experiment just described). For natural convection from a vertical flat plate, the transition to turbulence occurs at 4×10^8 (based on the distance up the plate and the temperature difference between the vertical wall and the undisturbed fluid at that point). However, the resulting temperature difference in the radial direction in this problem is smaller than T_H-T_L used to evaluate the Rayleigh number above. The Rayleigh number, based on T at the wall minus T at r=0 at the same height z and on z, would be less than 4×10^8 for the experiments seen. In addition as z increases, this ΔT will decrease. Turbulence was not observed in the experiments run with $T_H-T_L=50°C$.

To solve these equations and to resolve the boundary layer regime where the velocity is changing rapidly a varying grid system was used. An abrupt change in coordinate spacing can lead to large numerical errors.[4] Following Blottner, a coordinate stretching was used. A new coordinate, N, was introduced with a uniform interval ΔN and is related to the original coordinate by $r = r(N)$. This was done for both the r and z coordinates. An example of the transformation is

$$\left.\frac{\partial \omega}{\partial r}\right|_j = \frac{(\partial \omega/\partial N)_j}{(dr/dN)_j} = \frac{\omega_{j+1} - \omega_{j-1}}{r_{j+1} - r_{j-1}} + 0(\Delta N^2)$$

The coordinate derivatives were replaced by finite differences. This leads to a coordinate stretching once the transformation is defined. The relationship used to transform both r and z coordinates was of the form

$$x_j = 1 - \frac{K^{N_j/\Delta N_o} - 1}{K^{1/\Delta N_o} - 1}$$

where $N_j = (j-1)\Delta N$ and x = r or z. Both K and ΔN_o are constants that can be varied to achieve the desired grid spacing.

Using this transformation technique, Eqs. (1-6) were finite-differenced with central differences in space and forward differences in time. For the convection terms, a conserving upwind differencing method was used. To advance the scheme in time, a modified alternating direction implicit technique was used, i.e. the convection terms always were evaluated at the old time level, the diffusion terms were split in the traditional manner of ADI.[5] This method is not unconditionally stable but it does retain more of the physics. Evaluating the convection terms at the old time level maintains the transportive nature better than using an implicit method for these terms. However, the use of ADI for the diffusion terms increases the stability of the numerical solution over a totally explicit scheme.

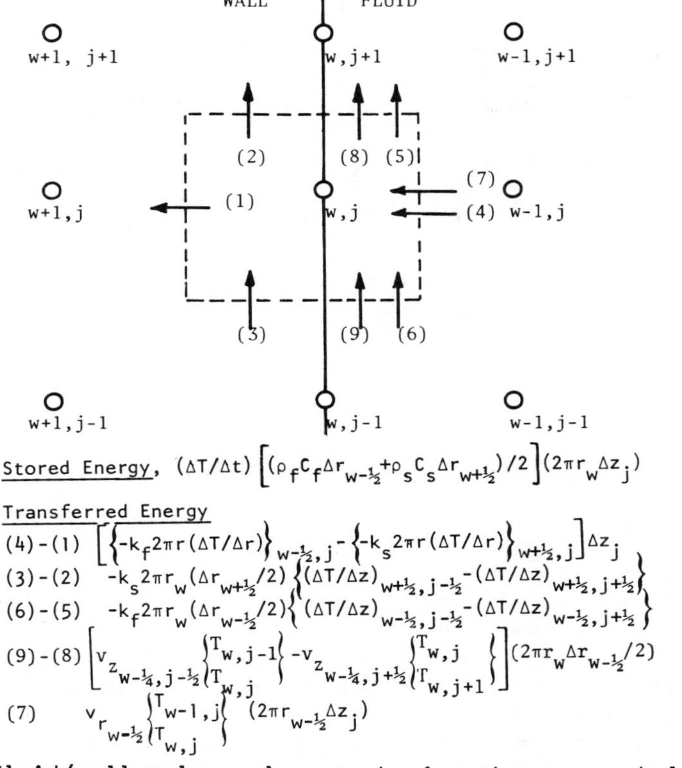

<u>Stored Energy</u>, $(\Delta T/\Delta t)\left[(\rho_f C_f \Delta r_{w-\frac{1}{2}} + \rho_s C_s \Delta r_{w+\frac{1}{2}})/2\right](2\pi r_w \Delta z_j)$

<u>Transferred Energy</u>

$(4)-(1)$ $\left[\left\{-k_f 2\pi r(\Delta T/\Delta r)\right\}_{w-\frac{1}{2},j} - \left\{-k_s 2\pi r(\Delta T/\Delta r)\right\}_{w+\frac{1}{2},j}\right]\Delta z_j$

$(3)-(2)$ $-k_s 2\pi r_w (\Delta r_{w+\frac{1}{2}}/2)\left\{(\Delta T/\Delta z)_{w+\frac{1}{2},j-\frac{1}{2}} - (\Delta T/\Delta z)_{w+\frac{1}{2},j+\frac{1}{2}}\right\}$

$(6)-(5)$ $-k_f 2\pi r_w (\Delta r_{w-\frac{1}{2}}/2)\left\{(\Delta T/\Delta z)_{w-\frac{1}{2},j-\frac{1}{2}} - (\Delta T/\Delta z)_{w-\frac{1}{2},j+\frac{1}{2}}\right\}$

$(9)-(8)$ $\left[v_{z_{w-\frac{1}{4},j-\frac{1}{2}}}\begin{Bmatrix}T_{w,j-1}\\T_{w,j}\end{Bmatrix} - v_{z_{w-\frac{1}{4},j+\frac{1}{2}}}\begin{Bmatrix}T_{w,j}\\T_{w,j+1}\end{Bmatrix}\right](2\pi r_w \Delta r_{w-\frac{1}{2}}/2)$

(7) $v_{r_{w-\frac{1}{2}}}\begin{Bmatrix}T_{w-1,j}\\T_{w,j}\end{Bmatrix}(2\pi r_{w-\frac{1}{2}}\Delta z_j)$

Fig. 4 Fluid/wall volume element showing the energy balance. The notation { } means to take the upper value if v>0 and the bottom if v<0.

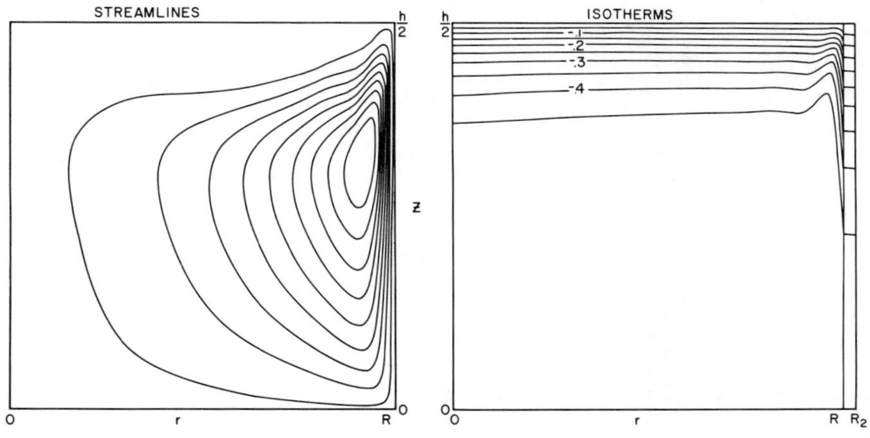

Fig. 5a Equally spaced streamlines and isotherms for Ra = 10^8 and t* = 0.003.

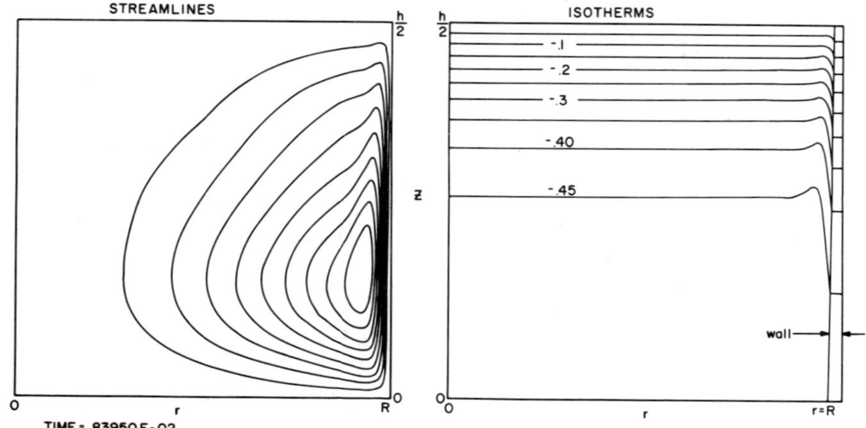

Fig. 5b   Equally spaced streamlines and isotherms for Ra = 10^8 and t* = 0.0084.

As stated the energy equation in the wall and fluid were solved separately. This allowed for the use of different time steps and different grid spacings between the two regions. However, the heat transfer at the fluid/wall interface must be matched. The energy balance at this boundary is shown in Fig. 4. The subscript w denotes the node at the fluid wall interface, w+1 is the first node inside the wall, etc. By doing a balance of the energy terms as shown in the figure, and by dividing out by the volume of the element, the desired equation is

$$(\overline{\rho C})_w \left(\frac{\Delta T}{\Delta t}\right)_{w,j} = \frac{1}{r_w \Delta r_w} \left\{ k_s \left(r \frac{\Delta T}{\Delta r}\right)_{w+\frac{1}{2},j} - k_f \left(r \frac{\Delta T}{\Delta r}\right)_{w-\frac{1}{2},j} \right\}$$

$$+ \frac{\overline{k}_w}{\Delta z_j} \left\{ \left(\frac{\Delta T}{\Delta z}\right)_{w,j+\frac{1}{2}} - \left(\frac{\Delta T}{\Delta z}\right)_{w,j-\frac{1}{2}} \right\} + \frac{r_{w-\frac{1}{2}}}{r_w \Delta r_w} (v_r T')_{w-\frac{1}{2},j}$$

$$- \frac{1}{\Delta z_j} \left\{ (v_z T')_{w-\frac{1}{4},j+\frac{1}{2}} - (v_z T')_{w-\frac{1}{4},j-\frac{1}{2}} \right\} \frac{\Delta r_{w-\frac{1}{2}}}{2\Delta r_w}$$

The term $(\overline{\rho C})_w$ is the average heat capacity of the volume element

$$\left\{ \rho_f C_f \, \Delta r_{w-\frac{1}{2}} + \rho_s C_s \, \Delta r_{w+\frac{1}{2}} \right\} / \, 2 \, \Delta r_w \, .$$

Similarly the effective thermal conductivity is

$$(k_s \, \Delta r_{w+\frac{1}{2}} + k_f \, \Delta r_{w-\frac{1}{2}}) / 2 \, \Delta r_w$$

Also the term T' means to use the conserving upwind technique to evaluate the temperature at this point.

It should also be noted that, for axisymmetric flow, the fluid velocity at the half-node points, needed for the convection terms, must be evaluated from a non-uniform weighting of the velocities at the node points. A volume source error is generated if an equal weighting is done near r=0. This was pointed out by Parmentier and Torrence[6] who then used the circulation theorem to correct this error. A weighting of ψ at the node points to obtain ψ at the half-node points was done so $\int_\ell \vec{v} \cdot d\vec{\ell} = \int_A \vec{\omega} \cdot d\vec{A}$ was satisfied. However, the same weighting of the streamfunction is obtained by considering an inverse weighting of ψ by the appropriate fluid volumes. An important point, though, is that no matter what the radial spacing is, the weighting factor of ψ must be used near r=0. To obtain ψ at $\Delta r/2$, the weighting values are ¼ of ψ at r = Δr and 3/4 of ψ at r = 0. This is independent of Δr. However, these weightings quickly drop to ½ and ½ a couple of grid spacings away from the centerline.

The boundary conditions on the temperature were: (1) the cylinder is insulated or $\partial T/\partial z = 0$ at z = 0 and $\partial T/\partial r = 0$

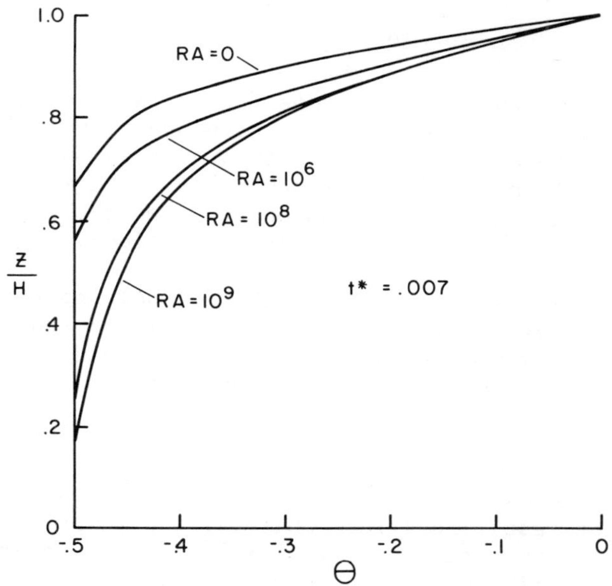

Fig. 6 Numerical solution (including convection terms) of the centerline temperature profile for Rayleigh numbers up to 10^9 (t* = 0.007).

at $r = R_2$; (2) at $z = h/2 = H$, the temperature will always be T_M (there is no loss in energy from the tank and the flow is symmetric); (3) because of symmetry, $\partial T/\partial r = 0$ at $r = 0$. The boundary condition on the velocity at the solid surfaces was that there was no slip. The vorticity at the boundary was found in terms of the stream function. At $z = 0$, $v_z = 0$, so $\omega = \partial v_r/\partial z = -(1/r)\partial^2\psi/\partial z^2$. Because of the varying grid system, a first order expansion involving ψ at the boundary and at one grid point away was used. This was done to avoid using information too far from the boundary as the grid spacing was large here. At $r = 0$, $v_r = 0$ and $\partial v_z/\partial r = 0$, so $\omega = 0$. At $z = h/2$, both the fluid at the top and bottom are flowing toward $r = 0$, so $v_z = 0$, $\partial v_r/\partial z = 0$ and $\omega = 0$.

At the fluid/wall interface, $\omega = -\partial v_z/\partial r$ and $\omega = -\partial(1/r)(\partial\psi/\partial r)/\partial r$. To evaluate ω at this point, an expansion in $\partial\psi/\partial N$ was used instead of in ψ. By letting $F = \partial\psi/\partial N = a+bN+cN^2$ and by evaluating F at the wall, at $\Delta N/2$ and at $3\Delta N/2$ in from the wall, the expression for the vorticity was found.

$$-r\omega = \frac{\partial^2\psi/\partial N^2}{(dr/dN)^2} = \frac{\partial F/\partial N}{(dr/dN)^2} = 9\,\frac{(-8F_w - F_{w-3/2} + 9F_{w-\frac{1}{2}})}{(-8r_w - r_{w-3/2} + 9r_{w-\frac{1}{2}})^2}$$

$$-r\omega = 27\,\frac{(\psi_w - \psi_{w-1}) - 3(\psi_{w-1} - \psi_{w-2})}{(-8r_w - r_{w-3/2} + 9r_{w-\frac{1}{2}})^2}$$

where a similar expansion was done for dr/dN. Actually this expression is still first order even though three points were used for the expansion. However, this expression has a smaller trancation error than expanding ψ in terms of just ψ_w and ψ_{w-1}. A second order expansion required a change in the velocity expression at w-1 and this was avoided.

This development gives the method of solving the problem of the recirculating flow with a boundary-layer regime and of matching the energy equation between a fluid and a solid. Results of the calculations are presented below.

Numerical Results

Several cases were run to illustrate the developing flow-field and the resulting effect on the fluid temperature. The intent is to determine the effect of the fluid properties and the characteristics of the enclosure itself on the rate of temperature stratification in the fluid. In general, it is desirable to predict the temperature profile in the fluid as a function of the nondimensional parameters of the problem. What is presented here is the effect of the Rayleigh number, aspect ratio of the enclosure and the enclosure wall thickness on the heat transfer. However, comparisons are made between different enclosures on the dimensional scale in addition to the nondimensional scale.

For the cases considered here, a grid size of 30 nodes in the z-direction and 50 nodes in the r-direction was used. The radial spacing varied from 0.0028 near the fluid/wall interface to 0.067 at r = 0. In the z-direction, the variation was from 0.0063 at the top to 0.095 at z = 0. In the enclosure wall itself, 4 grid points were used in the radial direction with a uniform spacing, but for the z-direction the grid spacing was the same as in the fluid. This simplified the matching of the heat transfer at the fluid/wall interface. The nondimensional time step used was 10^{-5} in the fluid and 10^{-6} in the wall, except for a Rayleigh number of 10^9 where a smaller time step was needed when solving the fluid equations to insure stability. The cases run were for different Rayleigh numbers, different aspect ratios, and different wall thicknesses. The ratio of the wall thermal conductivity to that in the fluid was not changed, and a value of 85 was used. This ratio corresponds to a fluid similar to water in an enclosure constructed to steel. The Prandtl number was kept equal to 1. For water, the Prandtl number is 3.5 at a mean temperature of 50°C but drops to 1 at higher temperatures. This is due to the decrease in fluid viscosity. The Prandtl number dependence is generally not large in natural convection, but higher Prandtl numbers will be investigated later. However, a Prandtl number of 1 does correspond to the thermal storage of water at high temperatures (200°C).

Figures 5a, 5b show the developing flowfield and the temperature changes in the fluid and the wall for Ra = 10^8 at two different times. The other conditions are as stated above. The equally spaced streamlines indicate the flow starts near the wall at z = h/2. At later times, the recirculation zone moves down the enclosure but stays near the wall. The temperature difference between the wall and the fluid decreases at the top at later times, because of the large convective heat transfer there. However, as this temperature difference does decrease, the fluid velocity also decreases and the heat transfer coefficient is reduced. There is a thin boundary layer type regime near the wall where the fluid is being heated up. For this particular case, the boundary layer regime is about 5% of the radius. For the numerical solution, there are 15 node points in the radial direction in this region, and good resolution of both the velocity and

temperature profiles is possible. As the fluid moves up, the
isotherms are distorted toward the top. The recirculating
fluid, though, causes the isotherms to move down in the
enclosure at a faster rate than if only conduction of the
fluid were important. As seen in the figure, the isotherms
are almost uniform in the radial direction except near the
wall. This is the region where the velocity is changing.
The net overall effect of the convection generated at the
wall is an effectively higher fluid conductivity. What is
also important to notice is that the convection effects are
not limited to the wall region, but involve the entire
enclosure. The changing temperature of the fluid is characterized though by the centerline temperature profile, so the
effect of different enclosures can be considered by comparing
the centerline temperature profiles for the different cases.

Figure 6 shows the centerline temperature profile for
different values of the Rayleigh number. The plot is given
for t* = 0.007. Again the other variables are as specified
above. As the Rayleigh number increases (from 10^6 to 10^9),
the convection effects become more important. This means
that, for the same size enclosure, but for a larger temperature difference, mixing will occur faster as expected. However, for the same temperature difference, but a larger
volume, the comparisons are not as simple, because the profiles plotted are for different real times. For a factor of
10 increase in the Rayleigh number, the actual volume would
increase by 10 for the same imposed temperature difference.
However, the real time corresponding to the same t* would
increase by only $4\frac{1}{2}$. For a thermal storage tank, it would
seem that a 10-fold increase in volume storage might have to
be stored 10 times as long. Or the question might be, is it
better to store the fluid in one tank or in several smaller
tanks? These comparisons should be done at the same real
time instead of the same t*. Figure 7 does the comparison
of the temperature degradation for different possible storage
tanks at real times.

Several comparisons are made in the figure. One possible
comparison is between two enclosures, one 10 times larger
than the other, at the same t. Profiles a and b make this
comparison. Profile a is for the larger tank whereas b is for
the smaller one. For these two cases, the same ℓ/R ratio of
0.03 was used, as well as the same aspect ratio. At this
time, the larger tank is only about 5% degraded while the
smaller one is about 15% mixed. If a comparison were made
for the same actual wall thickness, the ℓ/R ratio for the
larger tank would be smaller and the degradation would be
less than 5%. In either case, the larger tank mixes less

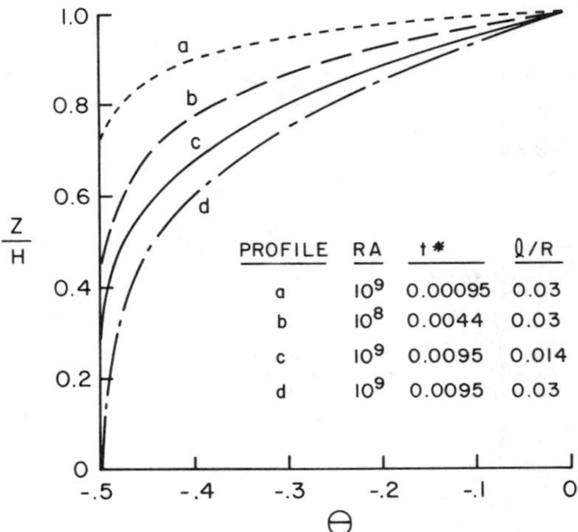

Fig. 7 Comparison of centerline temperature profiles for different sized tanks at the same real time.

rapidly for the same imposed temperature difference. However, if a larger storage tank must be able to store the fluid longer, the comparison should be between Profiles b and c for the same value of ℓ and between b and d for the same ℓ/R ratio. Profiles c and d are for the larger tank at a time 10 times as long, instead of at the same t. For the Profile d, the percent mixing is 25% and for Profile c (ℓ/R = 0.014), it is 20% mixed. This says that the percent mixing in the tank is not a linear function, i.e., a storage tank 10 times as great as another tank will not store the fluid 10 times as long for the same degradation of the thermal energy.

A third way of comparing the tanks is whether to use one large tank or a series of ten smaller ones, but having the same net volume. This comparison would again be for the same t. However, for several tanks, the storage of both hot and cold fluid would be in only one of the tanks. When the large tank was mixed 5%, the series of smaller tanks would be mixed 15%/10 or 1.5%. However to do an accurate comparison it would be necessary to look at the change in the thermocline while the filling or discharging the fluid took place. For the smaller series of tanks, the velocity at the wall, due to the forced fluid motion, would be increased by 10 causing increased heat transfer at the walls. This 1½% mixing would probably increase substantially, and it is likely the one large tank would appear better than a series of ten tanks.

Another point of interest is that the percent of the fluid that has heated up is approximately the same for the

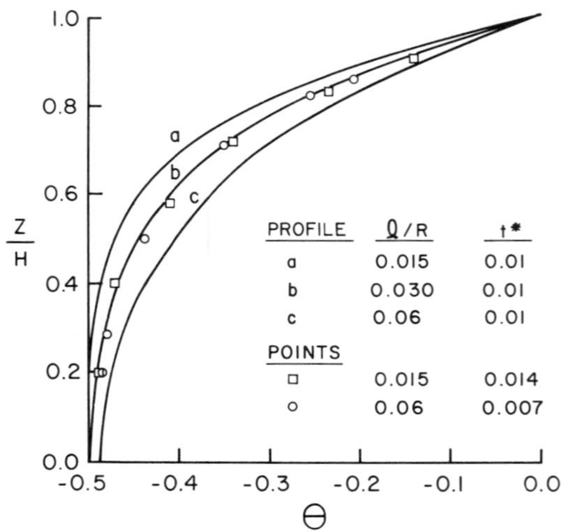

Fig. 8 Numerical solution of centerline temperature profile for different ℓ/R ratios. All other parameters are the same.

different Rayleigh numbers near the top. For the lower Rayleigh numbers, the heat initially is transferred slowly away from the wall, so the wall itself heats up. As the temperature difference between the wall and the fluid increases, the flow accelerates and the heat transfer increases. For larger Rayleigh numbers, the heat transfer initially is large between the wall and the fluid. As the mixing increases, the temperature difference between the wall and the fluid decreases, so the flow decelerates. The centerline temperature profiles tend to be about the same after a short time in this region. At the bottom of the enclosure, the heat transfer is still larger for increased Rayleigh numbers. At the wall itself, the wall remains colder for larger Rayleigh numbers since the fluid is so effective in transferring energy away from the wall.

Since the convection generated at the wall produces this effectively higher fluid conductivity, it is of interest to change the characteristics of the tank to determine their effect on the heat transfer. Figure 8 shows the temperature profiles when the wall thickness is increased but all other parameters are kept constant. The wall thickness to radius ratios considered were ℓ/R = 0.015, 0.03 and 0.06. The Profile 8-c corresponds to all three ℓ/R ratios but at different times. For ℓ/R = 0.015, this profile was obtained at t* = 0.014; for ℓ/R = 0.03, the profile was at t* = 0.01, and for a ℓ/R = 0.06, it is at t* = 0.07. Profile a and c

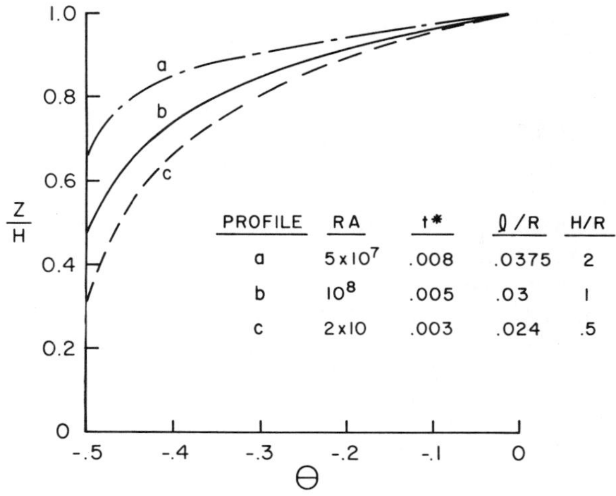

Fig. 9 Numerical solution of centerline temperature profile for different aspect ratios. All other parameters are the same.

are for $t^* = 0.01$ and for $\ell/R = 0.015$ and 0.06, respectively. A rough estimate of the wall dependency is that it is square root power, i.e., for the same value of $t^*(\ell/R)^{\frac{1}{2}}$, the percent mixing is the same.

Figure 9 is a graph of the effect of changing the aspect ratio from $H/R = 1$ to $H/R = 0.5$ and $H/R = 2$ while keeping ΔT, the volume, and the wall thickness constant. From the aspect ratios considered, the heat transfer decreases for the increased aspect ratio, because the area of the dividing surface decreases with respect to the side walls. The area available for heat transfer at the wall is $(2\pi R)\ell$. As R decreases, so does this area, and in turn the heat transfer. However, for a thermal storage tank, this effect would be countered by the fact that, fluid actually circulates through the tank instead of being at rest. For a larger aspect ratio, but the same volume, the fluid would come in contact with a larger wall area, $(2\pi RH = 2\pi(\text{volume})/R$ which increases as R decreases), and the heat transfer would increase. To determine the best aspect ratio for a thermal storage tank, the rate of fluid flow through the tank would have to be considered in more detail. An estimate would be that $H/R = 1$ would be best considering both these effects.

Conclusion

The heat transfer from the top surface of an enclosure to the fluid contained in it is increased when the enclosure wall has a larger thermal conductivity than the fluid. Convection currents, generated at the fluid/wall interface, result in an effectively higher thermal conductivity of the fluid. The flowfield of the fluid was shown to fill the entire tank and was not confined to the vicinity near the wall. To determine the flowfield numerically, it was necessary to resolve the boundary layer at the wall and to consider the interaction of the fluid and the wall. The resulting calculations gave increased heat transfer for higher Rayleigh numbers and for increased wall thickness. The heat transfer increased proportional to the square root power of the wall thickness to radius ratio. It also decreased as the area of the top surface decreased. The power of the Rayleigh number dependency was about -.14.

Acknowledgments

The author would like to thank Cecil Hess for performing the experimental measurements and the support of the National Science Foundation through Grant ENG 75-10294.

References

[1] Brumleve, T.D., "Sensible Heat Storage in Liquids," Sandia Lab., Livermore, Calif., SLL-73-0263, July 1974.

[2] Clakr, J.A., "A Review of Pressurization, Stratification and Interfacial Phenomena," Advances in Cryogenic Engineering, Vol. 5, 1968, pp 389-401.

[3] Barakat, H.Z. and Clark, J.A., "Analytical and Experimental Study of the Transient Laminar Natural Convection Flows in Partially-Filled Liquid Containers," International Heat Transfer Conference, Vol. 2, 1966, pp 152-162.

[4] Blottner, F.G., "Variable Grid Scheme Applied to Turbulent Boundary Layers," Computer Methods in Applied Mechanics and Engineering, Vol. 4, 1974, pp 179-185.

[5] Kee, R.J., "A Numerical Study of Natural Convection Inside a Horizontal Cylinder with Asymmetric Boundary Conditions," Ph.D. Thesis, Univ. of California, Davis, Calif., 1974.

[6] Parmentier, E.M. and Torrance, K.E., "Kinematically Consistent Velocity Fields for Hydrodynamic Calculations in Curvilinear Coordinates," Journal of Computational Physics, Vol. 19, Dec. 1975, pp 404-417.

EFFECTIVE CONDUCTIVITY OF REGULARLY
PACKED SPHERES: BASIC CELL MODEL WITH CONSTRICTION

Y. Ogniewicz[*]
University of California, Berkeley, California

and

M.M. Yovanovich[+]
University of Waterloo, Waterloo, Ontario

Abstract

Analytical solutions for conduction in basic cells of
regularly packed spheres are obtained. The basic cells are
shown to consist of a number of effective contact regions. The
influence of packing, mechanical load, and gas pressure is
determined. Experimental results for a single contact region,
and for an FCC basic cell, are in excellent agreement with the
analytical results at all mechanical loads and gas pressures
for air, argon, and helium.

Nomenclature

a	= contact radius
A	= cross-sectional area of a basic cell
c_p	= constant pressure specific heat
c_v	= constant volume specific heat
D	= sphere diameter
E	= Young's modulus
F	= mechanical load applied upon a basic cell $(P_a A)$
F_1, F_2, F_3	= constants defined by Eqs. (23-25)
h	= height of a basic cell
k_a	= apparent conductivity of a basic cell or of regularly packed spheres

---

Presented as Paper 77-188 at the AIAA 15th Aerospace
Sciences Meeting, Los Angeles, Calif., Jan. 24-26, 1977.

*Graduate Student.
+Professor, Thermal Engineering Group, Department of
Mechanical Engineering.

k_e = effective conductivity of a contact region
k_g = local gas conductivity
k_{ge} = effective gaseous conductivity associated with a contact region
k^*_{ge} = nondimensional effective gaseous conductivity associated with a contact region
k_s = solid conductivity
k_{se} = effective solid conductivity associated with a contact region
L = nondimensional contact parameter (D/2a)
M = modified Knudsen number ($2\alpha\beta\Lambda/D$)
N = normal load on a contact
P = gas pressure
Pa = apparent pressure due to mechanical loading
Po = standard gas pressure
Pr = Prandtl number
Q_g = total heat flow through the gaseous medium associated with a contact region
r = radial distance from the contact center
R_c = thermal constriction resistance
R_{cr} = thermal constriction region resistance
R_g = thermal resistance of the gaseous medium
R_r = radiation thermal resistance
R_s = thermal resistance of the solid medium
R_t = total resistance of a basic cell
T = mean gas temperature
T_o = standard temperature
ΔT_a = overall temperature difference of a contact region
x = nondimensional radial distance (r/a)

Greek Symbols

α = accommodation factor [$2(2 - \alpha_1)/\alpha_1$]
α_1 = accommodation coefficient
β = gaseous parameter {$2\gamma/[(\gamma + 1)Pr]$}
γ = ratio of specific heats (c_p/c_v)
δ = local spacing
δ^* = nondimensional local spacing (δ/a)
ϵ = emissivity
Δ = quantity defined by $\{1/[3P_a(1-\nu^2]/E\}^{1/3}$
ζ = outer radial limit of the gas region
ζ^* = nondimensional outer radial limit of the gas region
Λ = mean free path of a gas molecule
Λ_o = mean free path of a gas molecule at STP
ν = Poisson's ratio
ρ = solid fraction
σ = Stefan-Boltzmann constant

Introduction

Packed beds have a variety of applications in thermal systems. They provide a large ratio of solid surface area to volume. This property is useful in applications such as catalytic reactors, heat recovery processes, heat exchangers, heat storage systems, etc. Packed spheres also are used as an insulation material, acting as a radiation shield as well as a convection heat flow suppressant, at the expense of an increase of conduction heat transfer.

The heat-transfer characteristics of such beds, and particularly the conduction contribution, were the subject matter of a number of studies. The previous work on this subject may be divided among three groups of investigators. The first group [1-4] simplified the analysis by transforming the complex geometry of spherical particles into series and parallel arrangements of macroscopic rectangular volumes of gas and solid. One-dimensional heat-conduction analysis of such geometry then was performed. When the dependence of the gas conductivity on the Knudsen number was included, it was assumed uniform and was evaluated at some equivalent spacing obtained from geometry alone. A second group [5-8] has taken the problem further by retaining the actual geometry, i.e., spheres in contact, but the local gas conductivity was assumed uniform over the entire gas region, which actually varies in thickness. In both cases the solid constriction resistances, when incorporated, were added in parallel to the other resistances. The third approach [9,10] was to solve the problem numerically. The geometry was retained, and the gas conductivity was allowed to vary locally. The numerical solution was relatively successful, although mechanical load and packing were not considered.

In the present work, a detailed analysis of individual basic cells representing regularly packed spheres accompanied by experimental results is presented. This approach enables the isolation and analysis of the parameters, namely, mechanical loads, packing, gas pressure, and solid-gas-conductivity ratio. Constriction always occurs when the solid-to-gas-conductivity ratio is relatively large (which is true in most cases). It is obvious that the preceding condition is necessary for application of this model. The analysis of a single contact region includes the local variation of gas conductivity due to the variable thickness of the gas region, and the bending of the heat flow lines within the solid. A constriction of heat flow into the area in the vicinity of the solid-to-solid contacts both within the solid and the gas enables one to model the basic cell as an arrangement of thermal resistances associated with contact regions.

Analysis

A. Contact Region

Consider two identical spheres in contact, as shown in Fig. 1. The two solid surfaces are separated by a gaseous region varying in thickness from zero at the contact to some value δ_o at the outer adiabatic boundary (Fig. 1). The contact radius is a; the sphere diameter is D. T_{1a} and T_{2a} are temperatures of the two spheres far enough from the contact so as to be assumed relatively uniform. $T_1(r)$ and $T_2(r)$ are the temperature distributions at the solid gas interface. The foregoing is only a particular case of a contact region. The following analysis could be applied to different contact regions such as a sphere-flat contact (which is investigated subsequently), a cylinder-flat contact region, a contact region between a ball bearing and its race, etc.

Heat conduction is assumed to take place through the solid-to-solid contact region and through the gaseous layer in a parallel configuration. The validity of this assumption has been established previously by other investigators.[8,13]

For a relatively small contact-to-sphere-diameter ratio ($2a/D \ll 1$), the constriction resistance is well approximated by that of an isothermal circular contact on a half space [11] and is given by

$$R_c = 1/4k_s a \tag{1}$$

The contact radius a is evaluated by means of the well-known Hertz relation for elastic contacts, which, for two spheres, reduces to

$$a = [(3/8)ND(1-\nu^2)/E]^{1/3} \tag{2}$$

where N is the normal load, E is Young's modulus, ν is Poisson's ratio, and D is the diameter of the sphere.

The overall thermal resistance of the solid phase consists of two such constrictions and, hence, is given by

$$DR_s = L/k_s \tag{3}$$

where $L = D/2a$. The following result is obtained for the effective solid conductivity k_{se}, defined on the basis of a cube of side D:

$$k_{se} = k_s/L \tag{4}$$

Equation (4) is strictly valid only under vacuum conditions, but the contribution of the solid conduction to overall

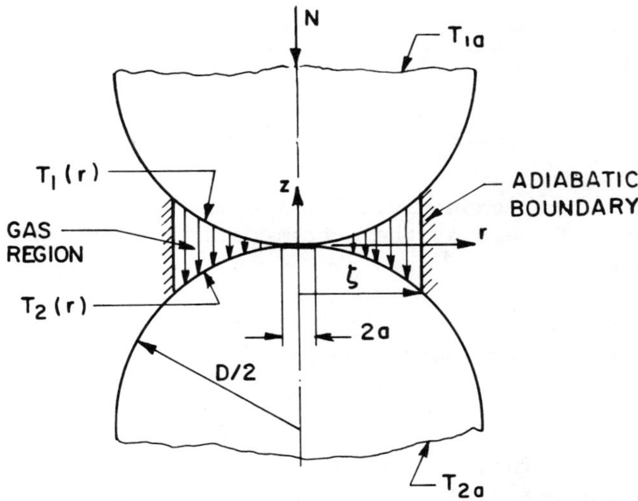

Fig. 1   Contact region between two identical spheres.

conduction diminishes as the gas pressure increases, and
hence the error incurred using Eq. (4) when calculating the
solid contribution to the overall effective conductivity is
small.   The thermal resistance of the gaseous phase is defined
based on the overall temperature difference (similarly to the
solid resistance)

$$R_g = \Delta T_a / Q_g \tag{5}$$

where $\Delta T_a = T_{1a} - T_{2a}$, and Q_g is the total heat flow through
the gaseous region.

The heat flow lines in the gaseous region are assumed to
be straight and perpendicular to the plane of contact.   This
assumption is valid in the proximity of the contact where the
two solid surfaces are nearly parallel to the contact plane,
and it is demonstrated later that the major portion of the
gaseous heat transfer occurs through this region.   The total
heat flow through the gas then is given by the following
integral:

$$Q_g = \int_a^\zeta k_g(r) \frac{\Delta T(r)}{\delta(r)} 2\pi r dr \tag{6}$$

where $k_g(r)$ is the local gas conductivity depending upon the
local spacing $\delta(r)$ and the interstitial gas mean free path Λ.
The local gas conductivity is assumed to be equal to the conductivity of a uniform layer of gas of thickness equal to the
local spacing $\delta(r)$, consistent with the previous assumption of

straight heat flow lines in the gaseous region. The local
conductivity of the gaseous region is given by [8]

$$k_g(r) = \frac{k_o}{1 + \alpha\beta\Lambda/\delta(r)} \tag{7}$$

where k_o is the conductivity of the gas under continuum conditions at STP, and α, the accommodation parameter, is defined
as follows:

$$\alpha = 2(2 - \alpha_1)/\alpha_1 \tag{8}$$

where α_1 is the accommodation coefficient, and β is defined by

$$\beta = (1/Pr)[2\gamma/(\gamma+1)] \tag{9}$$

where γ is the ratio of specific heats, and Pr is the Prandtl
number. The mean free path Λ is given in terms of the mean
free path at STP, Λ_o, as

$$\Lambda = \Lambda_o(T/T_o)(P_o/P) \tag{10}$$

The local spacing for two spheres in elastic contact is
given by the following [12]

$$\delta(r) = 2[\sqrt{(D/2)^2 - a^2} - \sqrt{(D/2)^2 - r^2} +$$

$$(2a^2/\pi D)\{[2-(r/D)^2]\sin^{-1}(a/r) + \sqrt{(r/a)^2 - 1} - \pi/2\}] \tag{11}$$

The local temperature difference $\Delta T(r)$ is taken as the
average of the following two cases: 1) The temperature distribution on the gas-solid interface is induced by the heat
flow through the solid-solid contact under vacuum conditions.
This was used by Yovanovich in his coupled model. [11] The
local temperature difference for this model is given by the
following expression:

$$\Delta T_1(r) = (2\Delta T_a/\pi)\tan^{-1}\sqrt{(r/a)^2-1} \tag{12}$$

2) The temperature distribution on the gas-solid interface is
induced by parallel heat flow through the gas as well as the
solid. This was used by Kunii and Smith [6] and Kaganer.[8]
Incorporating the local variation in the gaseous conductivity
given by Eq. (7), the following expression for local temperature difference is obtained:

$$\Delta T_2(r) = \frac{\Delta T_a}{[D/\delta(r)-1]k_g(r)/k_s+1} \tag{13}$$

The actual temperature distribution is somewhere between the two preceding. Since $\Delta T_1(r)$ and $\Delta T_2(r)$ do not differ substantially, the average is taken to yield

$$\Delta T(r) = (\Delta T_a/2)[(2/\pi)\ \tan^{-1}\sqrt{(r/a)^2 -1} +$$

$$\{[D/\delta(r)-1]k_g(r)/k_g + 1\}\{[D/\delta(r)-1]k_g(r)/k_g + 1\}^{-1}] \qquad (14)$$

Substituting Eqs. (7, 11, 14, and 6) into Eq. (5) and defining the effective gas conductivity in a similar manner to the solid effective conductivity (i.e., based upon a cubic cell of side D), there results [15]

$$k_{ge} = (k_o/L)I \qquad (15)$$

where the integral is defined as

$$I = \int_1^\zeta I_n\ dx \qquad (16)$$

and the integrand is

$$I_n = \frac{x\ \tan^{-1}\sqrt{x^2-1}}{\delta^*(x) + ML} + \frac{(\pi/2)x}{(k_o/k_s)[2L-\delta^*(x)] + [\delta^*(x)+ML]} \qquad (17)$$

The nondimensional parameters are defined as follows:

$$x = r/a; \quad \zeta^* = \zeta/a; \quad \delta^*(x) = \delta(r)/a \qquad (18)$$

and the modified Knudsen number is

$$M = 2\alpha\beta\Lambda/D \qquad (19)$$

The value of the integral was obtained numerically.[15]

The integrand I_n may be viewed as a nondimensional heat flow per unit length in the radial direction r. Figure 2 shows the variation of I_n, as given by Eq. (17), with the parameter M.

For small M, corresponding to relatively high gas pressures, the main contribution to I is due to a very small region in the vicinity of the contact. This is observed in Fig. 2, where I_n reaches a maximum value when x is small. For this case, the gas conduction is of relative importance compared with solid conduction. At large values of M, I_n decreases and flattens, suggesting that the entire gas region contributes approximately equally to the overall gaseous heat

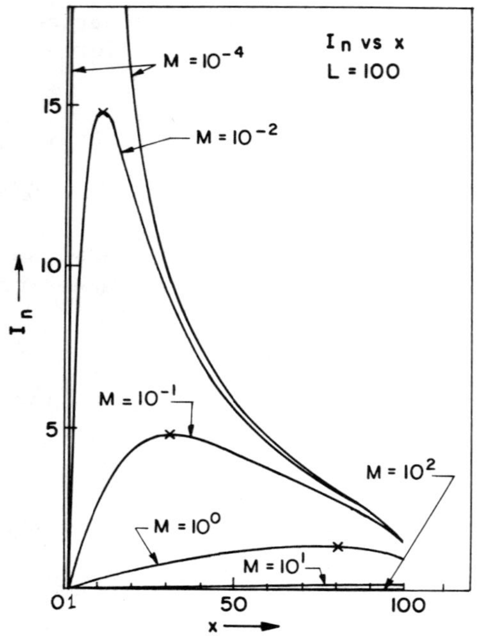

Fig. 2 Local variation of heat flow per unit length of radius.

flow I but this occurs at lower gas pressures where the solid conduction dominates the heat transfer. This observation suggests that the conduction through the interstitial gas of variable thickness should be modeled as constriction heat flow through a small gas region associated with each contact.

Figure 3 compares the effective gaseous conductivity obtained from Eq. (15) with the conductivity obtained from an analysis by Kaganer,[8] in which the conductivity of the gas was evaluated at the constant spacing of 2D/3. The present model improves upon the previous analysis by allowing for the region where the gas is rarefied ($M \leq 10^{-1}$) to move toward the contact area (x=1) with increasing gas pressure. This results in a monotonic increase in conductivity over a very wide range of M. It differs substantially from the behavior of a uniform gas layer of 2D/3 thickness, in which the variation of k^{*}_{ge} is more rapid in the range $10^{-2} < M < 10$ and furthermore does not allow for increase in the gaseous conductivity beyond a value of M approximately equal to 10^{-2} (Fig. 3).

Figures 4 and 5 demonstrate the effect of the solid-to-gas-conductivity ratio k_s/k_o and the sphere-to-contact-diameter ratio L. Because the temperature difference across the gaseous region decreases with the ratio k_s/k_o, the k^{*}_{ge}

curves do not collapse into one, which would mean that k_{ge} is directly proportional to k_o. The effective gas conductivity k_{ge}^* at small values of M is larger than unity because the relatively highly conducting solid allows the heat to flow through the very thin gas region adjacent to the contact area. The effective gaseous conductivity increases with a decrease in the contact-to-sphere-diameter ratio because the maximum value of I_n increases and moves inward.

Since the solid and gaseous effective conductivities were based on the same cell dimensions and are thermally in parallel, the overall effective conductivity k_e is simply the sum of the two, and therefore

$$k_e = k_{se} + k_{ge} = (1/L)(k_s + k_o I) \qquad (20)$$

B. Basic Cells

Simple cubic packing (CP), body-centered close packing (BCC), and face-centered close packing (FCC) are particular

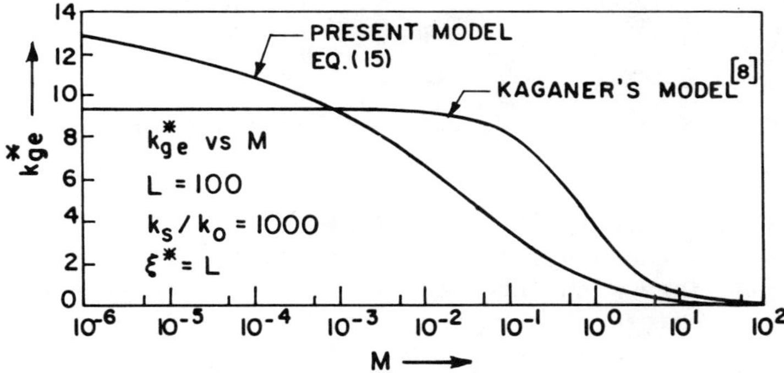

Fig. 3 Comparison between the present model and one employing a constant local gas spacing[8].

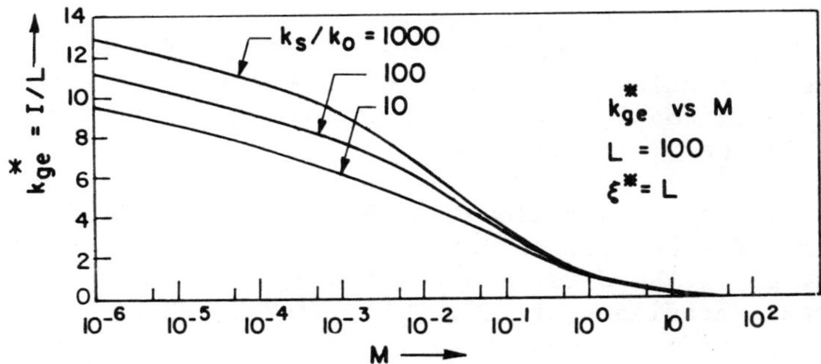

Fig. 4 Effect of solid-to-gas-conductivity ratio k_s/k_o on effective gas conductivity.

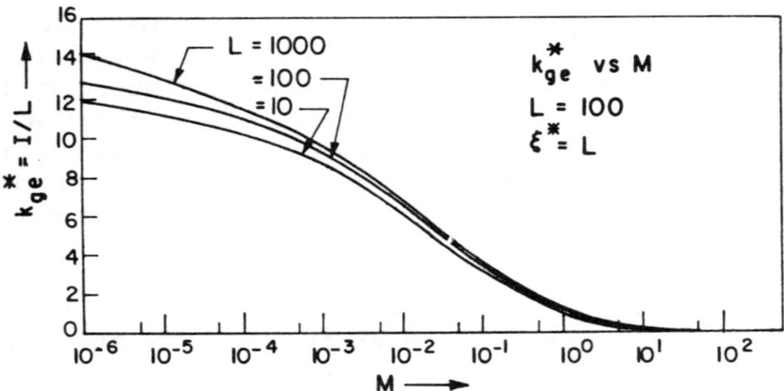

Fig. 5 Effect of sphere-to-contact-diameter ratio L on effective gas conductivity.

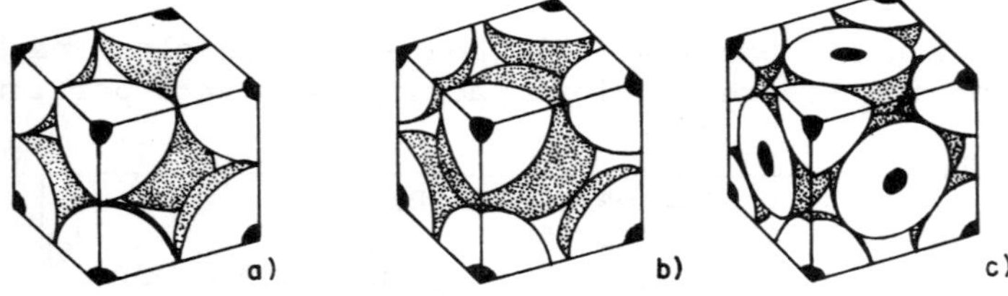

Fig. 6 Basic cells of CP, BCC, and FCC packings in orientation A.

examples of regularly packed spheres. These three packings are chosen as the subject of consideration, although other regular packings with intermediate contact numbers and solid fractions ρ exist.

Analysis of basic cells representing CP, BCC, and FCC packings was presented by Chan and Tien.[14] Their analysis considered evacuated beds only and, furthermore, only a single orientation of each packing with respect to the macroscopic temperature gradient. Figures 6a – 6c correspond to CP, BCC, and FCC basic cells, respectively. These basic cells represent the particular orientation of each packing, here designated orientation A, where the macroscopic temperature gradient is in the z direction.

Three such orientations, A, B, and C, for the three packings under consideration are shown in Figs. 7a – 7c, respectively. The direction cosines of the macroscopic temperature gradients are (0, 0, 1), ($\sqrt{2}/2$, $\sqrt{2}/2$, 0), and ($\sqrt{3}/3$, $\sqrt{3}/3$, $\sqrt{3}/3$), respectively. These orientations enable the construction of basic cells, which consist of two parallel isotherms and perpendicular adiabats.

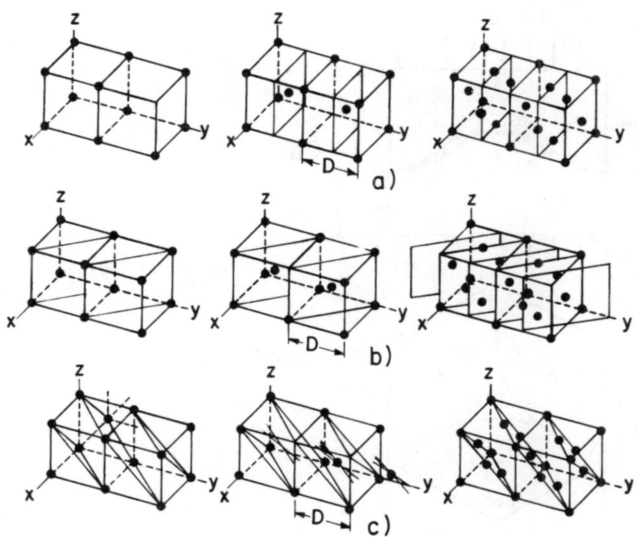

Fig. 7 Three orientations, A, B, and C, for packings CP, BCC, and FCC.

An example of a cell at an orientation different from A is shown in Fig. 8. This cell represents an FCC packing in orientation C, denoted FCC(C). The two horizontal boundaries are at uniform temperatures, T_1 and T_2, whereas the six vertical boundaries are adiabatic.

Each basic cell consists of a number of contact regions, each encompassing a circular solid-solid contact and an associated volume of gas of a particular shape, depending upon the specific packing structure encountered. The heat flow lines within each gas region become increasingly perpendicular to each contact plane as the solid-solid contact area is approached. Since the major gaseous contribution to the heat flow occurs relatively near the solid-solid contact, the area through which the gaseous heat flow occurs is transformed into an equivalent circular area, and, in addition, the heat flow lines are assumed to be perpendicular to the contact plane. With this simplification, the problem of heat conduction in a basic cell can be solved directly in terms of the results presented earlier for a single contact region.

The analysis of a basic cell proceeds as follows:

1) The apparent pressure P_a on a basic cell is related through geometry to the normal load N on a contact. The Hertz relation, Eq. (2), is used to calculate the sphere-to-contact-

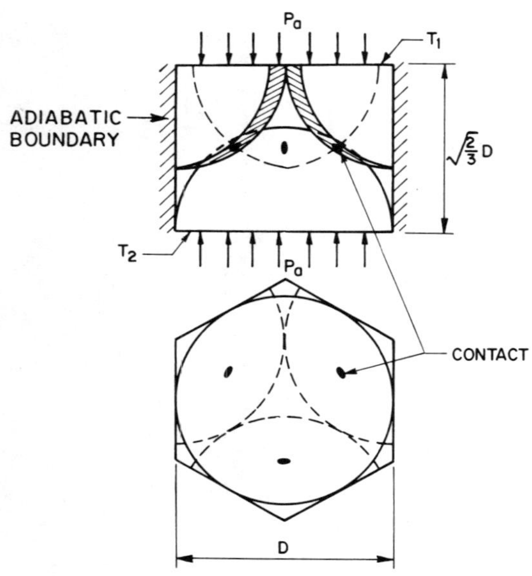

Fig. 8   FCC(C) basic cell.

diameter ratio L as follows:

$$L = \frac{F_1}{[\,3P_a(1-\nu^2)/E\,]^{1/3}} \tag{21}$$

where F_1 is obtained from the geometry of the basic cell alone
Defining

$$\Delta = \{[\,3P_a(1-\nu^2)/E\,]^{1/3}\}^{-1} \tag{22}$$

L then may be written as:

$$L = F_1\Delta \tag{23}$$

2) The gaseous region area is transformed to a circular one,
and its radius ζ^* is evaluated, resulting in

$$\zeta^* = F_3\Delta \tag{24}$$

The gaseous region area is found by dividing the sphere surface area by the number of participating contacts. It should
be noted that contacts perpendicular to the macroscopic
temperature field do not participate in conduction and therefore are not considered.

3) The overall resistance R_t is determined from the thermal
resistance network for each basic cell.

4) The apparent conductivity of a basic cell k_a is obtained
incorporating the overall dimensions, i.e., the height h and

Table 1 F_1, F_2, and F_3 for various basic cells

Basic cell (ρ)	Orientation	F_1	F_2	F_3
CP(0.524)	A	1	1	1.00
	B	1	1	0.87
	C	1	1	0.75
BCC(0.680)	A	$3^{1/6}$	$3^{1/3}$	0.79
	B	$3^{1/6}$	$3^{1/3}$	1.04
	C[a]	...	...	...
FCC(0.740)	A	$2^{1/2}$	2	0.94
	B	$2^{1/2}$	2	0.85
	C	$2^{1/2}$	2	1.05

[a]Not analyzed because of difficulty in constructing a thermal basic cell.

the area A, and is written as follows [15]

$$k_a = (F_2/\Delta)[k_s + k_o I(M,\Delta,k_s/k_o,F_1,F_3)] \tag{25}$$

The procedure was employed successively for each of the basic cells [except BCC(C)], and the results are summarized in Table 1.

C. Regularly Packed Spheres

The preceding analysis yielded apparent thermal conductivities for three discrete solid fractions, $\rho = 0.524$, 0.680, 0.740. The dependency of the apparent conductivity on the various packings and orientations is manifested through the three constants F_1, F_2, and F_3. In the light of the relatively small range of F_3 and its weak influence on the gaseous conduction, one may use an overall average value of F_3 ($\overline{F_3} = 0.91$) for all packings and orientations. Hence the packings are nearly isotropic, and only the constants F_1 and F_2 are required to evaluate the apparent conductivity.

The apparent conductivity of regularly packed spheres with intermediate solid fractions may be evaluated by interpolation of the results of the three discrete solid fractions. Second-order polynomials were fitted to the values of F_1 and F_2 given in Table 1, yielding

$$F_1(\rho) = 10.53\rho^2 - 11.39\rho + 4.077 \tag{26}$$

$$F_2(\rho) = 29.95\rho^2 - 33.22\rho + 10.18 \qquad (27)$$

The preceding functional relations for F_1 and F_2 in terms of the solid fraction in conjunction with Eq. (25) provide the means of determining the apparent conductivity of regularly packed spheres with corresponding solid fraction ρ.

Experimental Results and Discussion

A. Sphere-Flat Contact Region

Experimental data for a sphere-flat contact region were obtained by Kitscha and Yovanovich [13] for varying mechanical load and gas pressure for air and argon. For a sphere-flat contact region, a special case of the model presented here, one can determine 1) the contact radius

$$a = [\,(3/4)ND(1-\nu^2)/E\,]^{1/3} \qquad (28)$$

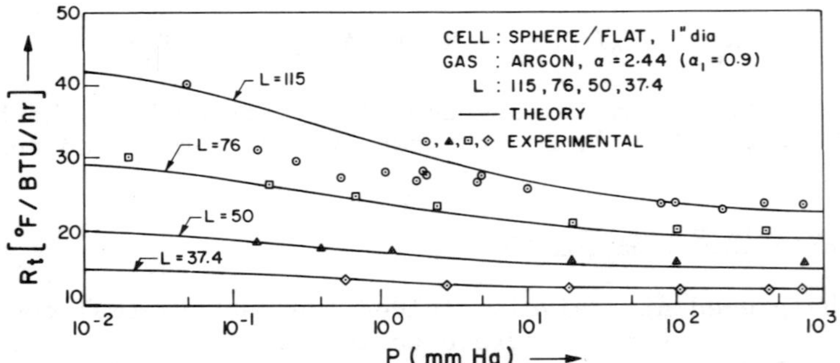

Fig. 9   Thermal resistance of a sphere-flat contact region with air at various loads and gas pressures.

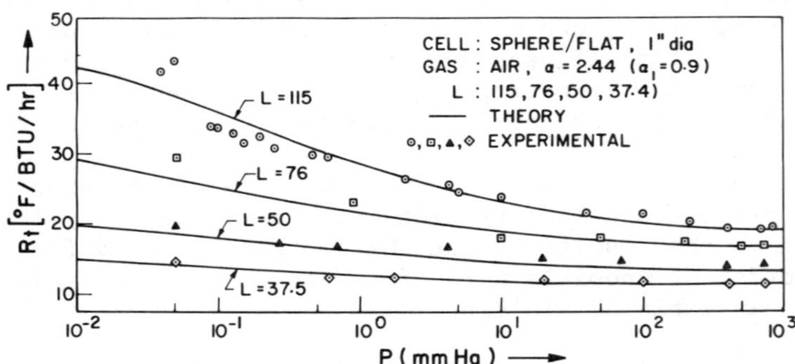

Fig. 10   Thermal resistance of a sphere-flat contact region with argon at various loads and gas pressures.

and 2) the nondimensional local spacing:

$$\delta^*(x) = \sqrt{L^2-1} - \sqrt{L^2-x^2} + (2/\pi L)[\,(2-x^2)\sin^{-1}(1/x) +$$

$$\sqrt{x^2-1} - (\pi/2)\,] \tag{29}$$

A small radiative contribution to the overall heat transfer was observed during the test program. To account for his heat transfer, the following thermal resistance was added in parallel to the gaseous and solid conduction resistances:

$$R_r = (4\varepsilon_1 A_1 F_{12}\sigma T_m^3)^{-1} \tag{30}$$

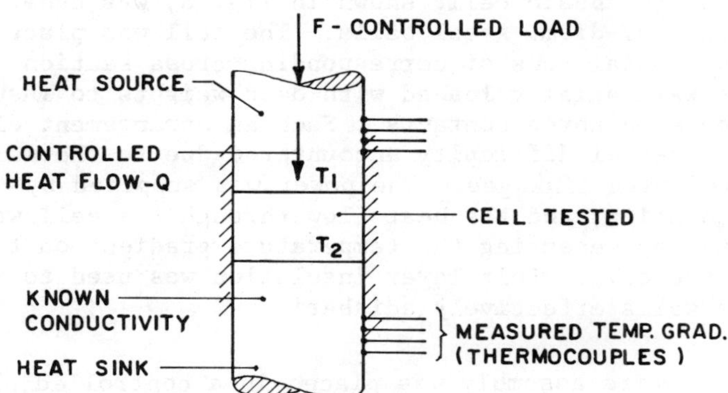

Fig. 11 Experimental setup of a basic cell.

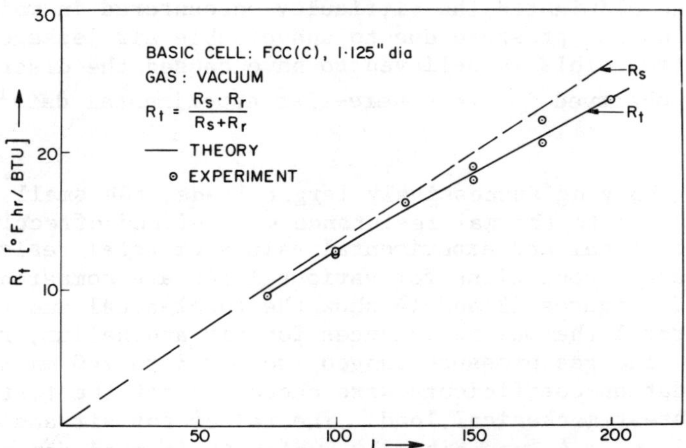

Fig. 12 Thermal resistance of an FCC(C) cell vs mechanical load in vacuum.

where ε_1, A_1, F_{12}, σ, and T_m are the emissivity, surface area, radiative view factor, Stefan-Boltzman constant, and mean temperature, respectively. Free convection was demonstrated to be negligible and is not considered in the present model.

The experimental and theoretically calculated values are presented in Figs. 9 and 10. The agreement is excellent except for relatively small loads, when L = 115 for air and argon. The discrepancy is believed to be caused by experimental error. Such results demonstrate that the effects of the mechanical load and gas properties are well modeled by the present analysis.

B. FCC(C) Basic Model

An FCC(C) basic cell, shown in Fig. 8, was constructed using 1.25 in.-diam. steel balls. The cell was placed between two coaxial rods of corresponding cross section (Fig. 11) The rods were axially loaded with dead weights to insure equal loads on three contacts. Such an arrangement eliminates the experimental difficulty encountered due to thermal stresses associated with linkages. The power was supplied by Joulean heating, and the heat flow through the cell was determined by measuring the temperature gradient on the sink side of the cell. Multilayer insulation was used to render the side walls effectively adiabatic.

The entire assembly was placed in a controlled pressure environment. Low vacuums (10^{-2} to 1 mm Hg) were obtained by establishing pressure equilibrium with a bleeder valve and constant evacuation by a mechanical vacuum pump. This technique eliminated the difficulty encountered in maintaining a constant pressure due to unavoidable air leakage into the system. This is believed to have caused the discrepencies observed in the sphere-flat experimental data[13] when L = 115.

By applying successively larger loads, the small radiant contribution to thermal resistance was defined effectively. The theoretical and experimental values of total resistance under vacuum conditions for various loads are compared in Fig. 12. Figures 13 and 14 show the theoretical and experimental total thermal resistances for air and helium, respectively. The gas pressure ranged from 10^{-1} to 740 mm Hg. The accommodation coefficients were chosen to fit the test data at the lightest mechanical load. The values for air and helium were 0.9 and 0.4, respectively, which is in good agreement with values reported in the literature.

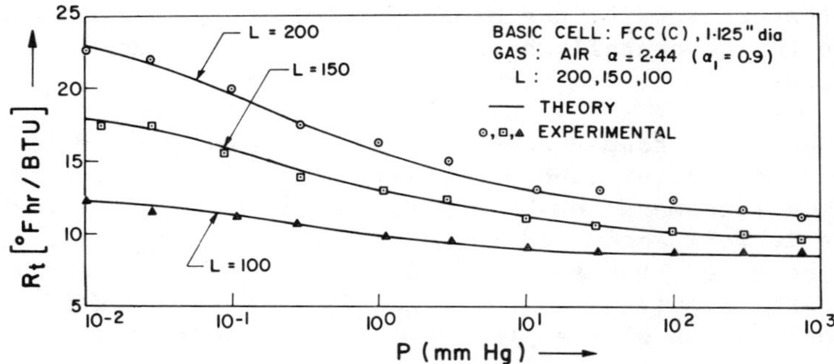

Fig. 13 Thermal resistance of an FCC(C) cell with air as a function of load and gas pressure.

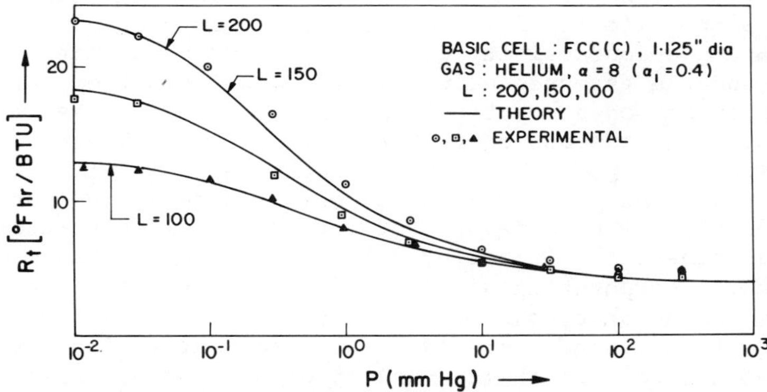

Fig. 14 Thermal resistance of an FCC(C) cell with helium as a function of load and gas pressure.

The solid thermal resistance varies linearly with L, as seen in Fig. 12. The experimental results for thermal resistance span a substantial load range (75<L<200). The excellent agreement of the experimental results with the predicted values (incorporating radiation) over the entire range effectively verifies the effect of load in the present theoretical results, as given by Eq. (3).

The total thermal resistance varies gradually over a wide range of M , as seen in Figs. 13 and 14. It continues to do so beyond $M = 10^{-2}$ (corresponding to P $\backsim$ 1.5 mm Hg for air, and p $\backsim$ 5 mm for helium). This trend is not predicted by the previous models, which did not allow for local variation in gas conductivity (see Fig. 3). The experimental total resistance varies substantially from vacuum to atmospheric pressure, particularly at large values of L. In light of the excellent agreement over the entire range of gas pressure (or M), the present model for gaseous contribution to the overall thermal resistance is demonstrated effectively.

The solid-to-gas-conductivity ratio k_s/k_o is different for air and helium (approximately 1800 and 300, respectively). The good agreement for both gases demonstrates that its effect is well accounted for by the present thermal model. It might be noted, though, that the values of k_s/k_o in both cases are relatively large, which is a necessary condition for application of the present constriction model.

Summary

Analytical results for the apparent conductivity of regularly packed spheres were obtained. The analysis incorporates the effects of mechanical load, gas pressure, packing, solid-to-gas-conductivity ratio, and other properties of the solid and the gas successfully. The results differ markedly from previous ones, particularly in the effect of gas pressure upon the gaseous contribution to the overall conductivity. The excellent agreement of the present analysis with experimental data obtained with basic cells indicates an improvement upon previous models insofar as regularly packed spheres are concerned. It is believed that future work, based on these results, incorporating additional parameters associated with random packed beds, should be successful. In addition, we may add that the present model for a contact region could be applied directly to a wide variety of problems consisting of contacts between curved surfaces submerged in a stagnant gas.

Acknowledgments

The authors acknowledge the financial support of the National Research Council. One of the authors (Y. Ogniewicz) thanks the National Research Council for an NRC scholarship. The advice and assistance of R. Kaptein during the experimental work are acknowledged gratefully.

References

[1] Beveridge, G.S.G. and Haughey, D.P., "Axial Heat Transfer in Packed Beds, Stagnant Beds Between 20 and 750°C," _International Journal of Heat and Mass Transfer_, Vol. 14, Aug. 1971, pp. 1093-1113.

[2] Luikov, A.V., Shashkov, A.G., Vasiliev, L.L., and Fraiman, YU.E., "Thermal Conductivity of Porous Systems," _International Journal of Heat and Mass Transfer_, Vol. 11, Feb. 1968, pp. 117-140.

[3]Yagi, S. and Kunii, D., "Studies on Heat Transfer Near Wall Surface in Packed Beds," A.I.Ch.E. Journal, Vol. 6, Jan. 1960, pp. 97-104.

[4]Krischer, O. and Kroll, K., "Die Wissenschaftlichen Grundlager der Trochnungstechnik," Berlin-Gottingen-Heidelberg, Bd. 1, 1956.

[5]Schotte, W., "Thermal Conductivities of Porous Solids," A.I.Ch.E. Journal, Vol. 6, Jan. 1960, pp. 63-67.

[6]Kunii, D. and Smith, J.M., "Heat Transfer Characteristics of Porous Rocks," A.I.Ch.E. Journal,Vol. 6, Jan. 1960, pp.71-78.

[7]Masamune, S. and Smith, J.M., "Thermal Conductivity of Spherical Particles," Industrial Engineering Chemistry Fundamentals, Vol. 2, May 1963, pp. 136-143.

[8]Kaganer, B.M.G., Thermal Insulation in Cryogenic Engineering, transl. by A. Moscona, Israel Program for Scientific Translation, Jerusalem, 1969.

[9]Wakao, N. and Vortmeyer, D., "Pressure Dependency of Effective Conductivity of Packed Beds," Chemical Engineering Science, Vol. 26, Oct. 1971, pp. 1753-1765.

[10]Wakao, N. and Kato, K., "Effective Thermal Conductivity of Packed Beds," Journal of Chemical Engineering of Japan, Vol. 2, Jan. 1969, pp. 24-33.

[11]Yovanovich, M.M. and Kitscha, W.W., "Modeling the Effect of Air and Oil Upon the Thermal Resistance of a Sphere-Flat Contact," AIAA Progress in Aeronautics and Astronautics: Thermophysics and Spacecraft Thermal Control, Vol. 35, edited by R.G. Hering, MIT Press, Cambridge, Mass., 1974, pp. 293-319.

[12]Cameron, A., Principles of Lubrication, Wiley, New York,1966.

[13]Kitscha, W.W. and Yovanovich, M.M., "Experimental Investigation on the Overall Thermal Resistance of Sphere-Flat Contacts," AIAA Progress in Aeronautics and Astronautics: Heat Transfer with Thermal Control Applications, Vol. 39, edited by M.M. Yovanovich, MIT Press, Cambridge, Mass., 1975, pp. 93-100.

[14] Chan, C.K. and Tien, C.L., "Conductance of Packed Spheres in Vacuum," *Journal of Heat Transfer*, Vol. 95, Aug. 1973, pp. 302-308.

[15] Ogniewicz, Y., "Conduction in Basic Cells of Packed Beds," M.A.Sc. Thesis, Dept. of Mechanical Engineering, Univ. of Waterloo, 1975.

CHANGES OF SURFACE SHAPES DUE TO CONTACT LOADING
WITH THERMAL AND PRESSURE CYCLING

A. Williams* and N. Idrus[+]
Monash University, Clayton, Victoria, Australia

Abstract

The thermal contact resistance of a metallic joint depends
critically on the shapes of the contact elements. Predictions
of thermal resistance rely on two-dimensional surface profiles
measured before the joint is assembled, whereas the shapes may
change considerably as a result of service conditions. This
paper describes briefly the development of a measuring system
to provide a three-dimensional description of surfaces and
presents results of tests to trace the changing characters of
finely ground aluminium surfaces following compression at high
pressures at room temperatures or thermal cycling at relatively
low pressures. The latter tests shed new light on the
"directional" effect of contact resistance.

I. Introduction

For the prediction of the thermal contact resistance of a
metallic joint, it is necessary to know, among other things,
the shapes of the contacting surfaces when loaded. The usual
source of information on these shapes is from magnified two-
dimensional profiles obtained from a Talysurf profilometer, or
any similar instrument, measured on the surfaces before they
are assembled together. However, when the assembled joint is
loaded, the contacts are deformed both elastically and plasti-
cally to an extent dependent on the applied load, the hardness-
es of the contact elements, and also on the shapes of the
microscopic asperities forming the spots of actual contact

Presented as Paper 77-791 at the AIAA 12th Thermophysics
Conference, Albuquerque, N. Mex., June 27-29, 1977. Copyright ©
American Institute of Aeronautics and Astronautics, Inc., 1977.
All rights reserved.
*Associate Professor, Mechanical Engineering Department.
[+]Graduate Scholar, Mechanical Engineering Department.

(which is usually a very small fraction of the apparent area
of contact). These very small regions of true metallic contact
may change in shape considerably if the joint is subjected to
any relative motion of the two contact elements. Such slipping
may occur as a result of conditions of service of a joint subjected to vibrations and side loads; hence the thermal resistance of the joint may vary with its history of service.

Very small lateral slipping movements also are believed to
be partially responsible for the so-called "thermal rectification" effect of a joint between dissimilar metals but are
superimposed upon distortions of the contacts due to other
differential expansions.[1] Some evidence for these changes of
shapes of contact spots was collected using a profilometer,[2]
but the measuring technique at that time was inadequate to
explore this in detail. The directional effects of dissimilar
metal contacts have received considerable attention in recent
years.[3-5] However, the aspect of surface damage has not been
examined in detail; hence this is one of the motivating factors
for this present investigation.

This paper describes work aimed at identifying and
measuring very small changes in the shapes of surfaces caused
by contact loading. This involved the development of a
measuring system and technique capable of three-dimensional
quantification of surfaces, as the present two-dimensional
information is inadequate. The objectives of these measurements include the real contact areas, the distributions of
actual contacts over the zone of apparent contact, the stat-
istical descriptions of surface shapes, etc. These objectives
have been achieved for a variety of surface finishes undergoing
contact loading in simple compression tests or after thermal
cycling.

II. Surface Measuring Equipment

Existing profilometers drag a stylus along a short
straight line across a surface and magnify the movement of the
stylus to obtain a cross-sectional profile for which the roughness [centerline average (CLA) or root mean square (rms)] is
computed automatically. This single parameter roughness value
commonly is used in engineering manufacture to compare
surfaces but gives little indication of the character of the
surface. Profiles having the same roughness may have very
different shapes, with different capabilities for specified
functions, e.g., to support load without large deformation, or
to resist abrasive wear, etc. This inadequacy prompted this
development toward a three-dimensional description, described
fully in Refs. 6 and 7, which follows similar recent developments in England.[8-10]

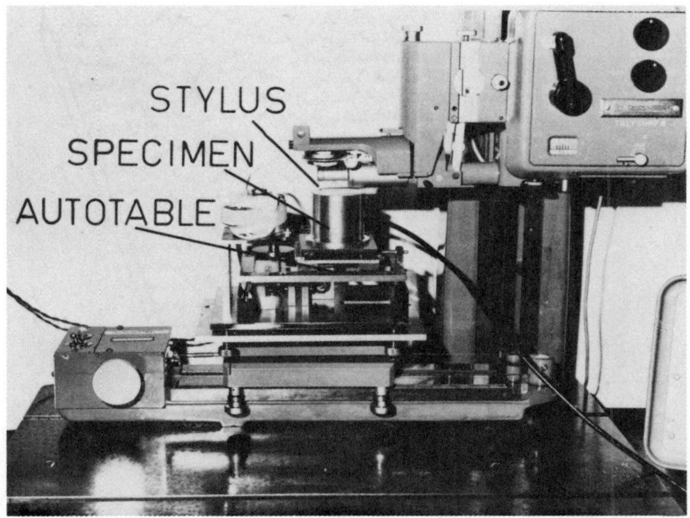

Fig. 1   Monash University "Autotable" in position on Talysurf profilometer.

The three-dimensional information is obtained by using a newly developed component, the "Autotable" (see Fig. 1), which moves the test specimen automatically under the fixed stylus of the Talysurf in a series of closely spaced parallel paths. At the end of each traversing motion, the specimen is indexed laterally, typically by 76μ (but this can be reduced to as low as 4μ if required), and the specimen is reversed along its next traverse. The motion thereby produces an automatic scanning of a rectangular area of the surface. Typical dimensions of the scan are 11 x 11 mm, subdivided into 168 traverses.

While the traversing occurs, the electrical output from the stylus is fed via an amplifier to an FM recorder (Electrodata 14-channel) to be stored on magnetic tape. One channel of the recorder is used to collect control signals from microswitches operated at each end of the Autotable's traverse; these are required later to recover proper sequencing of the profile signal.

At the end of a complete scanning, the magnetic tape is processed by a hybrid computer comprising a Nova 1200 digital computer and an AD 5 analog computer, with appropriate accessories. The outputs include roughness, skewness, and kurtosis of profile heights, a histogram of profile height frequency, correlation coefficients, profile asperity gradients, and an isometric map of the surface built up of parallel traces to present pictorially the features of the surface (see Figs. 2a and 2b).

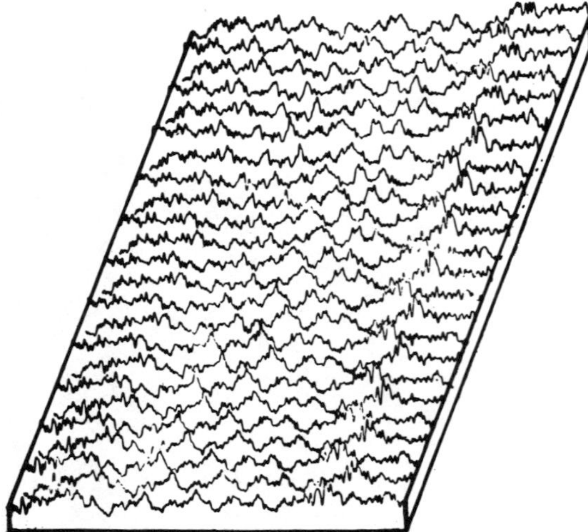

Fig. 2a   Isometric plot of a ground surface.   (Note: Alternate
traces are displayed laterally because of different response
characteristics of the limit switches.)

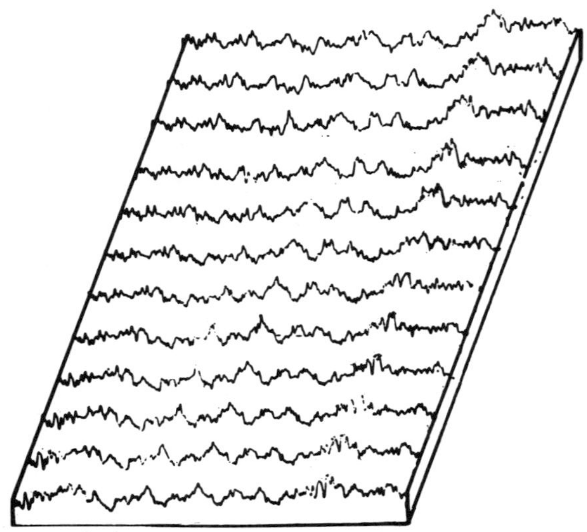

Fig. 2b   Same surface as Fig. 2a but with alternate traces
omitted to clarify topography.

     A vital feature of the Autotable is its facility to
reposition a test specimen after the specimen has been removed.
This repositioning is made by aligning hair lines on the
specimen with those on the holder, and with the Autotable
assembly mounted in precision Rank Taylor Hobson traversing

table. Traces made after repeated repositioning of a sample
were seen to be identical, thereby confirming the required
accuracy. This was checked over the entire area of scan and
for initial traverses in the two directions.

The purpose of this repositioning facility is to allow us
to examine the histories of deformation of recognizable
asperities after subjecting the surface to various loading
conditions, e.g., after the surface had been loaded during a
series of heat-transfer tests or by simple compression. The
isometric profile maps printed on tracing paper allow their
superimposition, to enable the changes of profiles to be identified and measured. These changes produce clear indications
of the patterns of the real contact spots and of contact areas.

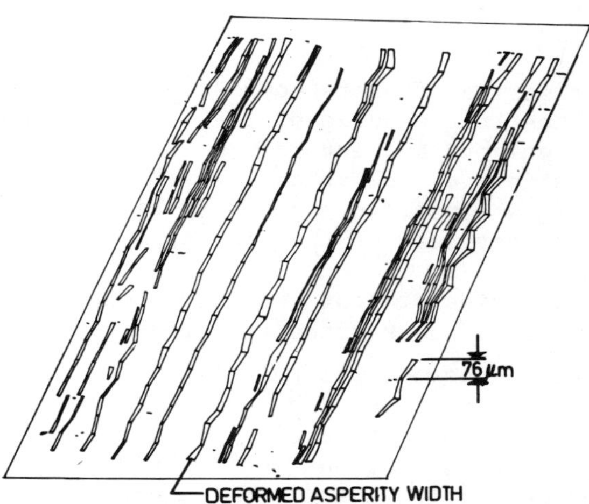

Fig. 3 Contact map of ground surface after compression against
rigid plane.

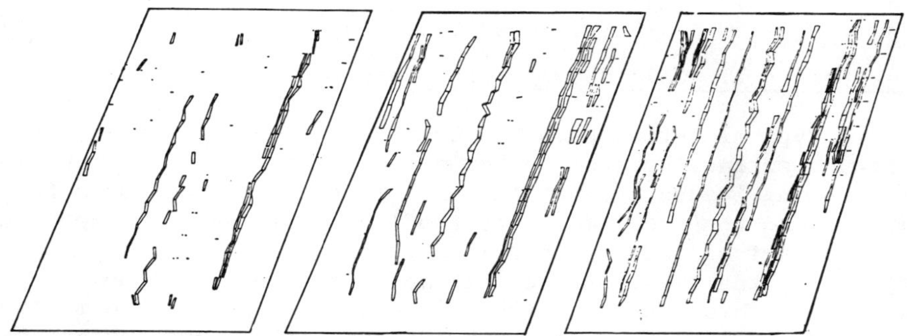

Fig. 4 Successive contact maps of ground surface after
compression against a rigid plane at contact pressures
of 50, 150 and 250 MPa.

III.   Test Procedure and Specimens

Cylindrical specimens of aluminium alloy A2011 were prepared, with their flat surfaces having ground finishes of controlled roughness over a range of CLA typically 1 to 4μ.   In
the first series of tests, these were of 5 cm diam. and were
compressed at room temperature against a very smooth flat surface (0.3 μ CLA) of a hardened steel specimen. The latter was
flat to within four interference fringes of yellow light when
compared with an optical flat. For the second series of tests,
the aluminium specimens were of 2.54 cm diam. and had surfaces
similar to the previous ones. Several of these specimens were
prepared by polishing them to almost optical flatness and then
stroking them against specified grades of emery paper to obtain
an accurately flat and uniformly rough surface with a single
direction of lay.

The first series of tests consisted of simple compressive
loadings in a test rig capable of loads up to 100 tons using a
system of dead weights and levers. This method of loading is
free from vibrations or pulsations, which may affect the contacts. This rig was used to apply very high pressures, comparable with the hardness of the aluminium alloy.

The second series of tests was performed in a rig designed
for investigations of heat transfer across joints. The maximum
available contact pressure was relatively low, e.g., $200 kN/m^2$
($\doteq$ 3000 psi). The joint could be subjected to heat flow in
either direction without disturbing the assembly, and mean
interface temperatures of about 200°C could be obtained. The
loading was by dead weights and levers, applied to a column
assembly that included the test joint, all contained within a
steel chamber in which a vacuum, e.g., 10^{-2}Torr, was maintained
for thermal insulating purposes.

IV.   Results of Tests

A.   Series 1: Cold Compression Tests

A typical set of parallel profiles is shown in Fig. 2 for
a ground surface before its being subjected to compression.
After being loaded against the smooth steel surface for a few
minutes and then replaced into the identical measuring position,
a second set of parallel profiles is produced, to be superimposed on the first set so that any differences can be
observed. This operation is performed manually, as the micro-
switch control signals cannot guarantee perfect phasing of the
forward and backward traverses of the Autotable, and the final
alignment therefore is done visually. Automation of this operation is being investigated.

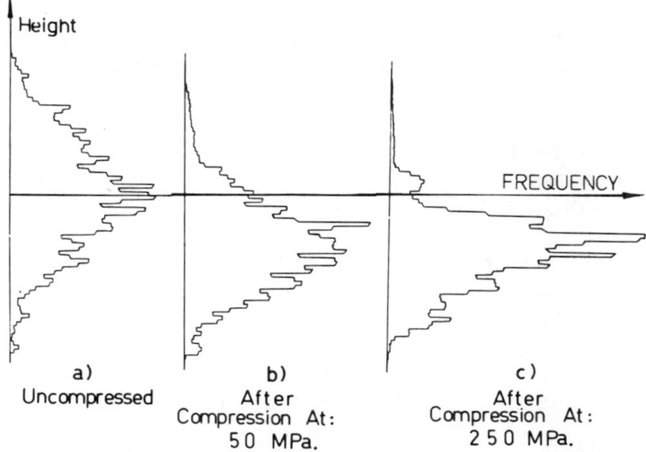

Fig. 5 Profile height frequency distributions of ground
surface before and after compression against a rigid plane.

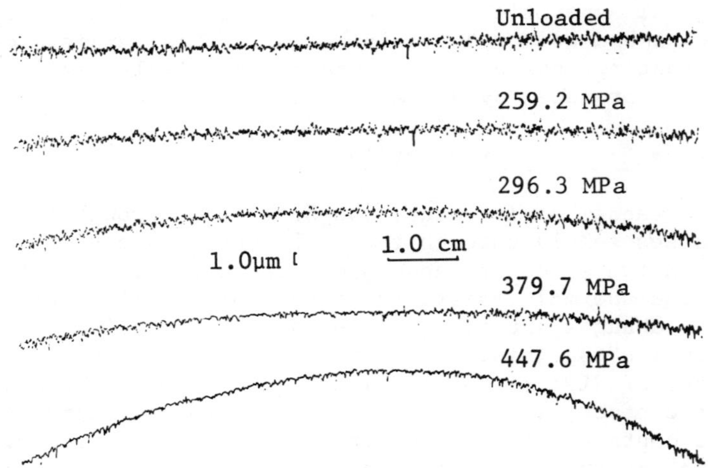

Fig. 6 Profiles showing large-scale deformations caused by very
high contact pressures. Note low horizontal magnification.

    Figure 3 shows a typical contact map obtained by the
superimposition process. It shows clearly those asperities
that suffered deformation and can be used for the measurement
of contact areas and spot distribution.

    Figure 4 shows three similarly produced maps for the same
specimen after incremental loadings at 50, 150, and 250 MPa.
The developing areas and numbers of actual contacts are identified easily.

The distribution of profile heights of a freshly ground surface is roughly gaussian, as can be seen in Fig. 5a, as expected. After compression against a plane to 50 MPa, the distribution is distinctly skewed, as shown in Fig. 5b, which indicates that the profile peaks are now much less frequent and that the maximum heights of these are reduced. The trend continues to 250 MPa (see Fig. 5c).

However, as the applied contact pressure approaches the yield stress of the material (296 MPa for the aluminium used herein), the test specimen begins to suffer bulk deformation. On removal of the specimen from the compression machine, there was obviously a partial elastic strain recovery, which produces a bowing of the previously flat surface, as shown in Fig. 6.

Repeated incremental compressions of this type alternately tended to increase and then reduce the severity of the bowing, with subsequent damage to the asperities around the highest part of the hump. For a maximum applied pressure of 448 MPa, the original asperities were still identifiable, although their peaks had been reduced in height. This persistence of surface features has been observed on relatively small-scale indentation produced by a steel ball on a rough surface.

B.   Series 2:   Heat-Transfer Tests

In a vacuum environment, the heat flow through a metallic joint passes solely through the real contact spots. Any changes in the sizes of these spots or in their distribution affect the thermal resistance of the joint. It is believed[2] that such changes are responsible for the so-called "thermal rectification" of joints of dissimilar metals. The two important types of distortion are 1) relative displacements in the plane of contact, and 2) relative displacements normal to the plane of contact. The former are believed to produce slight "smearing" of the softer of the two sets of mating asperities and are sought herein as evidence for the explanation of the rectification effect.

Three sets of tests were conducted, at comparatively light contact pressures (9.5 MPa, $\simeq$ 1300 lbf/in.^{2}) and with interface temperatures typically up to about 200°C. Even at these pressures, the surfaces were seen to become changed in character after only three cycles of thermal cycling (see Fig. 7).

The high peaks of the original surfaces disappeared as a result of the heat-transfer tests at contact pressures as low

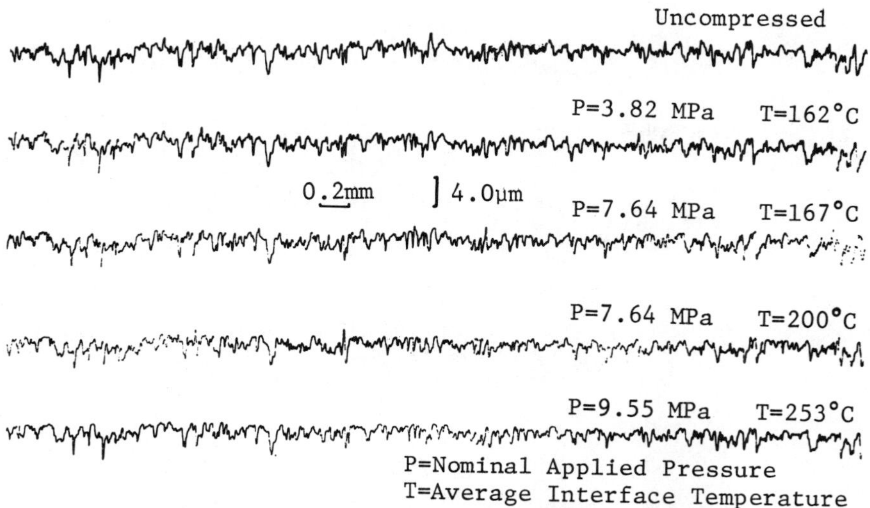

Fig. 7   History of profile of ground surface after thermal cycling at low contact pressures. Specimen No. 4 position 2 AL. A2011 yield stress = 296.5 MPa.

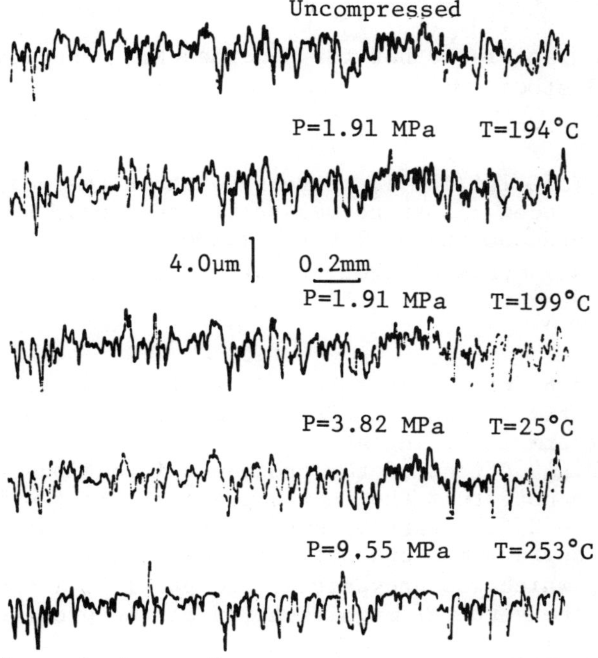

Fig. 8   Enlarged views of same portion of profile following thermal cycling at low contact pressures. Specimen No. 5 position 3 AL. A2011 yield stress = 296.5 MPa.

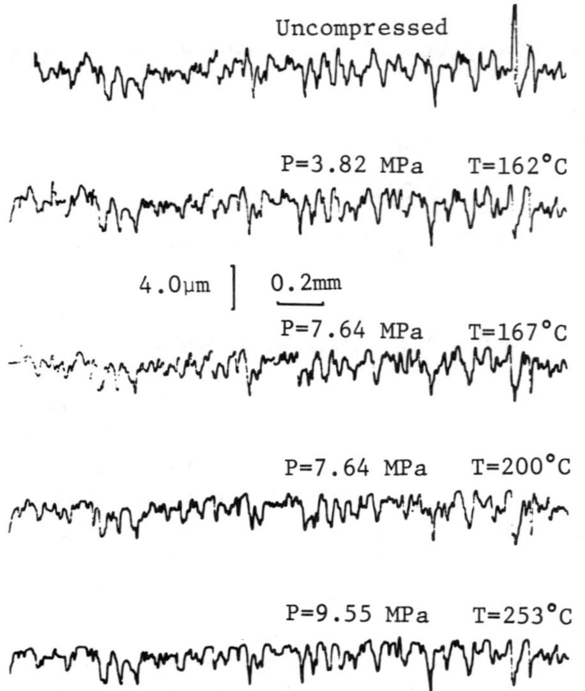

Fig. 9   Same as Fig. 8 but for Specimen No. 4 position 3 AL.
A2011 yield stress = 296.5 MPa.

as 3.8 MPa (550 lbf/in.^{2}), whereas, without the heating, this
pressure produced almost no measurable changes. For two cycles
of heating and cooling at 7.6 MPa (1100 lbf/in.^{2}), the surface
roughness had decreased from its original value by about 20%.
(This test was at an unusually high interface temperature of
253°C).

Figures 8 and 9 show enlarged sections of two different
specimens suffering substantial changes of shape when subjected to thermal cycling at moderately low contact pressures.
The "smoothing" effect of the procedure is obvious. It is in-
teresting to note that the fourth phase of the testing (p =
3.82 MPa, T = 25°C), carried out at low temperature, did not
change the surface measurably; the trace showed one additional
sharp peak, which was unexpected, and possibly a speck of dust.
(Note that the vertical magnification of the profile is 5000.)

V.   Discussion and Conclusions

The measuring equipment and techniques allow consistent
accumulation of data identifying very small changes in the

shapes of surfaces. The changing pattern of contacts of a ground aluminium surface loaded incrementally against a steel plate can be traced pictorially and quantitatively.

Predictions of thermal contact resistance of a joint all show that the resistance decreases as the contact pressure increases. However, the results of series 1 herein, for very high contact pressures, show that bulk deformations of the contact elements may occur when the load is removed and thereby make the thermal resistance much higher for subsequent loadings. This loading condition may represent that of a bolted joint, which requires occasional dismantling and reassembly during its service life. Great care must be taken if such a joint is required to have a consistent thermal behavior.

The heat-transfer tests of series 2 show that the surfaces suffered permanent deformations as a result of thermal cycling at low contact pressures. As a rough indication of the effect of the thermal applications, similar types of surfaces tested either cold or heated showed that, to produce comparable amounts of surface "damage," a contact pressure for cold tests needed to be about 40 times that of the heated tests. This comparison emphasizes the sensitivity of a joint to thermal distortions and adds information to the understanding of the phenomenon of thermal rectification of metallic contacts.

This equipment and its associated measurement techniques have been developed at Monash University during 1976; hence the amount of test data is limited. The almost two-dimensional nature of ground surfaces makes them relatively easy to handle, especially for the loading condition chosen. However, the technique is now well proven, and it is now our intention to explore truly three-dimensional random profiles in a variety of situations, including compression, thermal cycling, and abrasive wear conditions.

Acknowledgment

The authors are indebted to Monash University, which has given full support to this work.

References

[1]Williams, A., "Heat Transfer at the Interface of Dissimilar Metals," Aeronautical Research Council, Rept. ARC 23499, Strut 2409, Feb. 1962.

[2]Williams, A., "Heat Transfer Across Metallic Joints," PhD Thesis, Univ. of Manchester, United Kingdom, 1966.

[3]Clausing, A., "Heat Transfer at the Interface of Dissimilar Materials," _International Journal of Heat and Mass Transfer_, Vol. 9, August 1966, pp. 791-801.

[4]Thomas, T.R. and Probert, S.D., "Thermal Contact Resistance - the Directional Effect and Other Problems," _International Journal of Heat and Mass Transfer_, Vol. 13, May 1970, pp. 789-807.

[5]Veziroglu, T.N. and Chandra, S., "Directional Effect In Thermal Contact Conductance," _Heat Transfer_, Vol. 1, Elsevier, Holland, 1970.

[6]Idrus, N., Mitchell, R.W. and Williams, A., "The Roles of Computer in the Three Dimensional Analyses of Engineering Surfaces," _Proceedings of the Australian Conference on Computers in Engineering_, Sept. 1976.

[7]Idrus, N., "An Automatic Traversing Table for Three Dimensional Measurements of Surfaces," Monash Univ., Australia, MMEL Rept. 26, 1975.

[8]Sayles, R.S. and Thomas, T.R., "Microtopometry of Engineering Surfaces," _Paisley Tribology Conference_, Sept. 1975.

[9]Whitehouse, D.J. and Archard, J.P., "The Properties of Random Surfaces of Significance in their Contact," _Proceedings of the Royal Society (London)_, Vol. A316, March 1970, p. 97.

[10]Peklenik, J. and Kubo, M., "A Basic Study of a 3D Assessment of the Surface Generated in a Manufacturing Process," _Annalen College International Pour L'Etude Scientifique des Techniques de Production Mecanique (CIRP)_, Vol. XVI, 1968, pp. 257-263.

Chapter III—Thermal Control Systems

FLEXIBLE DEPLOYABLE-RETRACTABLE SPACE RADIATORS

J. W. Leach[*] and R. L. Cox[+]
Vought Corporation, Dallas, Texas

Abstract

A lightweight flexible radiator system for on-orbit
cooling of space payloads is described. The radiator is
packaged as a compact unit that can be attached to a vehicle
structure or hatch prior to or after launch. On-orbit, it is
deployed to provide the radiating surface needed for a
specific experiment. The unit is being developed and qualified independently as a heat rejection system that will be
ready for any spacecraft or experiment, and that will not
require significant structural and systems accommodation.
Design details and thermal vacuum test results are presented.

Introduction

Conventional space radiators such as those of the
Apollo and Gemini programs are structurally integral with
the vehicle skin, whereas the Space Shuttle Orbiter radiators
line the interior of the cargo bay door. Experiments or
spacecraft that need more radiating surface area than can be
provided by fixed radiators require deployable radiator
panels. Flexible radiators are designed to satisfy this
requirement, and have performance and weight advantages that
will make them attractive for other applications. Flexible
radiators are adapted easily to an existing vehicle, since
they can be stowed in compact units that are not susceptible
to damage by dynamic loads during launch. Since flexible
radiators do not require extensive structural support, they
are inherently lighter in weight than rigid panels.

Presented as Paper 77-764 at the AIAA 12th Thermophysics
Conference, Albuquerque, N. Mex., June 27-29, 1977. Copyright
© American Institute of Aeronautics and Astronautics, Inc.,
1977. All rights reserved.

*Thermodynamics Engineering Specialist, Propulsion and
Thermodynamic Technologies.
+Project Engineer of Advanced Space Thermal Systems.

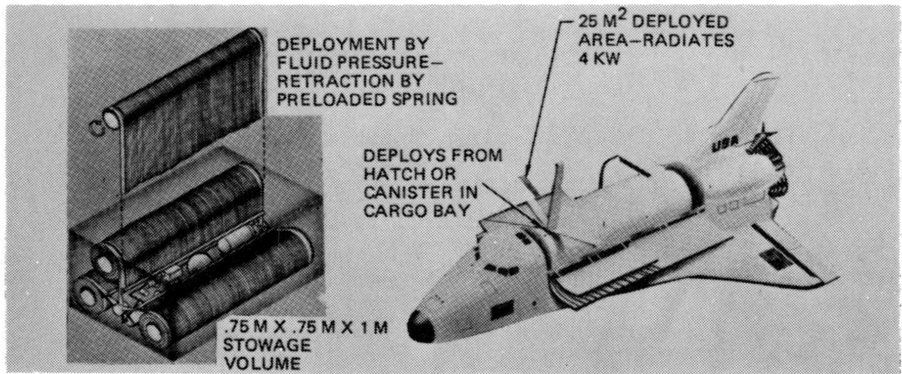

Fig. 1   Soft tube flexible system installation.

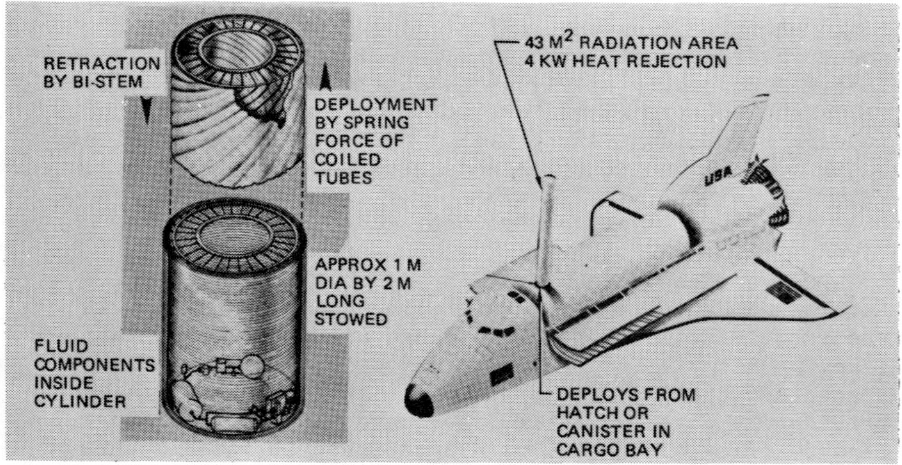

Fig. 2   Hard tube flexible system installation.

On-orbit, the flexible radiator is deployed as shown
in Figs. 1 and 2. Two alternate concepts are shown. The
first has plastic or elastomeric transport tubing and deploys
by unrolling like a party whistle, using a gas pressurant to
inflate two tubes on either side of the flexible panel.
Heavier deployment mechanisms such as storable tubular
extension members (STEM) may be substituted for the inflation
tubes to obtain more positive control of the radiator displacement. The teflon tubing design shown has three panels,
each 1 m wide x 8 m long, with a combined three-panel area of
24 m² and weight of 44 kg (including pumping power penalty
but exclusive of fluid loop components).

The second concept has aluminum radiator tubes. The
tubes are wound in a helical spring configuration, forming
a cylinder covered by the fin material. It deploys by the
inherent spring force, similar to a jack-in-the-box. The
concept shown is a single cylinder, 1.07 m diam and 12.8 m
long, with a surface area of 43 m^2 and a weight of 106 kg.

In this paper, we review the results of a test program
conducted to demonstrate the feasibility of the plastic and
metal tube concepts on test articles that have approximately
1/15 the area of the full-sized systems. In addition, the
design efforts on a subsequent program to fabricate and test
a full-scale prototype wing of the system shown in Fig. 1 are
discussed.

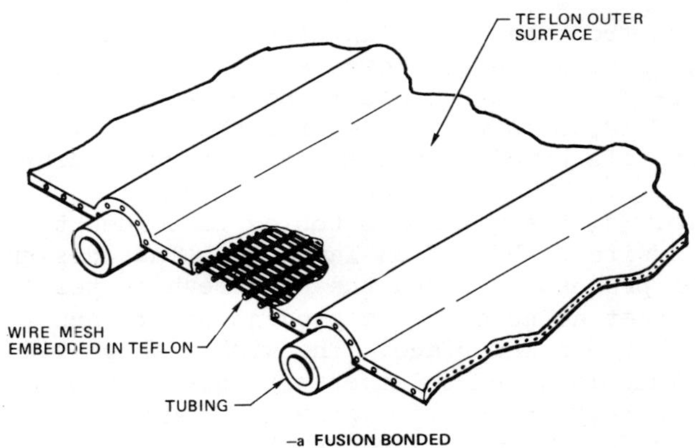

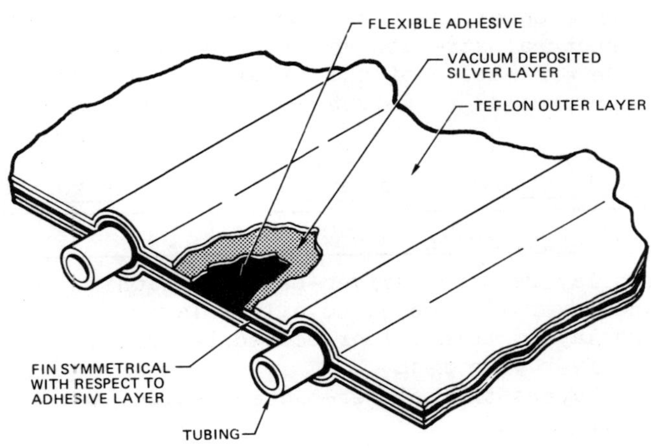

Fig. 3 Flexible radiator panel construction.

Design Details

Radiator Fin

A composite flexible fin material consisting of silver
and teflon has been developed which provides high thermal
conductance and emittance, resistance to degradation caused
by ultraviolet radiation, and strength and flexibility in a
cold environment. Figure 3 shows two alternate fin designs
evaluated during the feasibility demonstration program. In
Fig. 3b, the transport tubing is enclosed between the films
of teflon, each coated on the inner surface with a "thick"
layer of vapor-deposited silver, and with a flexible silicone
adhesive used to bond the laminate together. The silver layer
in this case provides three functions: 1) high lateral thermal
conduction, 2) low solar absorptance, and 3) protection of
the adhesive from the degrading effects of solar ultraviolet
radiation. Element tests showed a silver layer thickness
in the 6000- to 25,000-Å range to be obtainable; a nominal
thickness of 12,500 A was chosen for the feasibility demonstration article based on mechanical integrity considerations.

In Fig. 3a, the transport tubing is bonded to one side
of a silver wire mesh/ teflon laminate. This design uses a
heat fusion process to embed the wire mesh in teflon, with
the result that no adhesive is present other than that required at the tube interface. The wire mesh provides lateral
thermal conduction; and solar absorptance is minimized by
transmission between the wire strands. Table 1 lists all of the
radiator fin designs evaluated in this work.

The function of the outer fin layer of FEP teflon is to
provide structural strength and a high-emitting space stable
surface. Determination of the teflon film thickness is a
trade of performance vs weight because of the effect of thick-

Table 1 Comparison of radiator fin constructions

Fin construction
Two layers thick silver-backed teflon
One layer wire mesh, no silver layer
Two layers thin silver-backed teflon, one layer   silver wire mesh
Two layers thin silver-backed teflon, two layers   wire mesh
Two layers clear teflon, two layers wire mesh,   all fusion bonded.

ness on infrared emittance. Solar absorptance also increases
with thickness but to a lesser degree. Optimum film thickness
occurs in the 50-150-μ range, depending on operating temperature and area constraints for the radiator.

Transport Fluids and Tubing

The transport fluid and tubing for the flexible radiator
system were selected in an extensive materials evaluation
study, which surveyed the existing production and developmental materials. The transport tubing must provide for long
operating lifetime in a meteoroid environment, wide operating
temperature range, fluid retention, and flexibility and
strength consistent with the deployment/retraction system.
The characteristics of transport fluids which influence
fluid selection are boiling point (or vapor pressure), fire
point, pour point, toxicity, thermodynamic and transport
properties, and compatibility with the tubing material. The
study evaluated metal tubing and a great variety of flexible
materials including fluoroelastomers, perfluoroelastomers,
thermal and thermoplastic polyurethanes, polypropylenes,

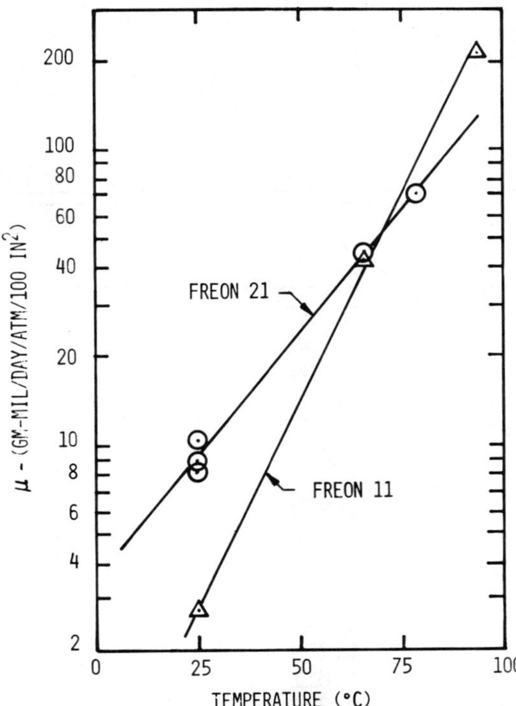

Fig. 4 Permeability of FEP teflon .

polyethylenes, polyester and silicone elastomers, and various
types of rubber and fluorinated polymers. Fluids surveyed
included fluorocarbons, silicate esters, and silicone fluids.

For the metal tube concept, the transport fluid may be
selected based on performance considerations alone. Freon 21
has been selected for its thermodynamic properties, pumping
power requirements, and operating temperature range.

The requirements for flexibility at low temperatures
and strength at high temperatures eliminate most tubing
materials for the soft-tube concept. MP-1880 ether-bond
polyurethane, ECD-006 perfluoroelastomer, Hytrel polyester
elastomer, and FEP teflon are materials that satisfy the
general requirements. Polyurethane tubing was used successfully in the feasibility/demonstration soft-tube flexible
radiator test article. FEP teflon has been selected for
fabricating the prototype radiator because it has superior
strength at high temperatures, and is compatible with a
greater variety of fluids.

When selecting the transport fluid for the soft-tube
radiator, the strength, permeability, and chemical compatibility of the transport tubing must be taken into account
before considering thermal performance. Freon 21 is
chemically incompatible with most flexible tubing materials.

Table 2 Properties of candidate fluids for flexible radiators

Fluid	ρ, g/cm^3	C_p, cal/g-°C	$K \times 10^4$, cal/sec-cm-°C
Oronite FC-100	0.897	0.45	2.35
Oronite 70	0.954	0.37	3.30
Dow Corning DC 331	0.939	0.36	2.07
General Electric SF 96	0.915	0.36	2.77
General Electric SR 81	0.971	0.36	3.60
Ethylene glycol-water (RS-89a)	1.075	0.73	9.05
DuPont Freon 21	1.367	0.25	2.93
DuPont Freon 11	1.476	0.208	2.07
DuPont Freon E-1	1.538	0.245	1.49
DuPont Freon E-2	1.683	0.240	1.69
3-M FC-88	1.618	0.244	1.53
3-M FC-75	1.779	0.24	3.26
3-M FC-77	1.811	0.24	1.57
Monsanto Coolanol 15	0.897	0.43	2.56

Teflon is the only plastic or elastomeric material that is
compatible with R-21 and has sufficient strength to contain
its vapor pressure at elevated temperatures. However, Freon
21 permeates teflon tubing at unacceptably high rates. Freon
11 is an alternate fluid with desirable thermal characteristics which permeates teflon. Figure 4 gives permeability
coefficients measured for these two fluids at temperatures
typical of space radiator fluids. The leakage rate of
Freon 21 at a nominal temperature of 27°C for the three-
panel system shown in Fig. 1 is approximately 0.23 kg/day.

Table 2 lists the transport properties of alternate
fluids considered in designing flexible radiators. An
optimum tube diameter and spacing exists for each fluid for
which the radiator panel and pumping system weight is
minimized.

Chemical compatibility and permeability tests showed
that some of the fluids of Table 2 cannot be used in the
soft-tube radiator. Coolanol 15 and RS-89A are typical of
fluids that are usable. These fluids have very low permeability coefficients and no measurable chemical incompatibi-
lities with the transport tubing.

Flow Instabilities

Space radiators with parallel flow transport tubes are
susceptible to flow instabilities that occur at operating
conditions where there are large differences between the fluid
inlet and outlet temperatures. When the temperature difference is too large, the flow distribution in the parallel tubes
changes abruptly from being uniform, where each tube carries
approximately the same flow, to nonuniform, where the flow in
one or more of the parallel tubes completely stagnates. Flow
instabilities are undesirable when accompanied by freezing
of the fluid in the nonflowing tubes, since this may prohibit
the flow from returning to the uniform distribution when the
operating conditions change. Also, metal tubing eventually
may fail due to stresses caused by the freezing and thawing
of the transport fluid.

The range of operating conditions for which the flow
distribution is stable depends on the properties of the
transport fluid. An equation from Ref. 1 gives the approximate stable operating limits.

Micrometeoroids

The radiator tubing wall thickness must be sufficient to
retain the transport fluid pressure after being struck by

micrometeoroids. The depth of the crater left by the most damaging meteoroid expected to strike the tubing during the designed operating life of the radiator is computed from a ballistic equation, which is based on ground-test data, and a meteoroid flux model derived from penetrations of metal foils in near-Earth orbits.

The mechanics of hypervelocity impacts of plastics are not completely understood. An equation given in Ref. 2 correlates the existing data reasonably well. The equation is

$$t = 0.65 \ (1/\varepsilon_t)^{1/8}(\rho_m/\rho_t)^{1/2}(V_m)^{7/8}(d_m)^{19/18} \tag{1}$$

where

t = thickness of material penetrated, cm

ε_t = percentage elongation of target material

ρ_t = mass density of sheet material, g/cm^3

ρ_m = mass density of meteoroid, g/cm^3

V_m = normal impact velocity, km/sec

d_m = meteoroid diameter, cm

The elongation term in Eq. (1) is much larger for plastics ($\varepsilon = 300$) than for metals ($\varepsilon = 3$). Limited experimental data in the literature substantiate the importance of this term.

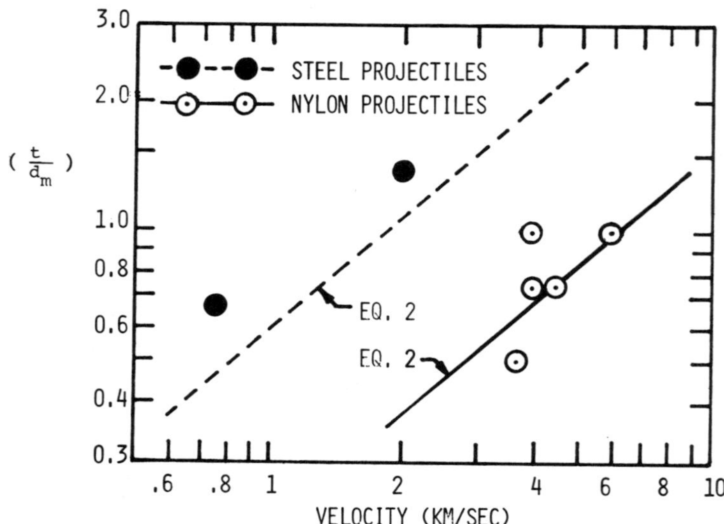

Fig. 5 Hypervelocity impact test of FEP teflon.

Qualitative data for plexiglas and polycarbonate[3] show
that polycarbonate (ρ = 1.2, ε = 115) is superior to plexiglas
(ρ = 1.2, ε = 5) for retarding meteoroids. Hypervelocity
test results for polyethylene (ρ = 0.9, ε = 500) in Ref. 4
are predicted conservatively in Eq. (1). The equation predicts slightly greater depths of penetration than are measured
at velocities below 12 km/sec and much greater depths of
penetration than Ref. 4 indicates will occur at velocities
typical of micrometeoroids, viz. 20 km/sec.

Tests were conducted at Texas A&M University[5] with
FEP Teflon to substantiate that Eq. (1) gives the correct
depth of penetration for the material selected for the prototype soft-tube flexible radiator. The results given in
Fig. 5 show that the penetration depths of nylon projectiles,
= 1.15, are predicted adequately by Eq. (1), whereas
steel projectiles, ρ = 8.0, penetrate greater depths than
the equation predicts. This indicates that the exponent of
the projectile density in Eq. (1) may be too small. If this
is the case, the equation is conservative for predicting
damage by micrometeoroids, ρ = 0.5.

Flexible Radiator Design Optimization

Given selected fin materials, radiator tubing, and
transport fluids, it is necessary to design the radiator for

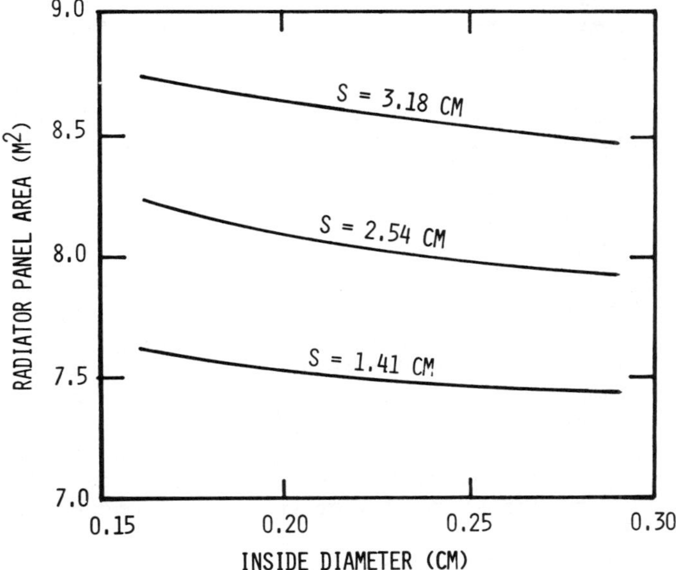

Fig. 6 Radiator area, coolanol 15 transport fluid.

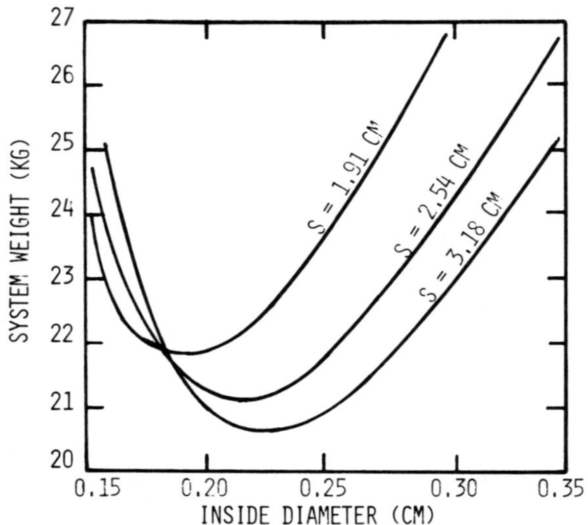

Fig. 7   Radiator system weight, RS-89a fluid.

minimum weight and area. The optimization analysis must
consider the requirements of the deployment/retraction system
and the fluid pumping system. The principal variables to be
considered in the analysis are tubing diameter and spacing.
The spacing of the tubes affects the radiator fin efficiency
and therefore the required surface area. It also affects
the stiffness and weight of the panel. Similarly, tubing
diameter influences the pumping power requirements, convective and conductive heat transfer, panel flexibility, and
system weight. Optimization analyses for the soft-tube
flexible radiator are described below to illustrate how the
tubing parameters influence the radiator design for ethylene
glycol-water and Coolanol 15 transport fluids. Similar computations have been performed for the hard-tube concept.

Figure 6 gives the influence of tube diameter and spacing
on required radiator panel area. The figure shows that the
required radiating surface area is much more dependent on
tube spacing than on tube diameter. The spacing of the tubes
affects the radiator fin effectiveness, whereas the tube
diameter influences the thermal conductance between the transport fluid and the base of the radiator fin. The radiator
is designed for laminar flow, so that tubing diameter has
no effect on the convective film temperature delta and only
a small effect on the shape factor governing conductance
through the tube wall.

Figure 7 shows how the tubing parameters affect the
radiator system weight. The panel weight increases in

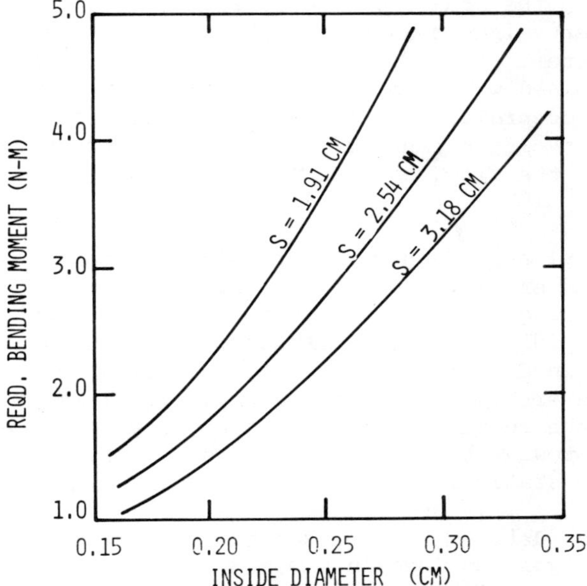

Fig. 8 Radiator panel flexibility.

Table 3 Comparison of flexible radiator designs

Design variable	RS-89a	Coolanol 15
Radiator panel length, m	7.35	7.83
Radiator panel area, m^2	7.14	7.62
Radiator panel width, m	0.965	0.965
Number of tubes	50	50
Tube spacing, cm	1.905	1.905
Tube outside diameter, cm	0.318	0.318
Tube inside diameter, cm	0.159	0.159
Relative weight,[a] kg	23.3	26.5
Pressure drop, bar	2.28	1.76
Bending moment for 25.4-cm-diam drum, N-m	1.6	1.6
Minimum outlet temperature (38°C inlet), °C	-29	-57
Radiator fin emissivity	0.71	0.71
Radiator fin efficiency	0.943	0.943

[a] The relative weight includes manifolds, the deployment drum, retraction springs, transport tubing and fittings, transport fluid, radiator fins, and weight penalty for fluid pressure drop.

proportion to the square of the tube diameter, whereas the
pumping power weight penalty is proportional to the inverse
of the diameter. Therefore, a minimum occurs in the curves
for the combined weight of the radiator and pumping systems.
The minimum occurs at relatively small tube diameters for
close tube spacing because of the number of tubes available
to transport the flow, and the corresponding low pressure
losses.

The final set of curves to consider in selecting the
tubing parameters is given in Fig. 8. The figure shows
that the bending moment required to stow the radiator panel
increases rapidly with tube diameter. The bending moment
also varies in direct proportion to the number of transport
tubes, or inversely with tube spacing. For large tube
diameters, the retraction springs that supply the bending
moment for stowing the radiator are relatively expensive and
create large static loads on the deployment system.

Based on data such as presented in Figs. 6-8, the
radiator designs outlined in Table 3 were obtained. The
selected panel configurations for Coolanol 15 and ethylene
glycol/water are approximately the same. The radiator panel
area is minimized with little weight penalty by designing
the radiator with close tube spacing. The minimum practical
spacing is approximately 1.91 cm. The system weight is
optimized for this spacing when the tubing inside diameter is
approximately 0.19 cm. Considering the structural requirements of the deployment system, it is desirable to select
tubing with a slightly smaller diameter. Standard-sized
0.0625-in.-i.d. tubing (0.16 cm) was selected because it
reduces the bending moment requirements by approximately 25%

Fig. 9 Hard-tube development test article.

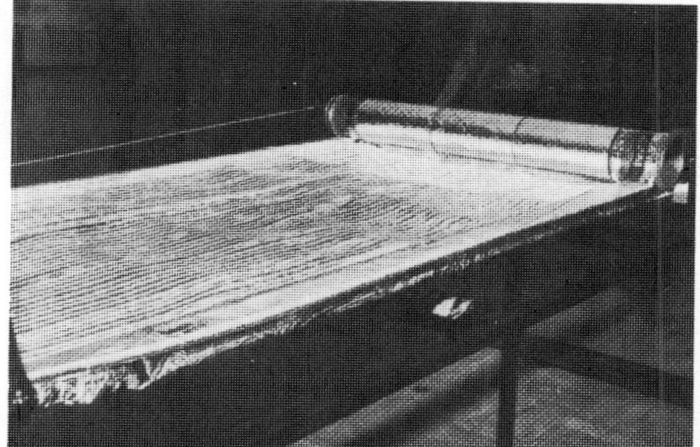

Fig. 10 Soft-tube development test article.

while increasing the system weight by about 1 kg. It also
eliminates cost for producing nonstandard tubing and connec-
tors.

Feasibility Demonstration Test Articles

Two feasibility demonstration radiators were fabricated
and tested in a vacuum environment to evaluate overall
system thermal performance and deployment concepts. The test
article shown in Fig. 9 is constructed with aluminum fluid
passages and deploys from the inherent spring force of the
coiled tubing: a motor-driven cable compresses the coils
to retract the radiator. The other test article, shown in
Fig. 10, has flexible tubing and is stored on a cylindrical
drum. Deployment forces are supplied by a gas pressurant,
which inflates two tubes or either side of the flexible panel.
Table 4 compares the construction and performance of the
two feasibility demonstration articles. The metal tube
concept has the wider operating range, whereas the flexible
tube concept has lighter weight.

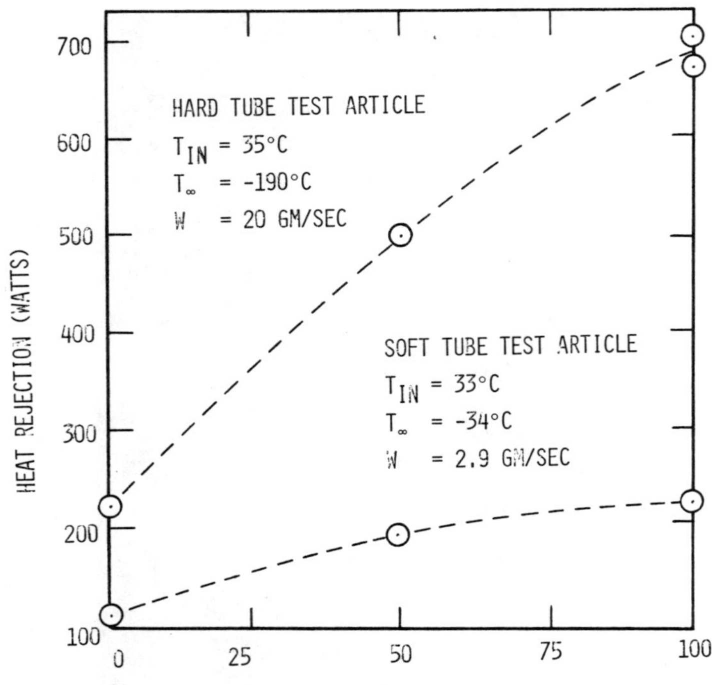

Fig. 11 Heat rejection for the development test articles.

The developmental model flexible radiators were subjected to thermal vacuum testing to evaluate properties such
as effective surface emittance, overall radiator fin conductance, joint conductance between the transport tubing
and radiator fin, and capacity for deployment and retraction
in a vacuum environment. The tests were conducted in the
Vought Space Environment Simulator and consisted of demonstration of deployment from a stowed configuration in a
25°C box into a cold vacuum environment and about 36 hr of
thermal vacuum testing on each test article. The test
objectives were as follows:

1) Demonstrate baseline heat rejection performance at
environmental sink temperatures of -190° and -18°C.

2) Demonstrate capability to deploy from a stowed environment to a -190°C space environment.

3) Demonstrate structural integrity over the range of
expected nominal and limit case operational environments under
steady-state and transient conditions.

4) Obtain experimental data on tube and fin temperature drops
and tube-to-fin and longitudinal gradients to evaluate
test article integrity and analytical procedures.

The -190°C environment represents a worst case condition
for testing the flexibility of radiator materials at low
temperatures; and, since it can be simulated precisely by
liquid nitrogen charged cold walls, gives an accurate measure
of the radiator performance. The -18°C case is representative
of actual flight conditions with IR exchange between the
radiator panels and the shuttle vehicle, the Earth Albedo,
and direct solar radiation.

Hard-Tube Model Test Results

Figure 11 gives typical heat rejection data for the
hard-tube test article. Table 5 compares the experimental
and predicted heat rejection for cold and warm environments.
Predictions are from a computer model, which accounts for
thermal interactions between the various components of the
test article and the wall of the environment simulation
chamber. The results for the cold environment agree almost
exactly, and the results for the warm environment agree

Table 4 Comparison of feasibility demonstration flexible
radiators

	Metal tube radiator	Flexible tube radiator
Tubing material	Aluminum	Polyurethane
Transport fluid	Freon 21	Coolanol 15
Fin material	Silver wire mesh/teflon	Thick layer silver/teflon
Deployment force	Coiled transport tubing	Inflation tubing
Model dimensions	0.7 m diam x 1.14 m	1 m diam x 1.85 m
Tube spacing	3.8 cm	2.54 cm
Radiator efficiency	0.85	0.72
Effective surface emittance	0.83	0.68
Upper operating temperature limit	300°F (450 K)	200°F (367 K)
Lower operating temperature	-140°F (178 K)	-20°F (245 K)
Weight	4.8 kg/m^2	1.9 kg/m^2

Table 5 Comparison of experimental and predicted thermal
performance of the hard-tube test article

Test point	T_{in}, °C	$\dot{w}$, kg/hr	T_∞, °C	Q(exp), W	Q(pred), W
1	35.7	74.5	-140	706	706
2	34.8	28.2	- 29	271	306

within what is believed to be experimental error. The small
discrepancy for the warm environment probably is caused by
inconsistent control or modeling of the environment. It is
difficult to maintain the test chamber wall at steady uniform
temperatures. At the higher ambient temperatures, small
variations in the cold-wall temperatures have a large effect
on the heat rejection from the radiator. Thus any error in
modeling or maintaining the environment would be reflected
in the test results. For cold ambient temperatures, errors
in representing the environment have less impact on the
radiator performance, and the predicted performance depends
more on the modeling of the radiator itself. The fact that
the experimental and predicted performances agree for this

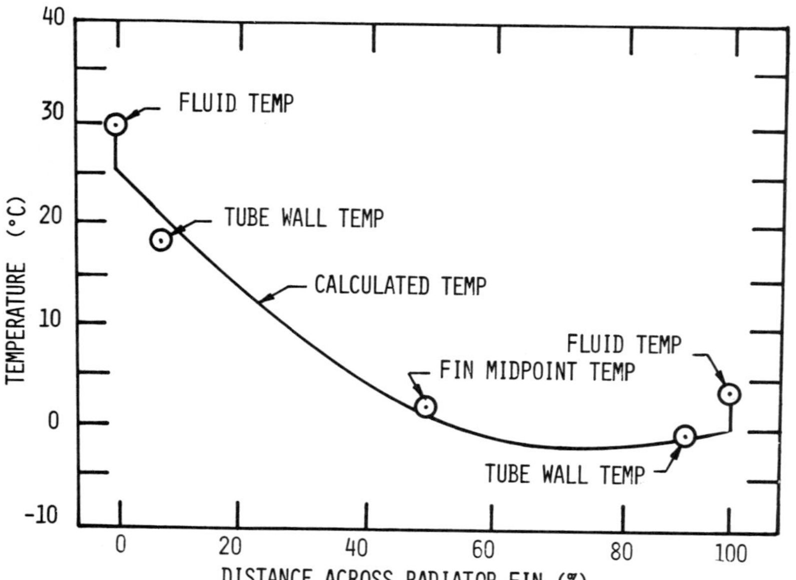

Fig. 12 Predicted and experimental temperatures for the hard-
tube development test article.

case indicates that the actual radiator construction and thermal properties are near the design values.

One area of uncertainty in the construction of the hard-tube radiator concerns the thermal contact between the tubes and the radiator find. The fins are glued to the tubes in such a way that it is difficult to predict the exact value of the joint conductance. Comparisons between predicted and experimental temperatures for the hard-tube radiator as in Fig. 12 indicate that the joint conductance is approximately the design value. Also, test points executed at the beginning of the thermal vacuum test were duplicated at the end of the test to determine whether the joint conductance had changed during the test. The measured heat rejection at the end of the test was found to be the same as at the beginning, thus demonstrating that the radiator performance had not degraded. The radiator showed no signs of wear as a result of over 100 deployment and retraction cycles and exposure to extremely cold vacuum environments.

Soft-Tube Model Test Results

Figure 11 gives typical heat rejection data for the soft-tube test article. Unlike the hard-tube model, the soft-tube radiator did not reject heat at the predicted rate. Table 6 shows that the experimental heat rejection is approximately 25% lower than predicted.

The low thermal performance of the soft-tube test article is believed to be caused by nonuniformities in the thickness of the silver conducting layer in the radiator fin. In practice, it is difficult to control the dimensions of the vacuum-deposited layer. Sample thicknesses selected randomly from the radiator fin stock were measured with a scanning electron microscope. The measured thickness ranged from 5000 to 18,000 A with an average value of 9000 Å. This is lower than the design value of 12,500 Å by an amount

Table 6   Comparison of experimental and predicted thermal performance of the soft-tube test article

Test point	T_{in}, °C	T_{∞}, °C	$\dot{w}$, kg/hr	Q(exp), W	Q(pred), W
1	35.0	−190	28.7	471	650
2	33.6	− 18	10.9	183	229

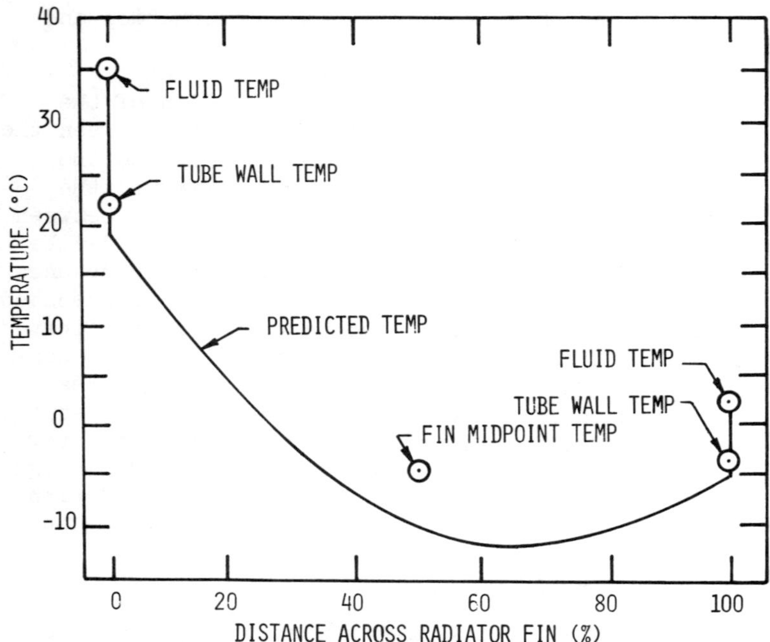

Fig. 13 Predicted and experimental temperatures for the soft-tube development test article.

sufficient to account for the low performance of the radiator. The fin surface emissivity also had a strong influence on heat rejection. However, measurements by NASA Johnson Space Center showed that the surface emissivity is actually higher than expected. Locally, the measured temperatures of the soft-tube model are predictable by a computer model based on design values of conductance and emittance. Figure 13 gives typical comparisons of predicted and experimental temperatures at locations where the results agree. At other locations the correlation is poor, giving further evidence of nonuniformities in the panel construction. The panel showed no evidence of wear as a result of testing, and the measured heat rejection at the end of the test was approximately the same as at the beginning of the test.

Prototype Soft-Tube Flexible Radiator

A prototype soft-tube flexible radiator is being fabricated and tested to evaluate fabrication techniques, thermal performance, and capacity for deployment and retraction on full-scale hardware. The prototype radiator panel consists of one wing of the three-panel system shown

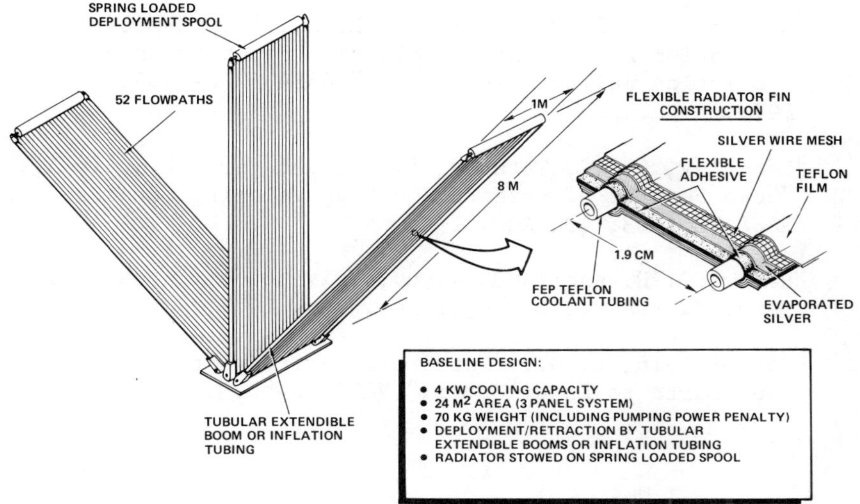

Fig. 14 Prototype soft tube flexible radiator.

in Fig. 14. The radiator is being tested with Coolanol 15
transport fluid and is based on the design outlined in Table
3. It is being fabricated with standard-sized 0.0625-in.-i.d.
(0.159 cm) FEP teflon tubing with wall thickness sized to
give a minimum expected life of 9.5 months in the near-Earth
micrometeoroid environment. The system is designed for fluid
inlet temperatures to approximately 100°C.

Acknowledgments

The authors wish to acknowledge H. F. Battaglia, W. E.
Ellis, and B. O. French of the NASA Johnson Space Center
for their many valuable suggestions and contractual support
under Contracts NAS9-13346 and NAS9-14476. We also wish to
express our appreciation to W. S. Slemp of the NASA Langley
Research Center for performing ultraviolet degradation tests
on flexible radiator materials. Thanks also are expressed
to the Vought Corporation for support of a significant portion
of the design studies.

References

[1]Leach, J. W., "Flow Instabilities in Spacecraft
Radiators," Vought, Dallas, Texas, Rept. T169-56,
Nov. 1974.

[2]Rittenhouse, J. B., "Meteoroids," Space Materials
Handbook, 3rd. ed., Addison-Wesley, Reading, Mass., 1968.

[3]Di Battista, J. D. and Humes, D. H., "Multi-material Lamination As A Means of Retarding Penetration and Spallation Failures in Plates," NASA TN D-6989, Nov. 1972.

[4]Leont'ev, L. V., Tarasov, A. V., and Tereshkin, I. A., "Some Results of Research on High Velocity Impacts," NASA TT F-13,740, Aug. 1971.

[5]Henry, G. R. and Rand, J. L., private communication, Texas A&M Univ., Bryan, Texas, Jan. 1977.

[6]Cour-Palais, B. G., "Meteoroid Environment Model, Near Earth to Lunar Space," NASA SP-8013, March 1969.

CONTAMINATION/DEGRADATION MEASUREMENTS ON OPERATIONAL SATELLITE THERMAL CONTROL SURFACES

D.G.T. Curran[*] and J.M. Millard[†]
Aerojet ElectroSystems Company, Azusa, Calif.

Abstract

The results of flight temperature measurements on geo-synchronous satellite thermal control surfaces are presented. The thermal data show long-term effects of contamination/degradation on radiator surfaces and calorimeter samples. The data relate to radiators designed for low-temperature cooling below 200 K using second-surface mirrors bonded to a metal substrate. Calorimeter data show environmental effects similar to that of the radiators on α_s/ε_H of selected thermal control coatings. Deposition data from quartz crystal microbalance monitors are shown and evaluated with respect to the measured effects of α_s/ε_H. Also, flight modeling techniques are discussed, together with details of on-going satellite thermal improvement program.

I.   Introduction

Radiators for passively cooled satellite systems require low values of solar absorptance (α_s) and high values of infrared emittance (ε_H). However, these thermal properties are subject to effects of the space/satellite environment which can cause an increase in α_s/ε_H, resulting in degradation of the system thermal performance. This degradation may involve the thermal control coatings through material damage and/or contamination deposition of particulate or outgassed materials.

Analysis of long-term temperature data of operational satellites has shown a gradual warming trend of the passive-

---

Presented as Paper 77-740 at the AIAA 12th Thermophysics Conference, Albuquerque, N.Mex., June 27-29, 1977.

[*]Specialist, Technical Staff.
[†]Member of Technical Staff.

ly cooled systems. Initially, temperature monitors gave
limited information such that thorough understanding of potential satellite-associated causes and their effects were
subject to conjecture. Therefore, means for providing
information for investigating the cause of the warming
trends were initiated.

Studies were conducted to provide more detailed thermal
data from flight systems combined with laboratory investigations of thermal control coatings subjected to simulated
launch loads, thermal vacuum temperature cycling, space
radiation, and particle effects.[1,2] Since contamination
of the radiator surfaces from satellite/spacecraft and
launch vehicle sources was a distinct possibility, studies
were conducted into potential deposition/degradation mechanisms. These included buildup of surface films and modification of this film due to photolysis and/or polymerization,
causing a subsequent increase in solar absorptance. Contamination models were postulated.[3] Deposition mechanisms
involved line-of-sight molecular impingement (outgassed
products) and electrostatic attraction of molecules due
to thrust, jet, plasma, and/or photoemission – induced
surface charge. Contamination from surface films due to
satellite ground handling also was investigated. Potential
sources included radiator fabrication, satellite testing,
storage, and prelaunch facilities. Launch environment
effects such as fairing jettison and booster blowback products also were studied.[4]

The results of these investigations showed the need
for revised flight instrumentation and improved ground-
handling procedures and satellite materials changes. A
program was initiated by Aerojet ElectroSystems Company
(AESC) under contract to the Space and Missile Systems
Organization of the Air Force Systems Command to follow
through with recommended changes to extend and improve
satellite thermal performance. Initial improvements involved revised ground cleaning (based on Auger Electron
Spectroscopy of cleaned thermal control coatings) of the
critical radiator surfaces, with cleaning verification
tests prior to launch.[5] Also, a lifetime monitoring instrumentation program was instituted to provide on-orbit
data to ascertain effects of hardware changes and to provide information on the thermal performance of contemplated
coatings for current and advanced systems. Temperature
monitors were relocated to give a finer grid of radiator
temperatures. As part of this effort, calorimeter experiments were flown to provide accurate measurement of α_s/ε_H
for several thermal control coatings untilized on the radi-

ator systems or contemplated as improved replacements for later satellites.

II. Calorimeter Analyses

Initially, four calorimeters were located on one of the main radiators of satellite A. The placement of these calorimeters with respect to radiator axial positions for each satellite is shown in Fig. 1. Several samples in each location were used to provide comparative environment effects between samples, whereas several locations were used to provide information on potential effects from exposure to satellite components.

A. Flight Sample Descriptions

The first calorimeter samples flown were typical of the thermal control coatings on the satellite radiators. These involved several types of second-surface mirrors using commercial and high-purity quartz. Subsequent calorimeter samples have been variations of these coatings using either silvered or aluminized second surfaces. Further changes have included ultraviolet (uv) filters to reduce α_s, as well as changes in the adhesives for bonding the mirrors to the magnesium substrate. Additional modifications have included grounding the mirror samples and

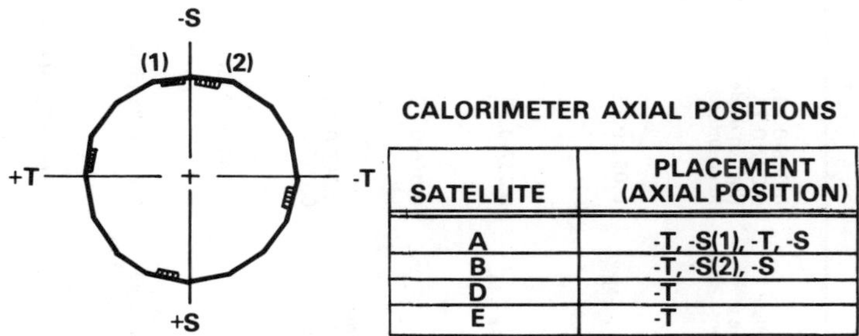

SATELLITE	PLACEMENT (AXIAL POSITION)
A	-T, -S(1), -T, -S
B	-T, -S(2), -S
D	-T
E	-T

Fig. 1   Calorimeter placement on satellite radiators.

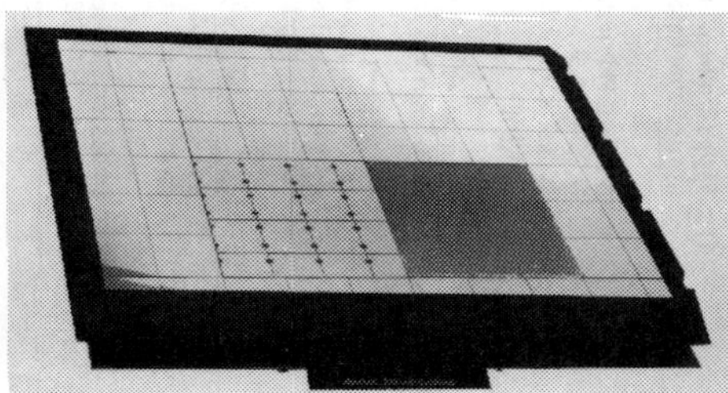

Fig. 2   Calorimeter with thermal control coating samples in place.

Table 1 Identification of thermal control coatings

Coating description	Solar absorptance/ emittance α_s/ϵ_H	Radiator location	Adhesive	Comments
1) Type I mirror, commercial quartz, dimensions 0.01 x 1 x 1 in., with and without indium oxide coating (with grounding tabs) on satellite E, dimensions 0.01 x 3 x 3 in.	0.066/0.75 0.082/0.75 (Indium oxide front surface)	Top and bottom one-third of radiator	GE RTV615 on satellite A, GE RTV566 on subsequent satellites	Used on satellite A, B, and E calorimeters, 9-mirror, 3 x 3 in. panel with 2-mil silvered FEP teflon gap seal (satellite E)
2) Type II mirror, commercial quartz, dimensions 0.01 x 1 x 1 in.	0.037/0.75	Middle one-third of radiator	GE RTV615 on satellite A, GE RTV566 on subsequent satellites	Used on satellite A and B calorimeters, ultraviolet filter to reduce α_s
3) Type III mirror, high-purity quartz, dimensions 0.01 x 1 x 1 in.	0.037/0.75	Used with type II in middle one-third of radiator	GE RTV615 on satellite A, GE RTV566 on subsequent satellites	Used on satellite A and B calorimeters, ultraviolet filter to reduce α_s
4) Type IV mirror, high-purity quartz, dimensions 0.01 x 1 x 1 in.	0.066/0.75	Not used	GE RTV615 on satellite A, GE RTV566 on subsequent satellites	Used on satellite A and B calorimeters
5) Silvered FEP teflon, dimensions 0.002 x 1 x 1 in. (satellite A), 0.002 x 3 x 3 in. on subsequent satellites	0.063/0.66 (300K) 0.063/0.56 (170K)	Used on sunshades and as gap seal between mirrors	3M 467, 3M Y-966 (satellite E)	Used on all satellites as referenced standard
6) Aluminized FEP teflon, dimensions 0.005 x 3 x 3 in.	0.144/0.80 (300K)	Not used	3M 467	Used on satellite D calorimeter
7) Silica cloth (GE Astroquartz 581) heat-laminated to 1-mil aluminized FEP teflon, dimensions 0.01 x 3 x 3 in.	0.197/0.86 (300K)	Not used	DC6-1104	Used on satellite D calorimeter
8) Silica cloth (GE Astroquartz 581) bonded to double-sided aluminized Kapton, dimensions 0.01 x 3 x 3 in.	0.165/0.86 (300K)	Not used	SR-585	Used on satellite D calorimeter
9) Embossed silvered FEP teflon, dimensions 0.005 x 3 x 3 in.	0.095/0.76 (228K)	Not used	Permacel P-223	Used on satellite E calorimeter

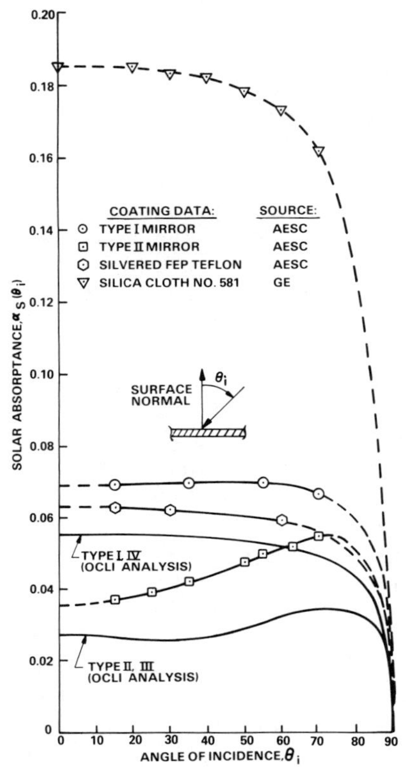

Fig. 3 Solar absorptance vs angle of incidence for thermal control coatings.

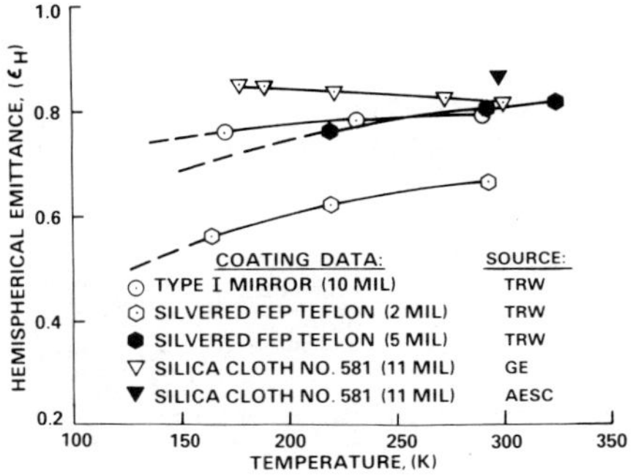

Fig. 4 Hemispherical emittance vs temperature for thermal control coatings.

using gap seals. All samples have been typically low α_S/ϵ_H coatings varying between 0.05 for second-surface quartz mirrors with uv filters to 0.23 for silica cloth (Astroquartz). A current calorimeter with samples installed is shown in Fig. 2. The identification of nine typical coatings flown on the calorimeters is given in Table 1. The table also indicates the use of several of the coatings on the main satellite radiator.

The variation of solar absorptance with angle of the incident radiation measured from the surface normal is shown in Fig. 3 for several representative coatings. Analytical predictions based on electromagnetic theory are given for second-surface mirrors as well as laboratory measurements. The analyses for type I and IV mirrors as well as type II and III mirrors were performed by the Optical Coating Laboratory, Inc. (OCLI), the manufacturer of the type I through IV mirrors used by AESC. The analyses show the predicted reduction of α_s vs angle of incidence (θ_i) for a uv filter optimized at normal incidence. The laboratory measurements indicate a similar reduction but significantly less at angles of incidence from the normal. Data also are presented for silvered teflon and silica cloth.

The temperature dependence of the hemispherical emittance (ε_H) for the thermal control coatings is shown in Fig. 4. The ε_H of 10-mil quartz mirrors (type I through type IV) is not affected significantly by a decrease in temperature in the 100 to 200 K temperature range. However, the thinner 2- and 5-mil teflon samples are affected significantly at low temperatures. The lower emittance of the silvered or aluminized second surface ($\varepsilon_H \gtrsim 0.1$) becomes important as the thickness of the teflon is decreased.[6] The 11-mil silica cloth shows a slight increase in ε_H as the temperature is decreased from ambient. A satisfactory explanation for this characteristic has not been given, but the weave of the cloth apparently plays a role in this behavior.

B. Phase I Calorimeter Analysis

A cross-sectional view of the calorimeter construction is shown in Fig. 5. This schematic illustrates the mounting

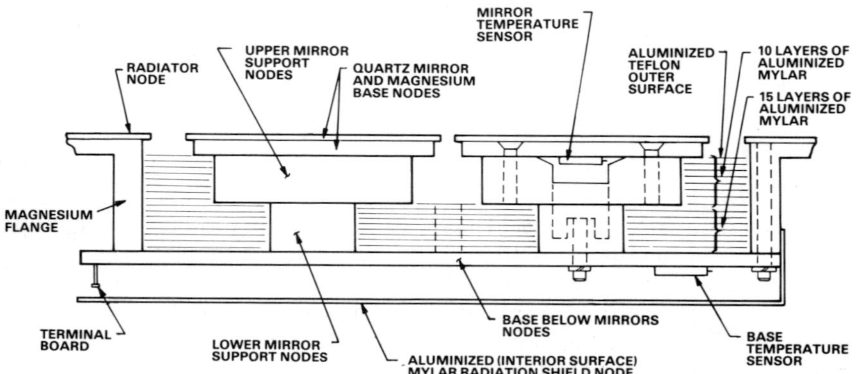

Fig. 5   Phase I calorimeter thermal model node schematic.

technique used for the quartz mirror samples. They were
bonded individually to a magnesium base similar to the
radiator construction. This base was, in turn, isolated
thermally from the radiator through fiberglass standoffs,
multilayer insulation, and a common fiberglass base. The
base was mounted to the radiator and shielded from internal
thermal radiation exchange with an aluminized mylar cover.
Temperature sensors of the platinum resistance type were
mounted to the magnesium base of each mirror sample and
to the module base for reference heat leak calculations.
Temperature monitors also were provided adjacent to each
calorimeter module to determine heat leak from the radiator.
Other temperature monitors on the satellite were available
to calculate radiation exchange, although the configuration
(view, shape) factors were small.

The thermal network used to represent the calorimeter
for orbital prediction of the sample thermal performance
also is referred to in Fig. 5. The thermal model consisted
of approximately 40 nodes. Satellite temperature data for
the radiator, module base, and external surfaces were used
to simulate the boundary nodes. Nodal network conductances
and radiation coefficients were calculated based on module
drawings and thermophysical property data. The amount of
absorbed solar flux was calculated by numerically integrating the expression

$$\frac{1}{2\pi} \int_{0}^{2\pi} \alpha_s\,(\theta_i)\,\cos\theta_i\,d\gamma \equiv \overline{\alpha_s F_s}, \quad -\pi/2 \le \theta_i \le \pi/2$$

where $\overline{\alpha_s F_s}$ represents the spin-averaged flux for a given position in orbit. The incident angle θ_i for a given angle

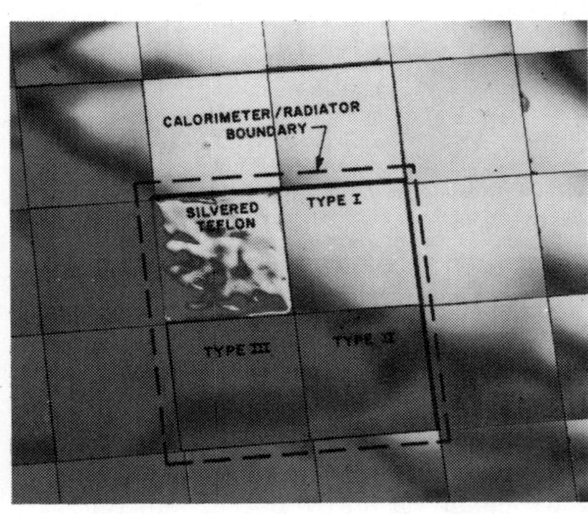

Fig. 6 Calorimeter loca-
tion on satellite A
radiator (-S axis).

of satellite rotation γ was calculated based on sample surface orientation with respect to satellite coordinates and orbital position with respect to the sun and Earth. An AESC computer program using Monte Carlo techniques was used to check these results, as well as for calculating satellite shading of the calorimeter samples and reflected solar irradiation from specular and diffuse satellite surfaces. The thermal model used the Systems Improved Numerical Differencing Analyzer (SINDA) computer program to calculate the orbital transient temperature diurnals for correlation with the measured temperatures. Telemetry data were available at 15-min intervals over a 24-hr period so that accurate temperature diurnals could be obtained for determination of α_s and ε_H.

The placement of one of the four calorimeters (-S axial position) on satellite A is shown in Fig. 6. The four samples for this calorimeter were silvered teflon, type I, type II, and type III mirrors. Note the gaps between the radia-

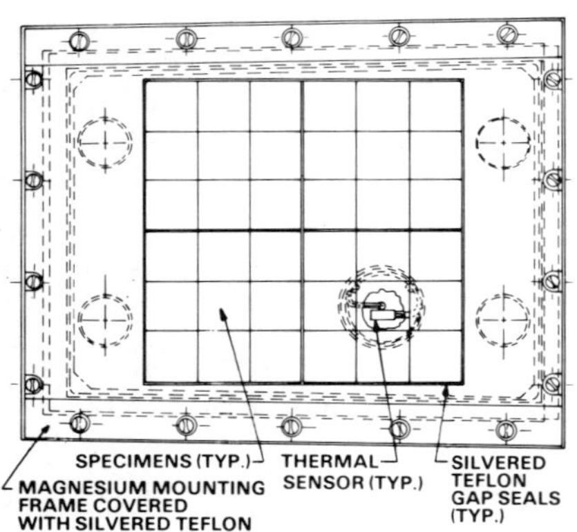

Fig. 7   Phase II calorimeter design.

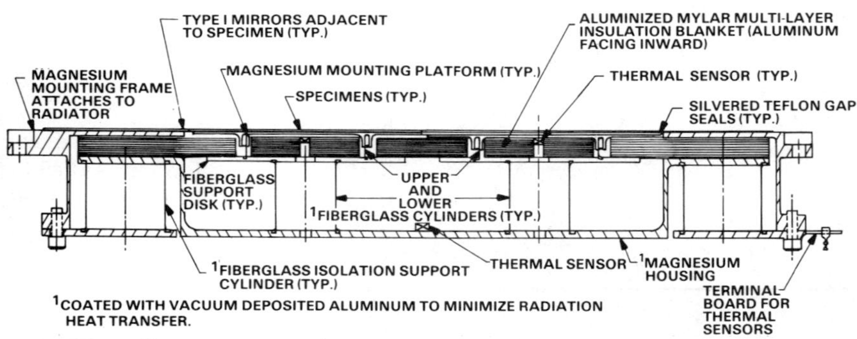

Fig. 8   Cutaway view of phase II calorimeter design.

tor mirrors, samples, and between the radiator mirrors and
calorimeter samples. The latter relatively large gaps were
due to installation of the calorimeter after the radiator
mirrors had been laid up. This problem later was alleviated
on the phase II calorimeter design. Modifications were
made on subsequent satellite radiators to minimize the
effect of the gaps. The mirror gaps are due to manufacturing tolerances and allowance for radiator thermal construction effects during cooldown in orbit. However, these gaps
were found to cause a significant uncertainty in the calculated solar absorptances for the calorimeter samples.

An error analysis was performed to determine the accuracy of α_s calculated from the thermal model correlation
of sample temperature data similar to that reported by
Millard.[7] The results indicated an error of ± 0.01 for an
α_s of 0.07 and ± 0.001 for a $\Delta\alpha_s$ of 0.03.

C.   Phase II Calorimeter Analysis

On all flights subsequent to satellite A, the redesigned
or phase II calorimeter was flown. This redesign was
based upon evaluation of the phase I calorimeter performance. Design modifications were employed in the phase II
calorimeter to increase its isolation from the radiator and
other components of the satellite, as well as to increase
the isolation of the individual samples from each other.
Also, the calorimeter sample area was increased from 1 to
9 in.^{2}, and the mounting design was changed to allow the
calorimeter module to be installed just prior to launch.
Figure 7 shows a top view of the calorimeter, whereas Fig. 8
displays a cutaway view of the module structure.

The samples are bonded to a magnesium mounting platform
that is bonded to a series of thin-walled (4 to 6 mil)
fiberglass support cylinders for improved thermal isolation.
These then are mounted to a magnesium baseplate in the module housing. The housing similarly is isolated with fiberglass support cylinders that are attached to the radiator

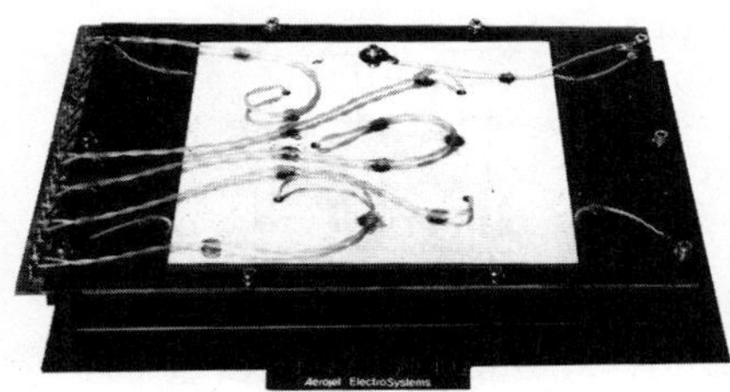

Fig. 9   Rear view of phase II
          calorimeter housing.

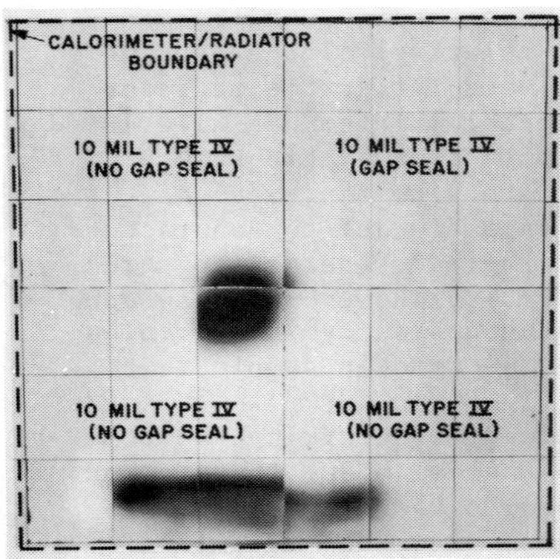

Fig. 10   Calorimeter location on satellite B radiator (-T axis).

mounting frame. All of the fiberglass support cylinders and disks along with the housing are coated on both sides with vacuum-deposited aluminum ($\varepsilon_H \leq 0.1$) to minimize radiation heat transfer between surfaces. Silvered teflon gap seals close the spaces between samples (and between sample mirrors) so that solar energy cannot reach the housing enclosure. Also, held in place between the support disks and mounting platforms, is a blanket of multilayer insulation designed to minimize the heat transfer between the samples and the housing enclosure. This blanket is composed of several thin aluminized mylar sheets and has an aluminized side facing the housing enclosure. The insulation blanket also extends outward and insulates the housing from the mounting frame, where it overlaps the housing adjacent to the samples. The top of the mounting frame in this region is covered with type I mirrors to minimize heat absorption adjacent to the samples. Where the mounting frame bolts to the radiator, a strip of silvered teflon serves the same purpose. A temperature status monitor is mounted beneath each sample mounting platform, from which insulated constantan leads travel through the fiberglass cylinders to the terminal board, as shown in Fig. 9. An additional temperature status monitor is mounted on the inside of the housing baseplate to provide a temperature history for heat leak calculations of the individual samples. The calorimeter position on the radiator of the satellite is similar to the phase I locations such that its view of exterior satellite components is minimized.

The phase II calorimeter thermal model for the -T-axis position consists of 23 nodes, of which seven are boundary nodes. As in the phase I calorimeter analysis,

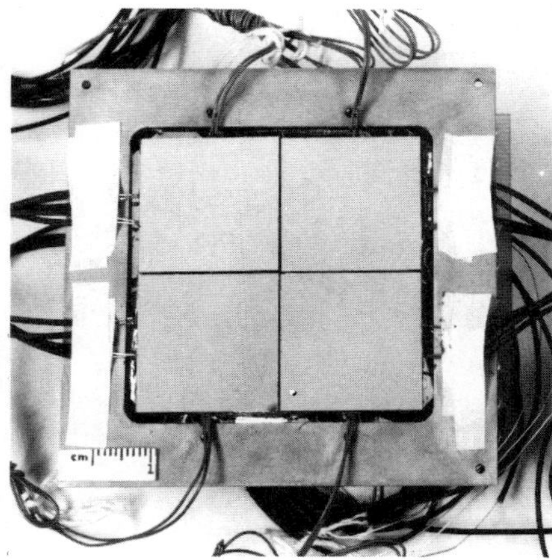

Fig. 11 Phase I calorimeter test module.

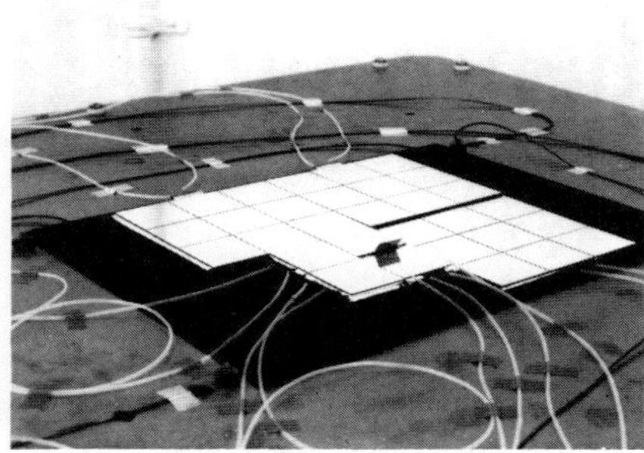

Fig. 12 Phase II calorimeter prior to installation in test fixture.

the SINDA program is used to compute temperature diurnals.
Each sample, combined with its adhesive, mounting platform,
and temperature monitor, forms a single node. The four
fiberglass support cylinders each are composed of two annu-
lar nodes, each for the lower cylinder and one for the support disk and upper cylinder. The insulation blanket is
quartered along gaps between the samples into four nodes.
The remaining seven nodes are boundary conditions, which
comprise the housing, mounting frame, space, and other
exterior satellite components.

The thermal model for each individual calorimeter
is constructed to take into consideration the actual
composition and thickness tolerances of individual materials
and components comprising each node. The thermal properties
of all nodes are inputed as functions of temperature, including ε_H and joule heating from the temperature monitors.
As in the phase I calorimeter thermal model, the heating
tables include the seasonal effects of solar intensity

variation, orbit inclination, and the effects of α_s (θ_i) variation for each sample due to satellite spin.

The $-T$-axis calorimeter on satellite B is shown installed on the radiator in Fig. 10. All four samples on this calorimeter were type IV mirrors. Each sample consisted of nine mirrors with gap seals only on the upper right-hand corner sample. The lower right-hand corner sample used nine 50-mil mirrors to test potential bulk property effects on degradation.

The visible effect of using the gap seal is evident in the figure. This arrangement was designed to test the effect of gap seals on initial solar absorptance and subsequent degradation effects. Unfortunately, temperature monitor failure on this calorimeter prevented comparisons. However, one sample (plus the base monitor) gave consistent data for four months before failure due, apparently, to improper bonding techniques of the sensor to the sample mounting plate. This apparent structural failure of the sensing element was reproduced on subsequent thermal vacuum tests. Later calorimeter sensors used a different adhesive combination (RTV in place of Stycast), which has proved successful on subsequent satellite calorimeters.

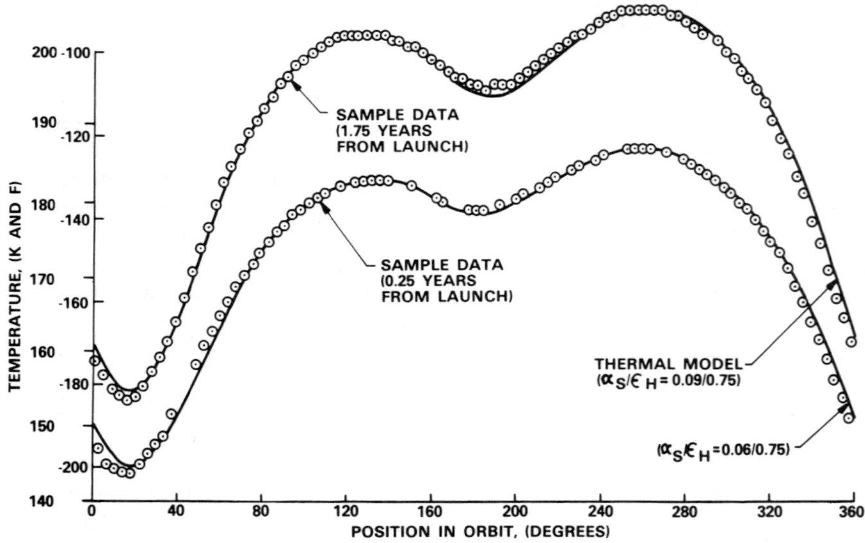

Fig. 13　Comparison of thermal model and calorimeter temperature data for the type I mirror (satellite A +S axis sample).

D. Calorimeter Thermal Vacuum Tests

Thermal vacuum tests were conducted on both the phase
I and II calorimeters. These calorimeter modules are shown
in Figs. 11 and 12, respectively, prior to test installation. Two tests were performed, namely, a boundary effects
test, in which the sensitivity of the mirror temperatures
to changes in the boundary temperature were established,
and a mirror sample cross-coupling test, in which the amount
of the thermal coupling between samples was measured. The
results confirmed thermal model calculations that indicated severe sensitivity of mirror temperature to boundary
temperature changes; i.e., 1° shift in boundary temperature
would result in a 0.5^o change in the mirror temperature.
However, the cross-coupling effects between mirrors were
found to be an order of magnitude lower. These results led
to the redesign effort for the phase II calorimeter to
improve sample thermal isolation for more accurate determination of environmental effects on α_s/ε_H.

The results of tests conducted on the phase II calorimeter justified the redesign effort. These results showed
at a 1° shift in boundary temperature would result in only
a 0.05^o change in the mirror temperature. The corresponding
mirror coupling tests indicated further improvement in
sample-to-sample isolation; i.e., a 1° shift in one mirror
sample temperature would cause an adjacent mirror to shift
by 0.02^o.

E. Results of Flight Measurements

Each calorimeter sample was analyzed for both solar
absorptance and infrared emittance changes independently
utilizing the dark and sunlit portions of the satellite

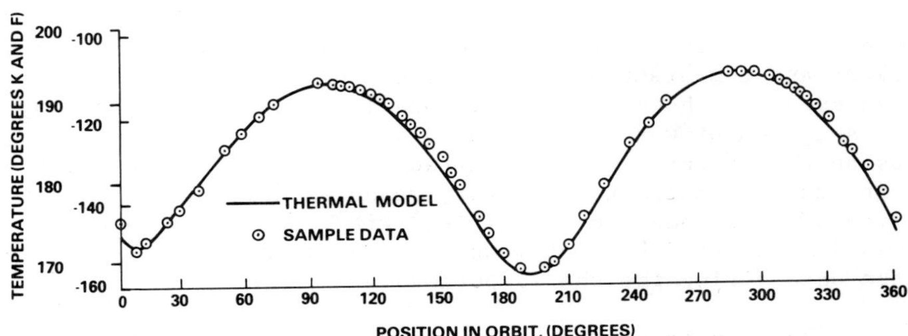

Fig. 14 Satellite D calorimeter 5-mil aluminized FEP teflon.

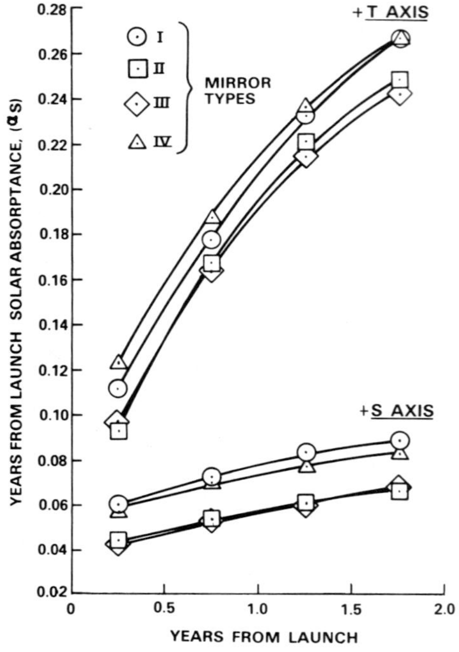

Fig. 15 Satellite A calorimeters sample degradation data.

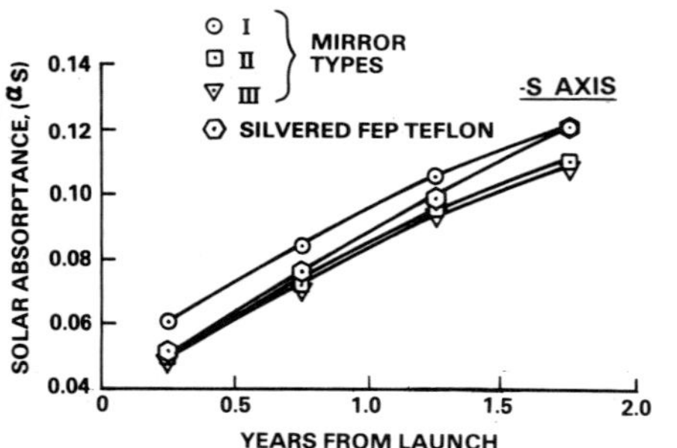

Fig. 16 Satellite A calorimeters sample degradation data.

orbit. Excellent agreement between thermal model calculations and temperature data for the diurnal cycle was achieved, as shown in Figs. 13 and 14. As seen in Fig. 13, this agreement was equally valid several seasons later using the assumption of α_s degradation ($\Delta\alpha_s$ increase) with constant ε_H based on laboratory measurements. The assumption of emittance decrease ($-\Delta\varepsilon_H$) with constant α_s also was investigated but was found not to correlate well with the data during the shaded periods.

The results of long-term orbital data analysis from the calorimeter experiments show substantial variation in

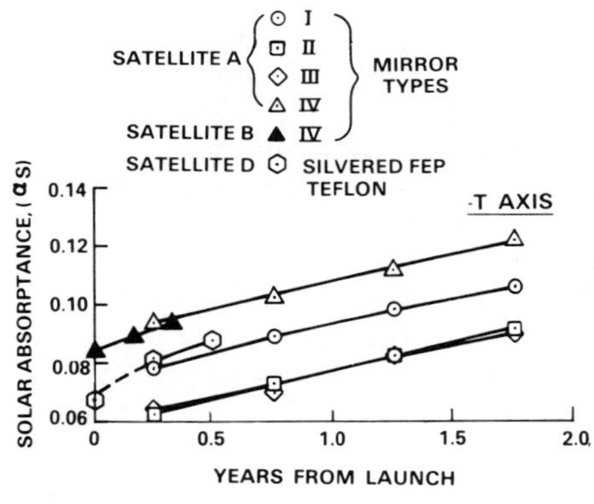

Fig. 17   Satellite A, B, and D calorimeters sample degradation data.

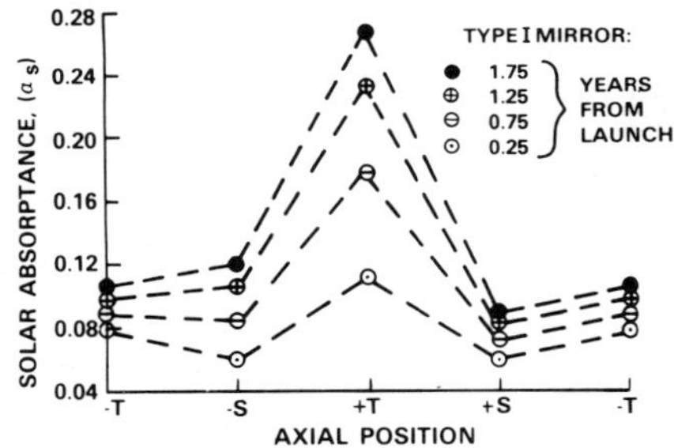

Fig. 18   Satellite A calorimeter circumferential degradation.

the degradation rates between calorimeter locations for satellite A, as shown in Figs. 15-17. The circumferential variation is shown more clearly in Fig. 18 as a function of axial position on the satellite. The dashed line indicates that the shape of the curve is not known, since the calorimeters represent only a fraction of the radiator area between axial positions. The data cover a period of 1.75 yr from launch for the type I mirror, used as a representative sample from the four calorimeters. The individual mirror samples on each calorimeter, however, indicate comparatively small differences on both an absolute and rate basis. Although the silvered teflon sample on the -S axis shows a higher degradation rate than the other quartz samples, the degree of circumferential variation in the degradation rates tends to rule out front surface material damage or delamination as the primary cause of the degradation.

A comparison of calorimeter sample data for satellites A, B, and D is given in Fig. 17. The data indicate that the degradation rates on the -T-axis calorimeter are quite similar between satellites and apparently again independent

of sample front surface material. The dashed line for
the silvered teflon sample data on satellite D indicates
that an initial rate change occurred during the first few
weeks after launch, and further analysis is required. Similar behavior occurred on the other three samples on this
calorimeter, as shown in Fig. 19. In this figure, the
two silica cloth samples appear to be degrading faster than
the silvered and aluminized teflon samples. This may be
due to intrinsic changes or from contaminants collecting
on the porous weave of the cloth samples. The weave has
a larger surface area available for contaminant deposition.

III. Radiator Analyses

As a result of the calorimeter findings, a study was
conducted to determine if the degradation rates for the
radiator, which housed the calorimeter units, would show
the same type of axial or circumferential variation. Previous analyses had assumed uniform degradation rates.

A. Flight Radiator Descriptions

Early satellite temperature status monitors were used
mainly to monitor nonradiator areas. After a warming trend

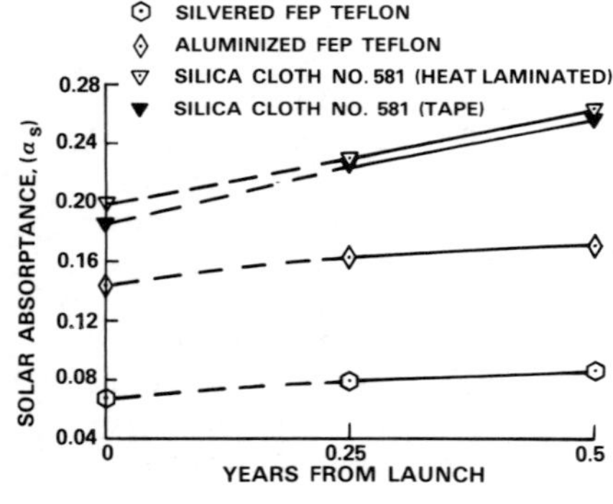

Fig. 19   Satellite D calorimeter sample
degradation data.

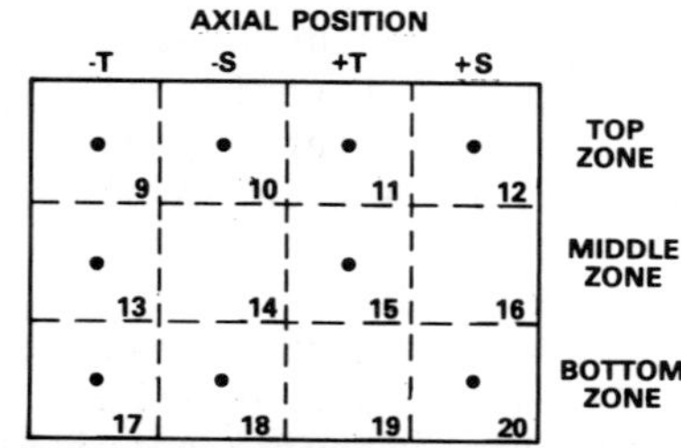

Fig. 20   Satellite radiator
nodal network and
temperature monitor
locations.

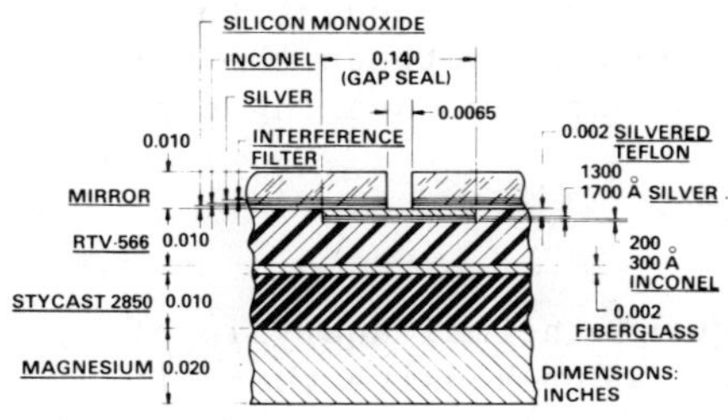

Fig. 21 Radiator thermal control surface composition.

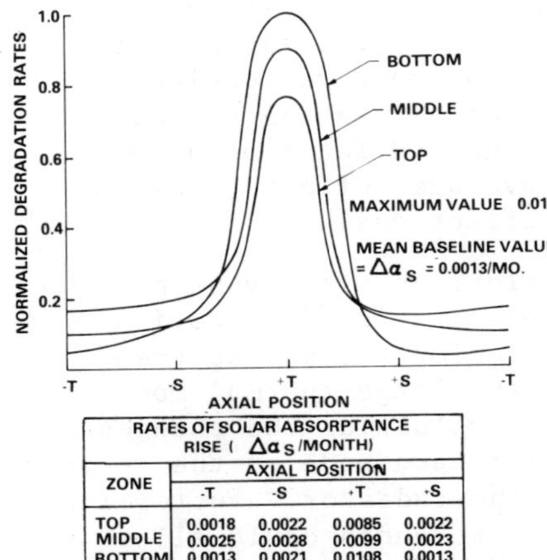

Fig. 22 Satellite A radiator solar absorptance degradation rates normalized to highest value (6 months from launch).

ZONE	RATES OF SOLAR ABSORPTANCE RISE ($\Delta\alpha_S$/MONTH) AXIAL POSITION			
	-T	-S	+T	+S
TOP	0.0018	0.0022	0.0085	0.0022
MIDDLE	0.0025	0.0028	0.0099	0.0023
BOTTOM	0.0013	0.0021	0.0108	0.0013

with time was observed on critical elements, attention turned to the potential contamination/degradation of the radiator and peripheral components which might cause an adverse change in the α_S/ε_H ratio of the radiators. Satellite B was the first to have sufficient temperature status monitoring of the radiator to allow extensive analysis to be performed. Figure 20 shows the positions of these monitors on the satellite B radiator with accompanying node and zone designations.

The radiator is composed of a magnesium shell support structure bonded to fiberglass panels with second-surface mirrors attached, as shown in Fig. 21. The top and bottom radiator zones of satellite A were covered with type I mirrors, whereas the middle zone was covered with type I and type II mirrors. All subsequent satellites had radiator top and bottom zones covered with type I and type II mirrors, whereas the middle zone was covered with type II and type III mirrors. In addition to these differences between satellite radiators, the silvered teflon gap seals

shown in Fig. 21 were used on satellites subsequent to
satellite A. The purpose of the gap seal between the mirrors was to reduce the solar absorptance by covering high-
absorptance RTV adhesive.

B. Flight Model Analyses

During the design and subsequent thermal analysis of
the satellites, the development of comprehensive thermal
models was evolved to serve as a design tool and predictor
of thermal performance. The basic 350-node model of each
satellite was modified to take into account individual
material and component changes. Effects of seasonal solar
intensity variation and orbit inclination were modeled as
in the calorimeter analyses. Similarly, heating tables
including the α_s variation of external surfaces as a function of incidence, spin-averaged effects, shadow-
ing, and solar specular/diffuse reflections.

Since the results of early calorimeter analyses on
satellite A indicated that the contamination/degradation
of radiator surfaces was responsible for the warmup trend
observed, it became apparent that the large thermal model
could provide an essential tool in evaluating the changes
in α_s/ε_H. Detrmination of α_s/ε_H for each node on the radiator could provide an α_s "map" of the radiator. Performing
such analyses for various times after launch could allow
a temporal pattern of contamination/degradation in terms
of α_s/ε_H to be developed.

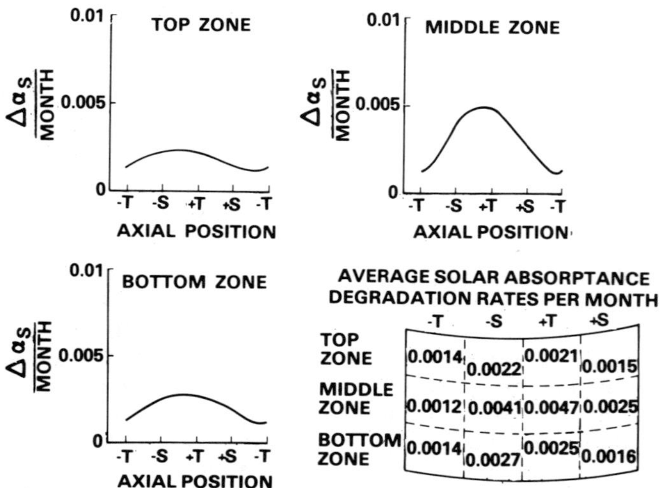

	-T	-S	+T	+S
TOP ZONE	0.0014	0.0022	0.0021	0.0015
MIDDLE ZONE	0.0012	0.0041	0.0047	0.0025
BOTTOM ZONE	0.0014	0.0027	0.0025	0.0016

Fig. 23 Average rates of change of solar absorptance
($\Delta \alpha_s$/month) on satellite B radiator nodes
(1.5 yr from launch).

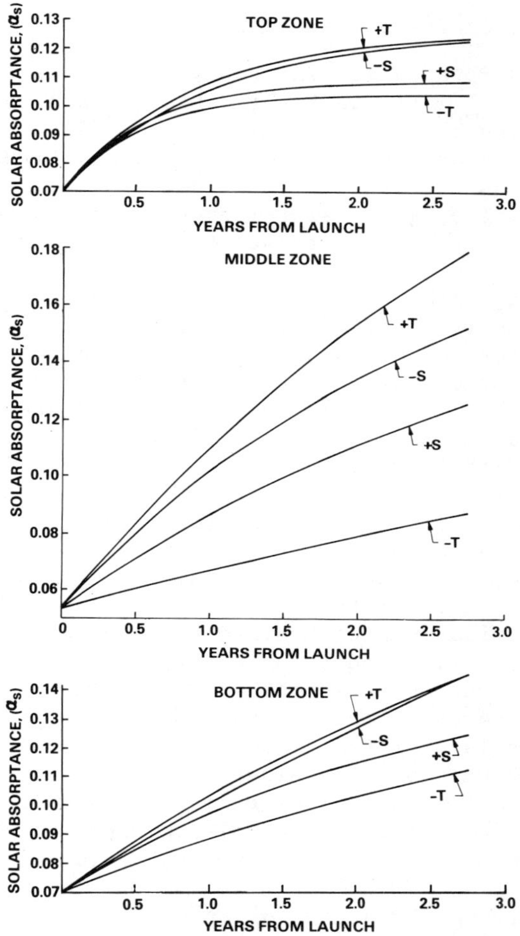

TYPE I RADIATOR ZONE	AXIS	NODE NO.	INITIAL SOLAR ABSORPTANCE, α_0	TIME CONSTANT, τ (DAYS)	ASYMPTOTIC SOLAR ABSORPTANCE, α_M
TOP	-T	9	0.07	192	0.1044
	-S	10	0.07	359	0.1263
	+T	11	0.07	318	0.1259
	+S	12	0.07	202	0.1088
MIDDLE	-T	13	0.053	3139	0.1792
	-S	14	0.053	949	0.2050
	+T	15	0.053	1350	0.2937
	+S	16	0.053	1284	0.1877
BOTTOM	-T	17	0.07	1671	0.1637
	-S	18	0.07	2888	0.3267
	+T	19	0.07	1504	0.2244
	+S	20	0.07	908	0.1519

Fig. 24   Solar absorptance exponential roll-off trends for the satellite B radiator (2.75 yr from launch).

C. Results of Flight Measurements

Solar absorptance mapping analyses for the satellite A radiator were limited to the first six months after launch because of temperature monitor failures. Also, with the scant monitoring on the radiator, many assumptions were required such as continuity of circumferential α_s distributions. Matching of internal temperature was used as a check to find feasible α_s map solutions.

Data on α_s gathered over the six-month period were used to compute rates of α_s rise. Statistically, only a linear rate of increase could be justified. Rates of rise were computed using a minimum least-squares average error criteria. Figure 22 displays a normalized distribution of these α_s rates of rise, together with a table presenting the absolute values. As can be seen in the figure, the circumferential distribution of α_s and corresponding α_s rates of rise for each zone are preferential in nature, in agreement with the calorimeter results. The highest rates are near the +T axial position, approximately five times the rates on the other axes. The magnitudes and distributions determined enabled temperatures on monitored exterior nodes and critical internal elements to be matched over the six-month period. Numerical solutions served, however, only as reasonable approximations because of the lack of temperature monitors on the radiator itself. Additionally, an evaluation was initiated to determine if a change in ε_H for the radiator could be justified. It was concluded that no detectable change could be observed, also in agreement with results from the calorimeters.

The satellite B radiator α_s mapping analyses covered a period of 2.75 yr. Early results showed linear α_s rates but with marked reductions in the magnitude of the distribution compared with satellite A. A 25% reduction in the overall rate of α_s rise was observed. The results for α_s rates of rise at 1.5 yr are presented in Fig. 23. A decrease in the peak preferential rates of α_s rise occurred, whereas minimum values were virtually identical to satellite A.

When the analysis had reached a point 2.75 yr after launch, data showed that a roll-off in rates of α_s rise clearly was evident and statistically justified. The exponential roll-off equation used is of the following mathematical form:

$$\alpha_s = \alpha_o + (\alpha_m - \alpha_o)\,(1 - e^{-t/P})$$

where

α_o = initial launch value of solar absorptance

α_m = asymptotic value of solar absorptance

p = time constant

t = time from launch

Figure 24 presents the constants derived from this equation for each node on the satellite B radiator, along with plots of the α_s trend correlation for each node. Figure 25 presents circumferential zonal plots of the trend correlations for every 100 days following launch. Figure 26 shows a typical α_s trend correlation.

IV. Quartz Crystal Contamination Monitor/Electrostatic
     Field Experiments

As part of corrective hardware changes for satellite B, instrumentation was flown to measure contaminant deposi-

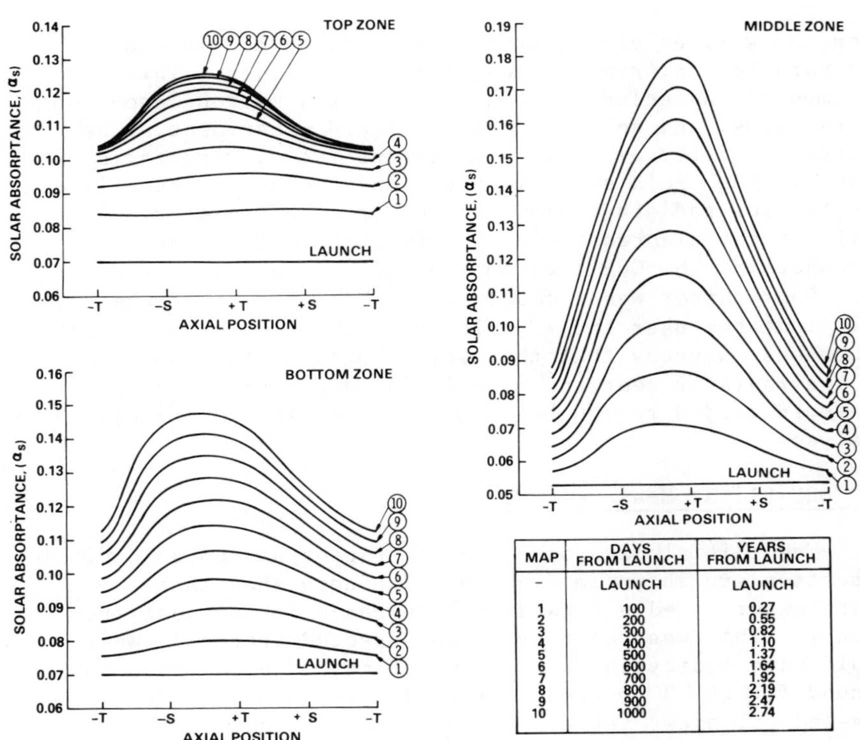

Fig. 25 Circumferential plots of satellite B solar absorp-
         tance trend correlations.

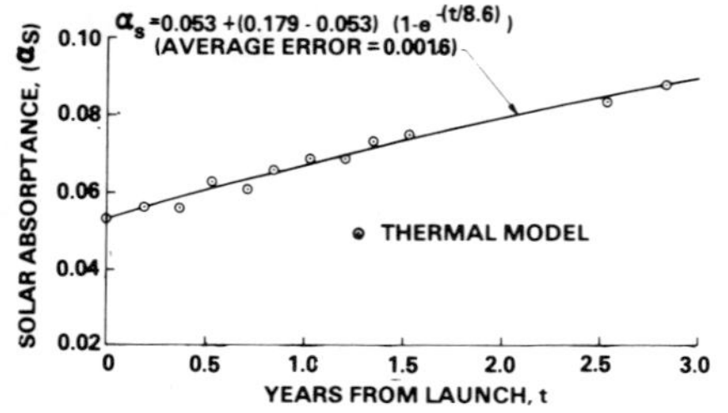

Fig. 26 Satellite B solar absorptance trend correlation (node 13).

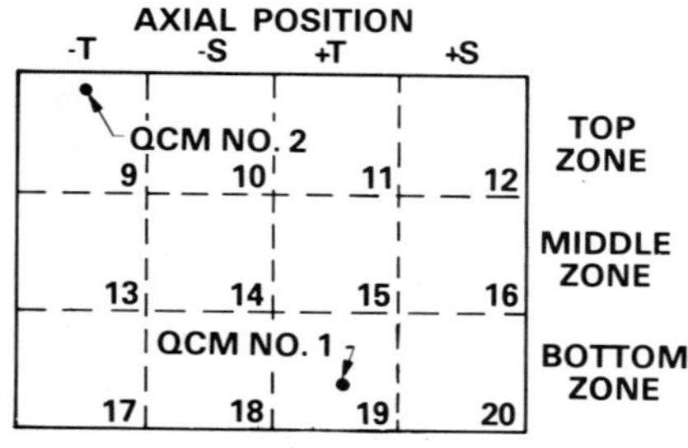

Fig. 27 Quartz crystal monitor locations on the satellite B radiator.

tion, as well as electrostatic effects, to aid in evaluating the parallel calorimeter and radiator analyses. This instrumentation included two quartz-crystal microbalance monitors (QCMs) and an electrostatic field experiment. The latter consisted of a field meter and two electrostatic probes. The field experiment was mounted on the satellite electronics radiator below the colder main radiator. Details of the electrostatic experiment have been reported elsewhere.[8] The QCM locations are shown in Figs. 27 and 28. One monitor was located on the bottom zone of the main radiator near the +T axis in view of several satellite sunshades, whereas the other was located on the top zone of the radiator near the −T axis, an apparently less contaminated/degraded region based on radiator and calorimeter analyses.

A. QCM Flight Measurements

The initial data obtained after geosynchronous orbital insertion are shown in Fig. 29. The data show that the units experienced a substantial decrease in the beat frequency. This was contrary to an expected increase, which would have indicated mass accumulation. The decrease continued for QCM 2 located on the −T axis; however, QCM 1 leveled off after several months in orbit and appeared to be increasing. These early data were not converted to a

Fig. 28   Placement of
          QCM 2 (-T
          axis).

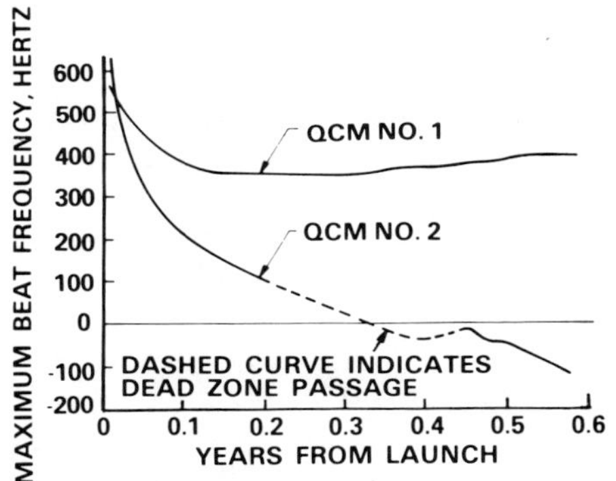

Fig. 29   Satellite B initial
          QCM data.

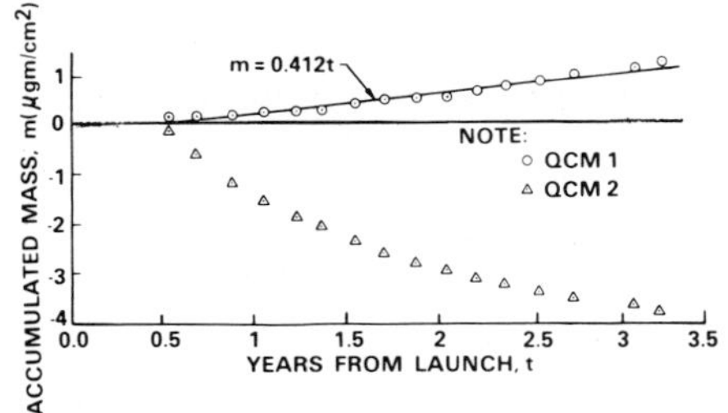

Fig. 30   Measured accumulated
          mass of QCMs 1 and
          2.

mass relationship because of their questionable validity.
Several theories were offered to explain this behavior,
including evaporation of initially condensed materials.
Other explanations of the QCM behavior included instrument
malfunction and insufficient calibration.

More recent data from QCM experiments flown on the
Skylab have, however, shown similar behavior.[9] Units exposed to a relatively clean environment, as is the case
for QCM 2, showed a continued decrease, whereas units exposed to definite contamination sources, as potentially

the case for QCM 1, showed mass increases. Subsequent QCM analyses, performed for several years of data (see Fig. 30), show that the QCM placed in the area of suspected major contamination indicates a linear increase in deposition with time (0.412 $\mu gm/cm^2$-yr) after the initial transient decrease. The QCM placed in the least-degraded radiator area continued, however, to show a decrease (1 to 2 $\mu gm/cm^2$-yr) but at a lower rate than the initial transient. These long-term trends indicate relative differences between QCM locations and, in this respect, are in agreement with the contamination theory.

B. Deposition and Irradiation Effects on α_s

Tests were conducted by Jones and Borson[10] on the effects of volatile condensable materials (VCM) on second-surface materials. The tests were performed by depositing Silastic 140 RTV while simulating the far uv (0.147 μm). The results showed that a contaminant thickness of 1000 Å would result in a $\Delta\alpha_s$ between 0.06 and 0.16 in a period of 1.5 yr. Assuming a contaminant density of 1 g/cm^3, an approximate sensitivity of $\Delta\alpha_s$ to Δm would be

$$\Delta\alpha_s/\Delta m = 0.01/100 \text{ Å} = 0.01/\mu g/cm^2$$

If the degradation rates for the radiator area in the vicinity of the QCM 1 location are compared with the QCM measurements, a sensitivity of $\Delta\alpha_s/\Delta m = 0.091/\mu g/cm^2$ is obtained. However, closer agreement is achieved if a correction is applied to the QCM 1 data based on the performance decrease of QCM 2, i.e., $\Delta\alpha_s/\Delta m = 0.19/\mu g/cm^2$. The latter correction assumes that both QCMs are subject to the same frequency decrease with time but that QCM 1 has sufficient material deposited to change this response.

V. Conclusions

The results of the thermal improvements on satellite B are evident from the reduced degradation rates. Corresponding temperatures and rise rates of the radiators are significantly lower than on previous satellites. This can be attributed to use of thermal control coatings with lower α_s values, radiator panel gap seals, refined cleaning procedures, improved quality control, and use of silvered teflon to replace S-13G white paint on satellite components such as sunshades. Also, as discussed elsewhere,[11] radiator phase-change materials further delayed the effects of the degradation, resulting in long-term improved thermal performance.

Contamination/degradation effects have been observed on other satellite/spacecraft systems.[12] The results of the analyses here have shown that contamination is apparently a major cause of the observed degradation:

1) Degradation rates with nonsymmetric distributions appear to be affected by satellite material changes and vent paths within line-of-sight of the observed degradation peaks.

2) The lowest level of degradation on several satellites appears to be independent of calorimeter thermal control coatings.

3) Ground through launch pad environments may affect degradation.

4) Launch vehicle environments may be a cause for the observed degradation, since recent QCM measurements[13] on a Titan III C payload indicate potential contamination for payloads during the ascent from retro-rocket plumes.

5) As reported recently,[14] tests have shown that thruster plumes have a wide mass spectrum in their exhaust products, which are potential contamination sources.

6) Calculated time constants from radiator degradation trends indicate short- and long-term contamination sources.

7) Radiator QCM measurements, although of questionable validity, support the contamination theory and appear to indicate extreme sensitivity of α_s to mass deposition rates.

8) Satellite-induced environments may affect initial and long-term degradation rates due to a) venting of satellite outgassed materials, b) thruster plume impingement and use rate, c) satellite spin effects on venting characteristics, apparently causing observed degradation rate shifts (see Fig. 25, top and bottom radiator zones), d) physical damage to coatings from electrostatic discharges, and e) additional electrostatic effects on non-line-of-sight contamination sources.

At present, analyses are continuing on several satellite experiments to quantify the effects of contamination/degradation on thermal control coatings. Proposed calorimeter/QCM experiments could investigate temperature level effects and protective cover concepts for selected coatings.

Current satellite programs will minimize further the con-
tamination/degradation effects using advanced thermal designs and selection of materials shown to be less susceptible to damage from effects of the space/satellite environment.

References

[1]Fogdall, L.B. and Cannaday, S.S., "Irradiation and Measurement of Second-Surface Mirrors," Boeing Co., Final Rept.
for Aerojet ElectroSystems Co., Azusa, Calif., Oct. 1972.

[2]Parkinson, J.B., "Second-Surface Mirror Degradation Study,"
Aerojet ElectroSystems Co., Azusa, Calif., Internal Rept.,
Nov. 1972.

[3]"Evolution of AESC Satellite Contamination Studies,"
Aerojet ElectroSystems Co., Rept. 4724, Feb. 1973.

[4]"Titan IIIc Payload Environmental Exhibit," Martin
Marietta Corp., Denver, Colo., Rept. MCR-75-157, Rev. B,
March 1976.

[5]"Cleaning of Second-Surface Mirrors and Lifetime Monitoring
Instrumentation (Phase II)," Aerojet ElectroSystems Co.,
April 1974.

[6]Heaney, J.B., "Suitability of Metalized FEP Teflon as a
Spacecraft Thermal Control Surface," American Society of
Mechanical Engineers, Paper 71-AV-35, San Francisco, Calif.,
July 12-14, 1971.

[7]Millard, J.P., "An Uncertainty Analysis For Satellite
Calorimetric Measurements," NASA TN D-4354, Feb. 1968.

[8]Nanevicz, J.E., Adamo, R.C., and Shaw, R.R., "Electrical
Discharges Caused By Satellite Charging At Synchronous Orbit
Altitudes," Conference on Lightning and Static Electricity,
Abingdon, Oxon, England, April 14-17, 1975.

[9]"Skylab Orbital Assembly Systems Mission Contamination
Evaluation Report-Assessment," NASA Marshall Space Flight
Center, Skylab Program Payload Integration, TR ED-2002-
1701, Rev. B, May 1974.

[10]Jones, P.F., and Borson, E.N., "The Effects of Deposition and Irradiance of Contaminants From the Outgassing
of Silastic 140 RTV," Aerospace, Rept. TOR-0059(6129-01)-
56, Feb. 1971.

[11]Keville, J.F., "Development of Phase Change Systems and Flight Experience On An Operational Satellite," _Progress in Astronautics and Aeronautics: Thermophysics of Spacecraft and Outer Planet Entry Probes_, Vol. 56, edited by A.M. Smith, AIAA New York, pp. 19-36.

[12]"Contamination and Degradation Of Spacecraft Systems: An Annotated Bibliography," Government-Industry Data Exchange Program. Corona, Calif., Rept. E053-0848, March 1976.

[13]Lynch, J.T., "Quartz Crystal Microbalance (QCM) Monitor of Contamination For LES-8/9," Massachusetts Inst. of Technology, Lincoln Lab., Project Rept. SC-33, June 1976.

[14]McCay, T.D. and Powell, H.M., "Direct Mass Spectrometric Measurements in a Highly Expanded Rocket Exhaust Plume," AIAA Paper 77-154, Los Angeles, Calif., Jan. 24-26, 1977.

EFFECTS OF SPACE RADIATION ON THIN POLYMERS AND NONMETALLICS

Lawrence B. Fogdall* and Sheridan S. Cannaday*
The Boeing Company, Seattle, Wash.

Abstract

Advanced materials for various spacecraft systems in the
1980s and 1990s have been evaluated in situ after exposure to
space radiation. Emphasis has been placed on materials having little or no previous base of combined environmental effects data. Application ranging from Earth orbit to near-sun
have been covered. High-temperature polymers and composite
coatings have been included. Silica composite coatings may
offer improved reflectance stability compared with metallized
fluorocarbons. Directional reflectance properties of FEP are
a function of charged particle energy and flux as well as
total exposure fluence and material characteristics. Data
obtained on polyimides and polyxylylenes under high-temperature radiation exposure conditions is discussed in the con-
text of near-sun solar sailing and rendezvousing.

I. Introduction

In a continuing program to evaluate and understand changes
in the thermophysical properties of materials for long-term
space use, an extended series of laboratory experiments has
been and is being conducted to simulate the effects of space
radiation. For various space systems in the 1980s and beyond,
various advanced aerospace materials are being considered;
some have little or no data base regarding the effects of
environmental exposure, and others that do are candidates for
uses during time periods up to 30 years. For these, available
short-term data must be extrapolated or interpreted or long-
term radiation exposure data obtained anew using state-of-
the-art techniques.

Presented as Paper 77-741 at the AIAA 12th Thermophysics
Conference, Albuquerque, N. Mex., June 27-29, 1977. Copyright
© American Institute of Aeronautics and Astronautics, Inc.,
1977. All rights reserved.
*Specialist Engineer.

This paper presents new information on the stability and
suitability of polymers and other nonmetallics for space uses
including large-area reflective surfaces, solar energy concentration, solar sail films, low-density structural members
or skins, as well as more conventional thermal control ap-
lications. Samples of applicable materials have been exposed
to an accurate simulation of the space radiation environment:
ultraviolet radiation,protons, and electrons accelerated
simultaneously onto samples that then are measured in situ.
Irradiation has taken place over real-time periods of thou-
sands of hours in a contamination-free test environment. The
ultra-high-vacuum environment has been obtained without organic pumping fluids. Sample temperatures have been regulated
continuously throughout the exposure time period and during
in situ measurements as well. Selective hot and cryogenic
surfaces and receptacles have been used to control the contamination that comes from test materials themselves. Exposure rates have been controlled to preclude rate effects
and to be applicable for the proposed use. Thus uv rates have
been held to a one-sun level at one a.u. in most cases, but
they have been increased more than an order of magnitude for
near-sun applications.

The other principal components of the space radiation environment, energetic electrons and protons, are highly variable
in intensity as a function of orbit and time. So electrons
and protons have been simulated in our laboratories over
approximately 3 orders of magnitude of flux (charged particles
per square centimeter-second), resulting in significant findings in such areas as specularity changes in reflective films
and survivability of solar radiation receivers.

II. Experimental Apparatus

These results have been obtained using portions of the
facilities in Boeing's combined radiation effects test (CRETC)
laboratories. There are presently three CRETC chamber systems, whose capabilities can be summarized as follows: 1) ultra-high-vacuum environment as refined as 2×10^{-9} Torr, using oil-free pumping subsystems; 2) space for test articles
up to 1 ft^3 (28,000 cm^3) in volume, irregular shapes included;
3) air-mass-zero ultraviolet radiation (uv) at rates from
less than one sun (1 a.u.) to approximately 16 suns and over
test array areas up to 0.5 ft^2 (465 cm^2); 4) vacuum ultraviolet radiation (Vuv) from a window-type source (no introduction of contamination) at rates $\sim$1 sun Lyman-α (1216 A)
over areas $\sim$20 cm^2; 5) protons to simulate positively charged
particles in the solar wind as well as Earth orbits (energies

<1 to ∿100 keV, fluxes $\leq$1 x 10^{10} p/cm^2-sec), proton beam area
up to 300 cm^2; 6) electrons <1 eV to >200 keV (flux up to 1 x
10^{11} e/cm^2-sec) simulating Earth orbital or neutralized solar
wind conditions, electron beam area generally ∿500 cm^2; 7) cw
or pulsed laser irradiation at HF, DF, and CO$_2$ wavelengths
under ambient conditions or 10^{-7} - Torr vacuum; 8) combinations of all of the preceding radiation sources, for simultaneous irradiations; 9) in situ measurement capabilities
for controlled sample temperature, spectral reflectance, spectral transmittance, semiconductor characteristics, dielectric
properties, and photography; future expansion to in situ emittance, Auger, and LEED measurements is planned[1,2]; 10) retention of in situ features during moves of portable chambers
to other radiation accelerators and sources; and 11) a high
degree of test automation and selective/protective shutdown
features for long-term combined radiation exposure at lowest
possible cost.

A representative array of instrumented test samples is
shown in Fig. 1, positioned in one of the CRETC vacuum chambers.
Another array is shown in Fig. 2 adjacent to an anticontamina-
tion baffle that traps outgassed sample adhesives and similar
volatile molecules.

Fig. 1 Spacecraft thermal control and solar array materials.

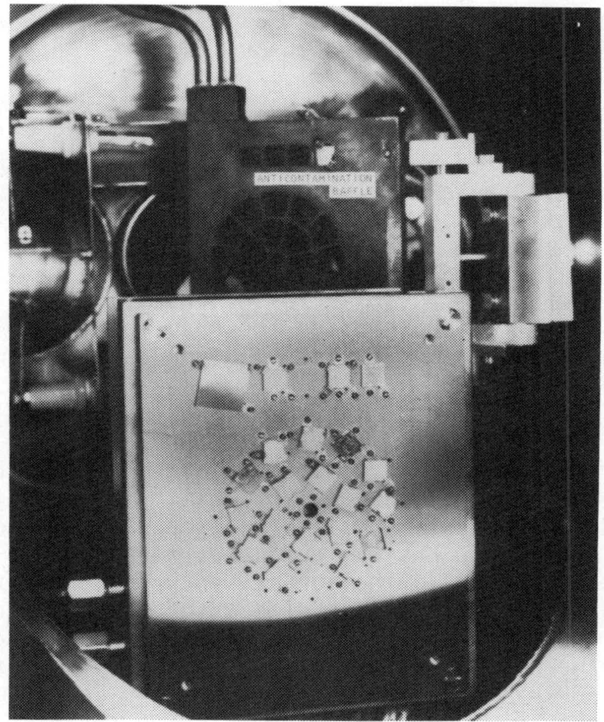

Fig. 2 Test sample array rotated away from anti-contamination
baffle.

III. Experiments and Results

Flexible, polymeric second-surface reflector materials
recently have displaced high-reflectance paints for many spacecraft thermal control applications. The metallic reflector
becomes the second surface by covering it with a transparent
polymer. The polymer greatly increases the laminate's emittance by means of its infrared absorption bands. One widely
studied (and used) laminate is transparent FEP teflon (a
fluorinated ethylene propylene copolymer), which is metallized
with vapor-deposited aluminum or silver on the back (second)
surface to produce an initially favorable absorptance-to-emit-
tance ratio material for many spacecraft and space system
exterior purposes. Tests for ultraviolet radiation-induced
reflectance changes began in the 1960s and generally showed
acceptable reflectance stability for Ag/FEP and Al/FEP. Irradiations with charged particles since have been conducted
in various radiation combinations.[3-9] Electron exposures soon
demonstrated the existence of the electrostatic charge and

field buildup problem that leads to material breakdown, and which since has been confirmed in orbit.[10-12]

In the most recent simulations of electrons in combination with protons and uv at actual orbital ratios, it is now clear that metallized FEP has definite limitations that are not uv-induced. These simulations were designed to yield data on spectral reflectance, $R(\lambda)$, and solar absorptance, α_s, that could be used as guidance in spacecraft thermal design. The definition of solar absorptance in opaque materials is $\alpha_s = 1 - R_s$, where R_s is solar reflectance, a coefficient spectrally weighted according to the shape of the solar spectral output or "Johnson curve." In equation form,

$$R_s = \int I_s(\lambda)\ R(\lambda)\ d\lambda / \int I_s(\lambda)\ d\lambda, \text{ where } I_s(\lambda)$$

is solar irradiance as a function of wavelength λ. Figure 3 shows the degree of $R(\lambda)$ and α_s instability of FEP after various long-duration tests simulating space radiation conditions. Simulation A for 4000 hrs was conducted to ascertain the effects of spacecraft illumination at synchronous altitude without the additional deleterious effects of magnetic substorm conditions, that is, uv irradiation at a one-sun rate (0.013 watt/cm^2 between 0.2 and 0.4 μm wavelength), plus 2 x 10^8 e/cm^2-sec (50-keV) and 2 x 10^8 p/cm^2-sec (40-keV) simultaneously. As curve A in Fig. 3 shows, reflectance changes in metallized FEP under nonsubstorm conditions, even after ap-

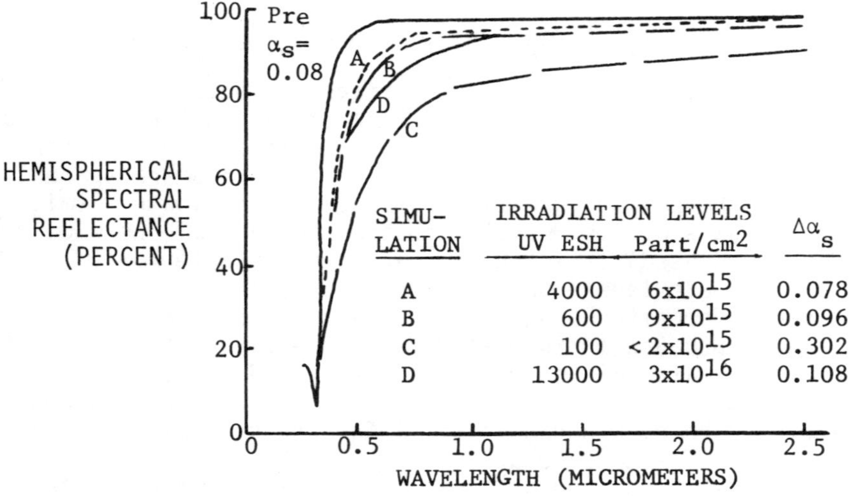

SIMU-LATION	IRRADIATION LEVELS		$\Delta\alpha_s$
	UV ESH	Part/cm^2	
A	4000	6x10^{15}	0.078
B	600	9x10^{15}	0.096
C	100	<2x10^{15}	0.302
D	13000	3x10^{16}	0.108

Fig. 3   Effects of uv, electrons, and protons on 5-mil FEP Teflon.

proximately 1 yr in orbit, are within acceptable tolerances of
all but the most critical thermal design. Moreover, the metallized polymer loses little of its specularity when exposed to
this relatively low flux of charged particles along with uv.

     Magnetic substorms can increase electron fuxes to 2 or
3×10^9 e/cm^2-sec. Simulation B considered a smaller fraction
of a year orbital time with magnetic substorm conditions included (uv, 1 sun; electrons, 2×10^9 e/cm^2-sec (115-keV); and
protons, 2×10^9 p/cm^2-sec (50-keV). Therefore, the ratio of
charged particles to uv electromagnetic energy increased by a
factor of 10, so that the charged particle fluences of simulations B and A were approximately the same. As shown in Fig.
3, the degradation of reflectance and increase in α_S is not
severe for simulation B. But post-test examinations of samples
show that dielectric breakdowns in the Polymer begin to cover
a significant fraction of the total material area, causing
some overall loss of specularity. This is of significance for
space systems and geometries that depend upon specular reflection for such functions as thermal balance or energy concentration. Figure 4 is a photomicrograph of metallized FEP after
exposure to simulation B. The induced discharge paths,
white in Fig. 4, are a scattering medium, resulting in diffuse
reflection.

     It also has been determined that the exact substorm char-
ged particle fluxes have a strong influence on the usable
lifetime of polymers such as FEP. During simulation C, char-

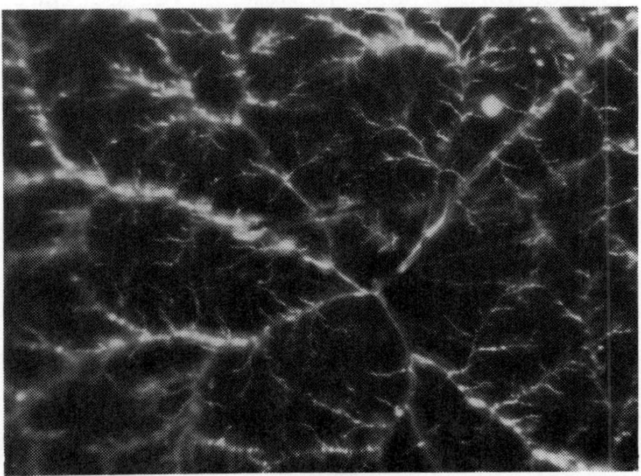

Fig. 4 Electrical discharge paths in FEP teflon after exposure
to simulated magnetic substorm charged particle fluxes
(magnification 100 X).

ged particle fluxes were increased 50% to peak orbital values of 3 x 10^9 e/cm^2-sec (115-keV) and 3 x 10^9 p/cm^2-sec (50-keV), with the uv rate set at 1.5 suns. Even though fluences were less than 1 x 10^{15} particles each sign/cm^2 (a 65-hr test period), these rates resulted in early dielectric failure from the internal electric field generated by trapped charges. Reflectance values and the α_S change are denoted "C" in Fig. 3. Silvered 5-mil FEP under these conditions rapidly becomes diffusely reflecting. Unmetallized (originally transparent) polymer becomes translucent or even opaque as irradiation is continued. Figure 5 depicts the visual degree of specularity loss after such a radiation exposure. Dark areas represent transparent FEP (still specularly reflecting if metallized), whereas white areas are a scattering polymer medium that is diffusely reflecting whether metallized or not.

The cases just cited are all for Earth orbit conditions (synchronous altitude), but similar results have been observed in a simulated solar wind. In simulation D, with a flux of 1 x 10^{10} 10-keV p/cm^2-sec combined simultaneously with ultraviolet radiation at a 16-sun rate (0.20 watt/cm^2), loss of specularity occurred to such an extent that FEP teflon's visual appearance was indistinguishable from that of white paint. Even before this point is reached, the material would be unsuitable for such uses as concentrators and reflectors in space, even though α_S changes are not themselves too great (see Fig. 3).

Most irradiation data have been obtained for 5-mil thicknesses of FEP teflon. There is some evidence that thinner

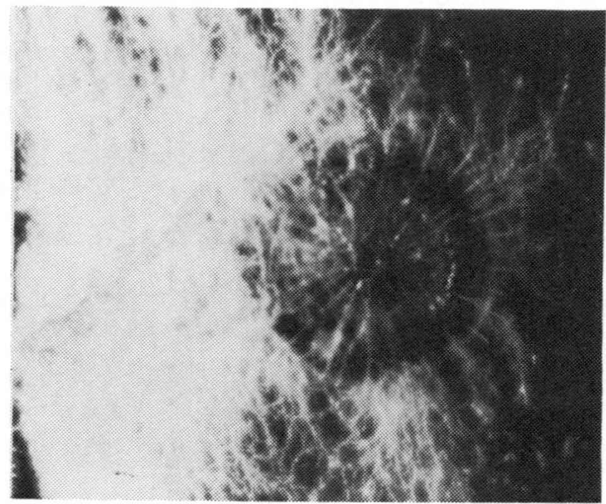

Fig. 5 Diffusely reflecting metallized FEP after extensive dielectric breakdown (magnification 140 X).

films may have more favorable properties (more resistance to
dielectric breakdown) if they are thin enough not to trap
energetic charged particles. Figure 6 (derived from Ref. 9)
compares solar absorptance changes in 2- and 5-mil FEP teflon
following irradiation by 115-keV electrons and 50-keV protons
simultaneously with ultraviolet radiation. The charged particle fluence values shown on the horizontal axis of Fig. 6 are
given for electrons and protons separately. The sum of the 2
was displayed within Fig. 3.

These differences do not occur when 2- and 5-mil metallized polymer films are irradiated using 2×10^9 fluxes of
lower-energy charged particles (30-keV electrons and 20-keV
protons) simultaneously with uv. Figure 7 compares solar
absorptance changes caused by reflectance losses after irradiation of 2- and 5-mil FEP by less energetic charged particles that have insufficient range to penetrate the 2-mil film.
In other words, 2- and 5-mil films appear more alike to lower-
energy charged particles that are trapped in the dielectric
FEP layer. Large though the α_s changes are (Fig. 7), these
films remain specularly reflecting following simultaneous
20-keV proton, 30-keV electron, and uv exposure.

More work needs to be done if the possibility is to be
confirmed that thinner polymer films have sufficient immunity
from dielectric breakdown to be usable on long-life space
systems. Data available on 1-mil films are scanty, and thick-

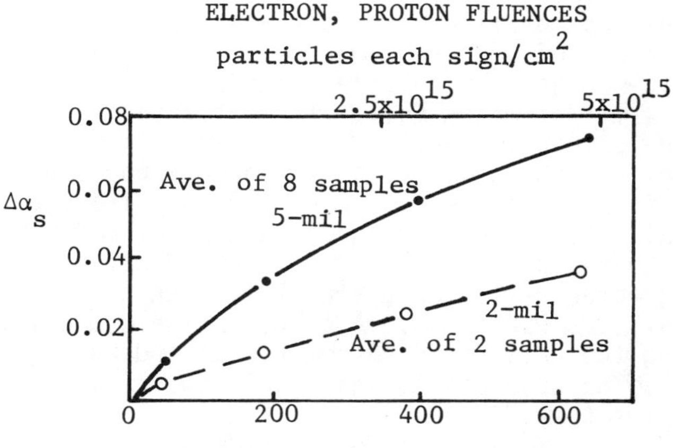

Fig. 6 Improved reflectance stability of 2-mil metallized FEP
compared to 5-mil FEP exposed to uv, 115-keV protons during
simulation B.

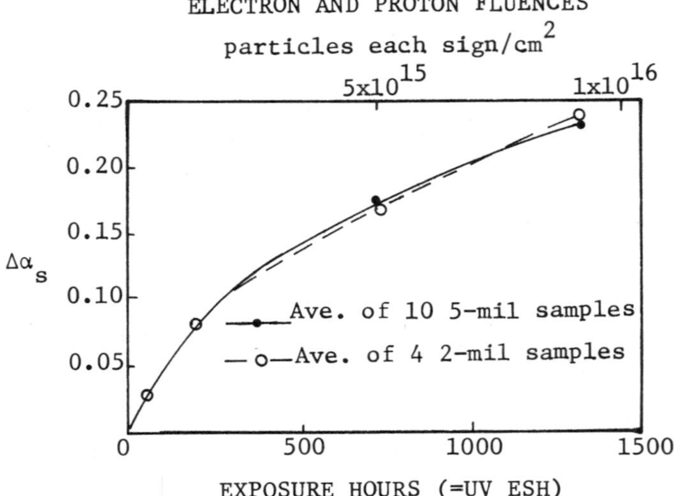

Fig. 7 Solar absorptance of metallized 2- and 5-mil FEP films irradiated by 30-keV electrons, 20-keV protons, and uv.

nesses of FEP less than 1 mil have not been investigated at all. It is these that will receive first consideration (due to lowest mass) for large-area advanced space systems.

Composite Materials

The possibility of extending the usable lifetime of space-craft exterior surfaces with composite coatings made of polymers and other nonmetallic materials also has been studied recently. In one of the programs in our CRETC laboratories, silica fabric was used to coat selected samples of 1-mil FEP backed by 0.4-mil aluminum foil. The reflectance characteristics of the composite coating is largely diffuse, with a contribution from the underlying aluminum mirror showing through the filaments of silica weave.[13] Simultaneous exposure of this silica/FEP/Al composite coating material to uv, 2 x 10^9 20-keV p/cm^2-sec, and 2 x 10^9 30-keV e/cm^2-sec results in greater reflectance stability than has been observed for metallized FEP alone. A quantitative comparison can be made using the data for FEP from Fig. 7 and adding to it the average solar absorptance changes in two silica composite coating samples (see Fig. 8).

Solar Radiation Receivers

One aspect of this topic has been covered previously: the case wherein metallized polymers are stretched or conform-

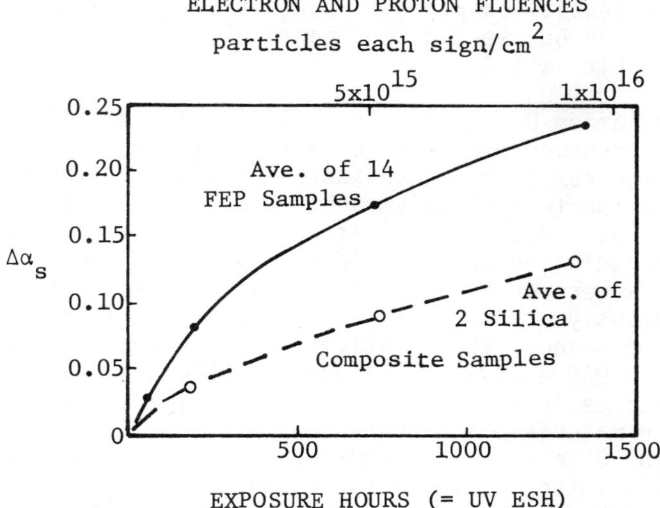

Fig. 8 Improved reflectance stability in silica/FEP/foil composite compared to metallized FEP.

ed to reflect solar energy elsewhere for thermal balance or concentration purposes. Polymer surfaces also are being considered for applications that involve optimal transfer of the radiation pressure of solar radiation for solar sailing. This pressure is greatest for the case of total reflection from such surfaces.

The equations related to radiation pressure of solar energy are understood most easily by first considering a totally absorbing surface or object. It can be shown that linear momentum p is delivered to a surface or object absorbing an amount of energy U according to the equation p = U/c, where c is the speed of light. The direction of the p vector is that of the incident solar beam. If solar energy instead is reflected normally from the surface, the momentum p undergoes twice the change, and transferred momentum is twice that indicated by the foregoing equation. Thus the maximum propulsive force (momentum per unit time) is F = p/t = 2U/ct. Material reflectivity between 1 and 0 will result in forces between 2U/ct and U/ct. In spacecraft applications, radiation pressure constitutes a very weak force, so that near-sun solar sailing for extended periods of time is required to obtain meaningful spacecraft velocity changes.

Solar orbits as close as 0.25 a.u. have been considered for several U.S. and European spacecraft. This translates to

16 suns intensity compared to Earth orbit. The maximum pressure that can be exerted on a perfectly reflecting surface in such an orbit is 1.5 x 10^{-3} dyne/cm^2 or 2.2 x 10^{-8} psi.

Simulation D included several thin-film candidates for use as solar pressure receivers or solar sailing films. Included were poly(paraxylylene) and Kapton polyimide. The latter was used as a quarter-mil polymer base, with reflection-enhancing overcoatings. Results indicating considerable stability after 812 hr irradiation are contained in Fig. 9. The irradiation conditions were a 16-sun level of ultraviolet radiation and simultaneously a simulated solar wind proton flux of 1 x 10^{10} 10-keV p/cm^2-sec. The relatively small changes shown in Fig. 9 after 13,000 equivalent uv exposure hours and a 3 x 10^{16} p/cm^2 fluence indicate that Kapton and similar polyimides may be usable materials for near-sun solar sail films. In the simulation D test, Kapton remained reflective as well as physically intact. It should be stated, though, that this overcoated Kapton material sample was bonded to a temperature-controlled metal substrate during simulation D. A solar sail film would be used in a free-standing mode to minimize mass per unit area. Its response to ionizing as well as thermal (including uv) radiation is the subject of follow-on investigations.

Poly(para-xylylene) was tested as an evaporated coating supported by a stainless-steel grid. A sample of this material, parylene N, was exposed during simulation D and underwent rapid degradation. Reflectance decreased to very low values in the wavelength region in which most of the sun's energy lies (Fig. 10). Moreover, the parylene could be observed during

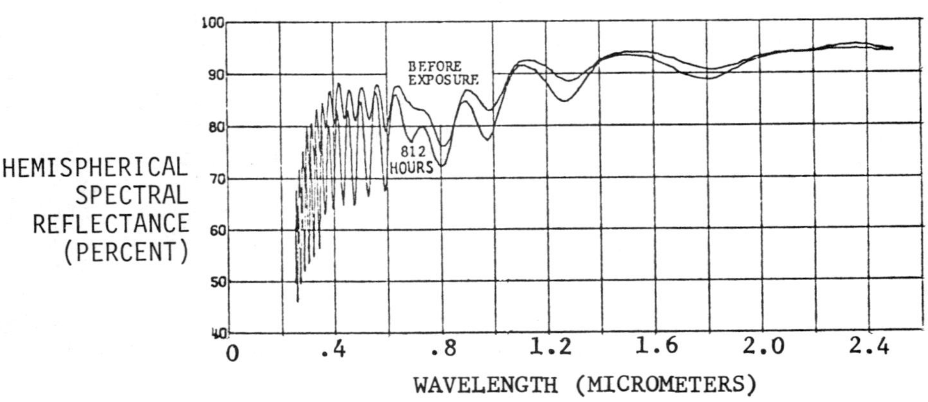

Fig. 9 Effect of solar wind and ultraviolet at 16-sun intensity on Al_2O_3-Al-overcoated 1/4-mil Kapton.

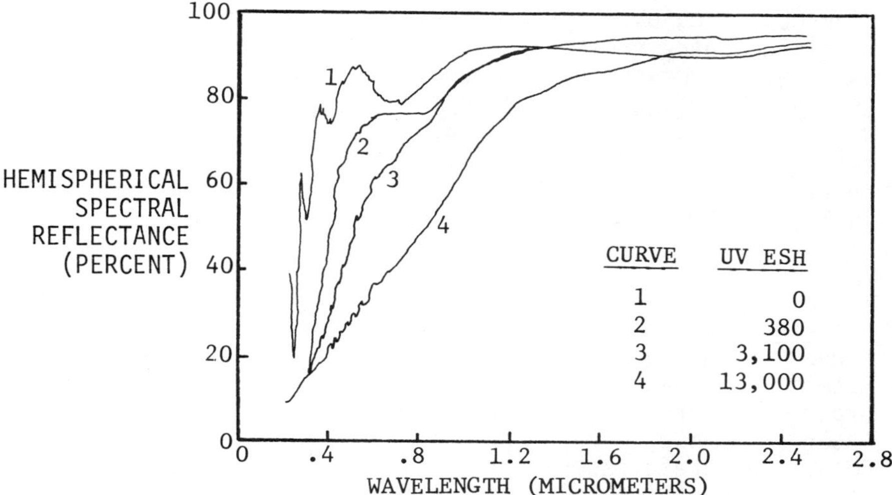

Fig. 10 Degradation of reflectance and physical condition of
SiO_x-Al_2O_3-Ag-overcoated aluminized parylene exposed to 16-sun
uv and solar wind.

and after the test (grid sample near the top of Fig. 2); it
was partly melted even early in the 812-hr test and severely
fragmented by the end of the test. Parylene would appear to
be unsuitable for near-sun solar sailing.

     The most suitable materials for near-sun solar sailing
films will be those with high emittance, as well as high front-
surface reflectance. The latter requirement means poor front-
surface emission. To improve back-surface emission may require
increased polymer thickness at the expense of mass per unit
area. Back-surface coatings may increase emittance more mass-
efficiently than making the base polymer thicker would. For
a large-area film that is at least approximately planar,
equilibrium temperature T can be determined by equating solar
power received and power radiated:

$$\alpha_s(SC)(SR) = \sigma(\varepsilon_f + \varepsilon_b)T^4$$

where σ is the Stefan-Boltzmann constant, ε coefficients are
determined separately for the film's front and back surfaces,
SC is the solar "constant," and SR is the sun rate. The product (SC)(SR), being power, is proportional to the energy U
referred to earlier. (For further treatment of this development, see Ref. 14.) Equilibrium temperatures approaching 400°
C have been calculated for an aluminized polyimide film without reflection-enhancing front-surface overcoatings or emis-

sion-enhancing back-surface coatings, even when exposed just
to 11-sun solar radiation. Substitution of silver for aluminum could decrease this temperature to 280°C. An equilibrium
temperature of 250°C under 16-sun solar radiation would re-
quire ε_b approaching 0.6 and $\alpha_s \approx 0.12$. Development of a viable solar sailing film may require quite advanced technology
in such areas as polymer science, vapor-deposited overcoatings, and space radiation simulation techniques.

IV Conclusions

The experimental data presented here represent state-of-
the-art knowledge concerning suitability of several candidate
materials for advanced space system applications. Further
experiments are underway in support of several projects of
interest to U.S. Government agencies and corporate laboratories.
Results of these experiments will allow expansion of current
conclusions, which are as follows:

1) Fluorinated ethylene propylene (FEP) copolymer is a
stable base material for thermal control, for solar energy
receiption, and for solar energy direction purposes, provided
that the energetic charged particle fluxes of magnetic sub-
storms at synchronous altitude are not encountered. A useful
lifetime of several years under these conditions is indicated.
When peak fluxes are encountered, this widely used material
can be expected to degrade, even rapidly under worst-case
conditions, so that substitutes must be identified through
further investigation.

2) In geosynchronous orbit, a composite coating of silica
fabric over FEP and metal foil may have better long-term reflectance stability than metallized polymer coatings during
combined uv/proton/electron radiation exposure to fluxes on
the order of 2×10^9 particles/cm^2-sec.

3) Overcoated polyimide film bonded to a metal substrate
retains reflectance and structural stability when irradiated
with near-sun radiation fluxes, whereas partially free-standing coated paraxylylene does not under conditions investigated
thus far.

V. Recommendations

Additional work should be done to evaluate materials in
the space radiation environment for applications in which they
would be utilized without substrates. Substrates and spacecraft interiors influence properties such as equilibrium

temperature and electrostatic charge trapping, so that results
obtained on nonmetallic materials on substrates are not
generally applicable to future uses involving free-standing
films. Both geosynchronous and near-sun applications need to
be considered in this regard:

1) Polymer films and metallized polymeric reflectors
should be exposed to uv, protons, and electrons simulating
geosynchronous altitude conditions, to ascertain via in situ
measurements their suitability for use without substrates on
large space structures.

2) Candidate solar sail film materials should continue
to be evaluated with realistic near-sun uv and solar wind
radiation intensities. Proposed front and back reflectances
and emittances need to be simulated closely to establish
equilibrium service temperatures and the corresponding suitable orbits.

Acknowledgments

This work was performed, in part, under NASA Contracts
NAS1-13530 and NAS1-14328, Langley Research Center, Hampton,
Wa., and under Contract 102178 A 36647 from ERNO Raumfahrtte-
chnik GmbH, Bremen, Germany. The authors appreciate the assistance of Loren D. Milliman with scientific data programs, E.
Russ Crutcher with Photomicrography, Richard W. Earle with
electronic test systems, and Paul W. Olson with mechanical
apparatus development.

References

[1] Fogdall, L. B. and Cannaday, S. S., "Solar Wind Studies on
Thermal Control Coatings and the Need for Further Simulation
and Thermophysical Measurement Improvement, AIAA Paper 70-833,
June 1970.

[2] Brown, R. R., Fogdall, L. B., and Cannaday, S. S., "Electron-
Ultraviolet Radiation Effects on Thermal Control Coatings,"
AIAA Progress in Astronautics and Aeronautics: Thermal Design
Principles of Spacecraft and Entry Bodies, Vol. 21, edited by
Jerry T. Bevans, Academic Press, New York, 1969, pp. 697-724.

[3] Fogdall, L. B. and Cannaday, S. S., "Dependence of Thermal
Control Coating Degradation Upon Electron Energy," Boeing
Doc. D2-126114-1 for NASA Goddard Contract NAS5-11164, May
1969.

[4] Fogdall, L. B. and Cannaday, S. S., "Proton and Electron Effects in Thermal Control Materials," Final Rept. to NASA Goddard for Contract NAS5-11219, May 1970.

[5] Cannaday, S. S. and Fogdall, L. B., "Irradiated Mirror Panel Discharge Tests," Boeing Test Report, Oct. 1972.

[6] Anonymous, "Natural Radiation Hardening Test," Boeing Test Report, March 1974.

[7] Fogdall, L. B. and Cannaday, S. S., "Proton and Ultraviolet Degradation Tests on HELIOS Spacecraft Materials," Boeing Doc. D180-17868-2, April 1974.

[8] Fogdall, L. B. and Cannaday, S. S., "Irradiation of Thermal Coatings," Boeing Test Report, July 1975.

[9] Fogdall, L. B. and Cannaday, S. S., "Effects of High Energy Simulated Space Radiation on Polymeric Second-Surface Mirrors," NASA CR-132725, Oct. 1975.

[10] Grard, R. J. L. (ed.), "Photon and Particle Interactions With Surfaces in Space, Dordrecht-Holland, D. Reidel Publishing Co., 1973.

[11] Rosen, A., "Large Discharges and Arcs on Spacecraft", Astronautics & Aeronautics," Vol. 13, June 1975, pp. 36-44.

[12] Cauffman, D. P. and Shaw, R. R., "Transient Currents Generated by Electrical Discharges," Aerospace Corp., SPL-74(4409-04)-1, June 1974.

[13] Fogdall, L. B. and Cannaday, S. S., "Polymer Modification and Simulated Space Radiation Exposure," NASA CR-145289, Feb. 1978.

[14] Van Vliet, R. M., Passive Temperature Control in the Space Environment, Macmillan, New York, 1965.

Bibliography

Halliday, D. and Resnick, R., Physics, Wiley, New York, 1961.

INSTRUMENT CANISTER THERMAL CONTROL

W. Harwell[*] and R. Haslett[†]
Grumman Aerospace Corporation, Bethpage, N. Y.

and

S. Ollendorf[‡]
NASA Goddard Space Flight Center, Greenbelt, Md.

Abstract

A transient thermal analysis and test of a thermal control canister is described. The 1- x 1- x 3-m canister
provides a uniform thermal environment for shuttle instrument payloads requiring fine temperature control, the design
goal being operation between 0° and 20°C, with a range of
±1°C at any selected set-point temperature. The canister
side walls are isothermalized by a system of longitudinal and
circumferential heat pipes rejecting heat through feedback-
controlled, variable-conductance heat pipes to side-mounted
radiators. A breadboard model of two side walls and two
radiators was tested in a thermal vacuum chamber. The breadboard was stable over a wide range of effective environments,
experiment dissipations, and control-point temperature
levels.

I. Introduction

The shuttle payload models being developed by NASA have
identified a class of experiments with a number of common
requirements. These are pallet-mounted experiments fitting
within a maximum envelope of 1 x 1 x 3 m. All require fine
pointing accuracy, and all operate over a narrow temperature
range.[1] To minimize premission thermal analysis and

Presented as Paper 77-761 at the AIAA 12th Thermophysics
Conference, Albuquerque, N. Mex., June 27-29, 1977. Copyright
© American Institute of Aeronautics and Astronautics, Inc.,
1977. All rights reserved.

[*]Senior Thermodynamics Engineer.
[†]Manager. Heat Pipe Programs.
[‡]Supervisory Engineer, Thermal Control.

eliminate some of the rigorous thermal vacuum acceptance testing normally associated with experiment checkout, it was proposed by NASA Goddard Space Flight Center (GSFC) that a thermal control canister be designed to accommodate these experiments. This canister would provide the individual experimenter with a standard environment within which his experiment would operate, effectively isolating the experiment from the external environment. The thermal specifications for the canister were set at maximum end-to-end and side-to-side temperature difference ±1°C, variable set-point control temperature within the range 0° to 20°C, control to be within ±1°C at any set point within the control range, maximum heat-rejection capability 300 W, and external thermal environment defined by the envelope of the shuttle orbital parameters.

Three thermal control concepts were proposed for the thermal control canister: a heated canister with thermostat control, a fluid loop with radiators, and a heat pipe system. Feasibility studies identified the heat pipe system as warranting further study. This paper briefly reviews these feasibility studies and reports the results of thermal vacuum tests performed on a heat pipe system thermal breadboard model to verify proof-of-concept.

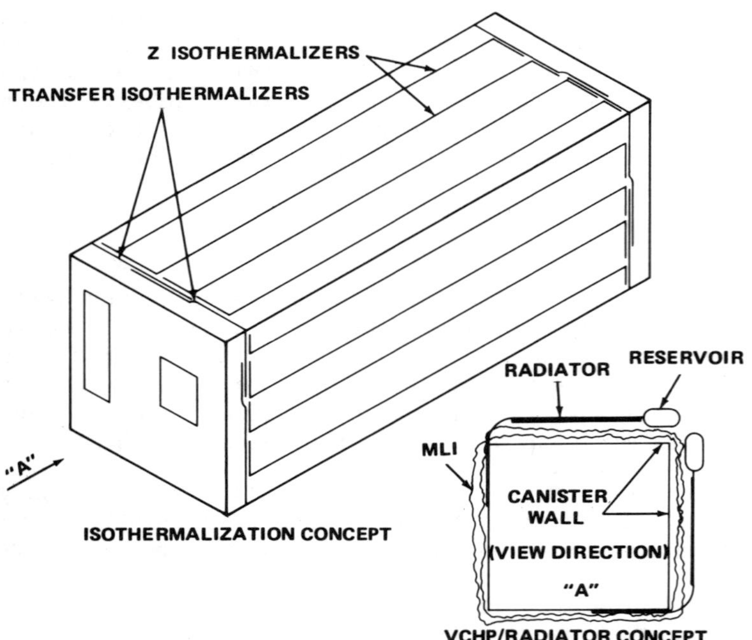

Fig. 1 Thermal control canister.

II. Design Concept

One concept for the proposed thermal control canister consists of a box with isothermalizer heat pipes built into the walls. The box is insulated externally with multilayer insulation (MLI), and the internal electronic dissipation is rejected to space through a system of variable-conductance heat pipes (VCHPs) mounted to radiators (Fig. 1). The system operates by virtue of the load-sharing capability of the radiators, with the VCHPs modulating the active area of the radiators. For tight temperature control of the canister, feeback-controlled VCHPs are required, with a sensor on the canister wall controlling the temperature of each VCHP gas reservoir. In its simplest form, the feedback control system consists of temperature sensors on the canister side walls, a controller unit, and an electric heater on the gas reservoirs of each VCHP. Each gas reservoir has an integral cooling fin coupling the reservoir to the external environment. These fins and the radiators are coated with silver-backed teflon.

The canister may be either midmounted or end-mounted. If midmounted, the canister would consist of three elements, a central frame and two boxlike ends, with all of the components thermally coupled to each other by isothermalizer heat pipes. The instrument would be mounted to the frame and the boxlike ends cantilevered off the frame. If end-mounted, the canister would consist of an end-mounting frame and one long boxlike section mounted to the frame. In either mounting scheme, access to the instrument would be by removal of the boxlike sections or removal of individual side panels.

Feasibility Studies

Parametric analyses were run to determine the required radiator area for different environments, experiment dissipation, radiator panel thickness, feeder heat-pipe spacing, feeder heat-pipe temperature, and radiator emittance. In this analysis, it was assumed that the radiator was insulated on one side.

The basic equation defining the heat-rejection capability (Q_{rej}) of a radiator can be defined in terms of the fin effectiveness (η_F), absorbed environment flux (Q_{abs}), and fin root temperature (T_R):

$$Q_{rej} = \eta_F \, (\varepsilon \, \sigma \, T_R^{\,4} - Q_{abs}) \, A$$

The fin effectiveness accounts for the temperature gradients across the radiator panel and, for small variations in absolute temperature, may be expressed as

$$\eta_F = \tanh (2\sqrt{\lambda}) / 2\sqrt{\lambda}$$

where

$$\lambda = \varepsilon \, \sigma \, (L_F^2 \, T_R^3 / k\delta)$$

and ε is emittance, σ is Stefan-Boltzman constant, A is area, L_F is half heat pipe pitch, k is thermal conductivity, and δ is radiator thickness.

Two methods of coupling the radiators to the canister were examined: VCHP headers between the canister and radiators with isothermalizer feeder pipes on the radiators, and VCHP feeder pipes coupled directly between the canister and radiators. The VCHP header design had a number of disadvantages compared with the VCHP feeder design, including the following:

1) Rapid freeze-out of the fluid in the isothermalizers (more rapid than diffusion freeze-out in feeder VCHPs).

2) Poor thermal coupling between canister and heat-rejection surfaces (important when attempting to control the canister to a low set point in a warm environment).

3) Poor redundancy. The VCHP header design would require only one VCHP, but for redundancy two would be necessary. If either of these VCHP's developed a small leak with loss of gas, then excessive radiator area would be active. In the VCHP feeder concept, if gas were lost from one VCHP, the other VCHPs would compensate. If there were enough leakage to result in loss of fluid, the failed header would result in decreased heat dump capability, whereas with the feeder pipe, the other VCHPs would compensate.

4) Poor ground testability. The VCHP header concept requires the feeder pipes to be at right angles to the header pipes, resulting in a difficult, if not impossible, configuration for ground test of the system in the heat-pipe mode.

The advantages of the VCHP header concept over the VCHP feeder concept are as follows:

1) Lower low-load heat leak. During parts of some possible missions, it is necessary to shut down a radiator by driving the gas/vapor interface into the adiabatic zone. There is some residual heat loss by conduction along the wall of each VCHP, and hence the greater number of pipes in the

feeder concept bridging the canister to the radiator results in a higher heat loss.

2) Lower reservoir control heater power. The dissipation of the reservoir control heaters is proportional to the number of reservoirs.

These factors were weighed carefully, and it was concluded that the ground testability requirement and the concern for freeze-out overrode all other requirements. Therefore, a feeder VCHP concept was adopted, the final design of which has three body-mounted radiators, each 10 ft^2 in area x 0.032-in.-thick aluminum plate with five VCHPs per radiator on 9-in. pitch.[2]

Similar tradeoff studies were performed on the canister side walls. The maximum design heat dump capability of a canister is 300 W. If this heat dissipation is assumed to be distributed evenly over the entire canister surface, then the flux is 25 W/m^2 (2.3 W/ft^2). For design purposes, an internal blackbody radiant flux of 82 W/m^2 (7.5 W/ft^2) was assumed, and the temperature distribution between two adjacent heat pipes was calculated for a one-dimensional conducting fin:

$$(T - T_S)/(T_R - T_S) = \cosh\left[2\sqrt{\lambda}\,(1-\chi/L_F)\right]/\cosh\left[2\sqrt{\lambda}\right]$$

where T_S is the effective black body source, T_R is fin root, T is local fin temperatures, L_F is half the tube pitch, and χ is measured from the fin root. As before,

$$\lambda = \varepsilon\,\sigma\,T_R^{\,3}\,L_F^{\,2}\,/\,k\delta$$

By solving for a given heat flux, the temperature distribution between two heat pipes can be calculated. Based upon the maximum allowable longitudinal and circumferential temperature differences of ±1°C and the 82-W/m^2 flux density, a panel thickness of 1.25 mm (0.05 in.) with an isothermalizer heat pipe spacing of 0.229 mm (9 in.) was selected. This is the thickness of the primary skin of the canister.

Assuming that the 300 W of electronic dissipation is rejected through one radiator (five VCHPs), then the maximum required transport capability of the VCHP (0.38-m evaporator, 0.30-m adiabatic, 0.97-m condenser) would be of the order of 59 W-m. This is well within the 120-W-m, zero-gravity capacity of the NASA GSFC axial grooved heat-pipe extrusion with ammonia as working fluid. With the 0.229-m (9-in.) pitch heat-pipe matrix proposed for the canister, the maximum transport required for any one isothermalizer pipe is less than 25 W-m., and again the NASA GSFC heat-pipe extrusion with ammonia working fluid was selected.

The temperature control system proposed for the canister depends upon the ability to modulate automatically the area of the heat-rejection system to account for changes in source dissipation and orbital variation in sink temperature. This modulation is achieved by the variable-conductance heat pipe coupling the canister to the radiator. Both dry- and wet-reservoir VCHP systems were examined. Dry-reservoir VCHPs are unpredictable because of vapor diffusing into the reservoir and displacing some noncondensable gas. In addition, once a dry-reservoir VCHP is charged, it is impossible to change its operating set point without changing the reservoir volume. Wet-reservoir VCHPs, on the other hand, depend upon the working fluid being present in the reservoir in both vapor and liquid phases so that the vapor partial pressure always equals saturation pressure. Then, by using a feedback control system, the set point can be changed over wide temperature ranges, the only limit being that the reservoir temperature required to maintain control must be higher than the equivalent environment temperature. A wet-reservoir VCHP with feedback control therefore was selected for the thermal control canister.

A transient thermal math model of a canister radiator system with feedback-controlled VCHPs was set up with simplified representation of an experiment package.[3] This model was used to examine the transient thermal stability of the canister and experiment, the variables being experiment mass and heat dissipation duty cycle, transient external environment, reservoir control heater dissipation, reservoir fin area, reservoir weight, and reservoir-to-condenser volume ratio. The VCHPs were modeled by flat-front theory with the gas, vapor interface location being updated after every temperature iteration of the transient thermal analyzer. The feedback logic for the VCHPs was modeled as a proportional controller, with a control band of $1^{\circ}C$, plus high- and low-temperature thermostats on the reservoirs. The controller and thermostats controlled the reservoir control heater. The effects of gas front disturbances and turbulances also were investigated by including time-lag equations in the VCHP model. This simplified math model demonstrated the inherent thermal stability of the system.[3]

Breadboard Configuration

For proof-of-concept, it was necessary to demonstrate that the system heat dump capability could be adjusted continuously

by the feedback-controlled VCHP radiator system to maintain
the canister temperature within a narrow temperature range.
Tradeoff studies were performed to select a breadboard configuration that would simulate the thermal characteristics of the
canister adequately. To minimize gravity effects associated
with ground testing of heat pipes, it was required that the
canister box geometry be replaced by a number of flat coplanar
plates, thus allowing all heat pipes to be tested in the heat-
pipe mode. For proof-of-concept hardware, it was decided that
two sides of the box needed to be simulated, and the breadboard shown in Fig. 2 was selected. The thermal breadboard[4]
consisted of two canister side walls and two radiators unfolded
to lie in one plane as shown. The side walls (Panels 1 and 2)
were 52- x 38- x 0.050-in. aluminum plates, and the radiators
(Radiators 1 and 2) were 47- x 30- x 0.032-in. aluminum
plates. The panels simulating two of the side walls were
isothermalized by a series of Z shaped longitudinal heat pipes
clamped to the surfaces, with thermal grease at all interfaces. The thermal mass of the other two side walls, which
would make up one-half of the canister, was simulated by
aluminum blocks coupled to the side panels by circumferential
heat pipes.

The radiator plates were preformed to conform to the
VCHP pipe profile and had to be sprung slightly to allow the
condenser length of the VCHP to fit into the groove. Nine of
the VCHPs were bonded in place using a silver-filled Chromerics
cement; the tenth pipe (VCHP 1) only was clamped in place
mechanically to allow easy removal and replacement. The
evaporator section of each VCHP was clamped to the side walls
with thermal grease at all interfaces. All isothermalizer and
VCHPs were fabricated from the NASA GSFC standard axial
grooved heat-pipe extrusion.

The breadboard was supported by a framework fabricated
from 3 x 1-1/2 in. aluminum channel. This heavy frame
allowed all isothermalizers to be lined up coplanar within
0.05 in., whereas all VCHP's were set up with their reservoir
end 1/8 in. below their evaporator end.

An instrument was simulated by a number of heated plates
and a patch heater (Fig. 2). These plates were insulated
from the support frame but were coupled conductively and/or
radiatively to the side walls. The conductive coupling was
approximately 1/2 W/°C. For the radiative coupling, the
canister side walls were painted with 3M Black Velvet paint
(emittance 0.86), and the simulated instrument plates were
taped with several layers of Kapton tape (emittance 0.85).
The patch heater consisted of two 6- x 6- x 1/16-in. aluminum
plates with a Nichrome ribbon heater sandwiched between them.

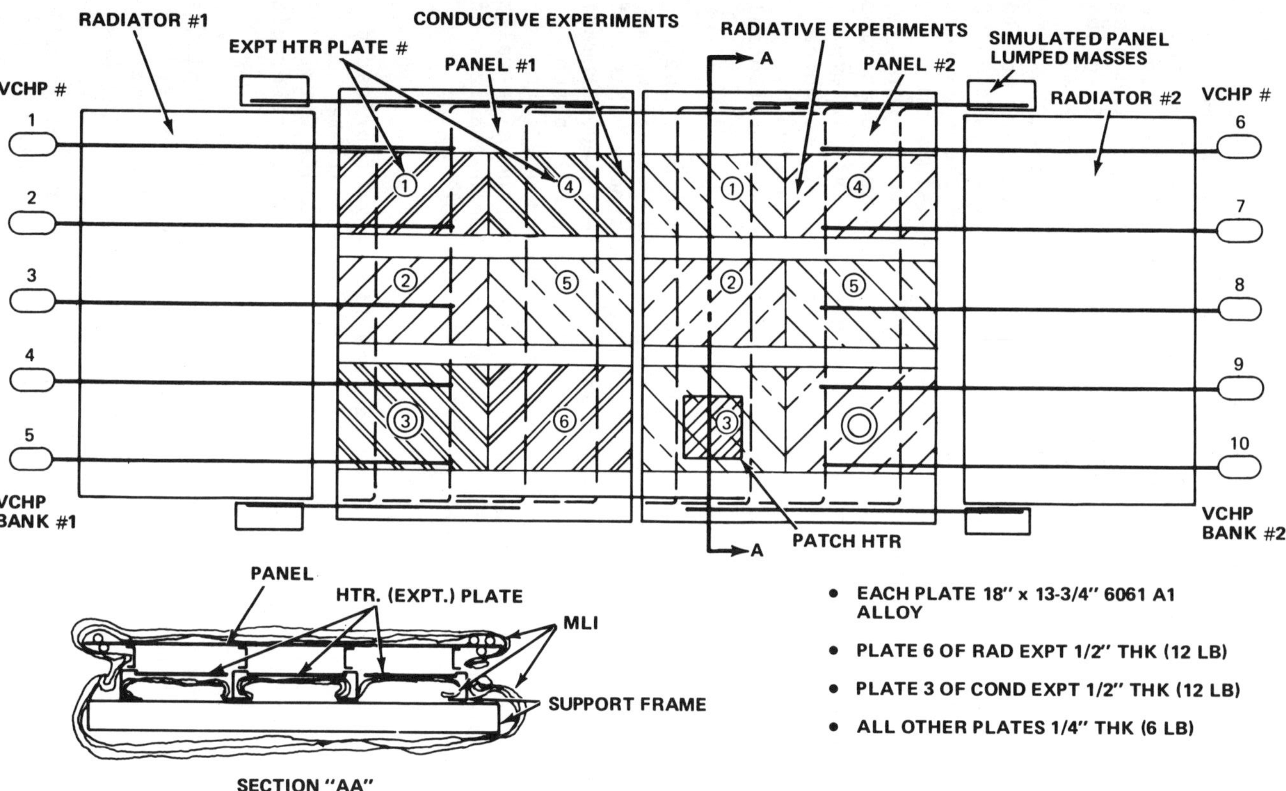

Fig. 2 Experiment simulation.

The patch heater was bolted directly to the side panel with thermal grease at the interface.

Three types of heaters were used, as shown below. All heater elements are fabricated from Nichrome ribbon laid down on double-sided Kapton tape and covered with single-sided Kapton tape:

1) Environment simulation heaters. These heaters were bonded to the underside of the two radiators and to the reservoir fins. They were used to simulate absorbed environment fluxes.

2) Instrument dissipation simulation heaters. The plates used to simulate the instrument and the lumped masses all had heaters bonded to them to simulate dissipation of experiment electronic boxes.

3) Control heaters. The control heaters of the feedback system were bonded to the underside of the gas reservoirs. These heaters provide heat to the reservoirs as required by the feedback controller logic.

A total of 180 thermocouples were distributed over the radiators and side walls, two platinum resistance thermometers were cemented to two of the reservoir fins, and six thermistors were cemented to the side panels and selected instrument heater plates. The thermistors were used as the sensor units for the feedback controller(s). Finally, the breadboard was insulated with a 25-layer aluminized Mylar blanket, the only exposed surfaces being the black-painted upper surfaces of the radiators and the reservoir fins.

The breadboard was tested in the 15- x 10-ft diam thermal vacuum chamber at NASA GSFC. This chamber has a nitrogen cold wall covering the entire inner surface. The breadboard was installed in the chamber with the framework resting on two goalposts with Teflon blocks at the framework/goalpost interfaces to thermally insulate the test item from the chamber wall and also to build in a 1/8-in. tilt to raise one side of the breadboard above the other. This tilt, together with the 1/8-in. tilt built into the VCHPs, insured that all heat-pipes were operating in the heat pipe mode.

The thermistor sensors for the feedback controllers were brought out to a selector box. This box allowed any single sensor or any pair of sensors to be fed as input signals to either proportional or on/off temperature controllers. The temperature controllers used were commercially available units operating from a 28-V supply, and all had variable set points within the range of -50° to $+50^\circ$C with a control band of $\pm\frac{1}{2}^\circ$C about the preselected control point.

III. Test Conditions and Objectives

The thermal breadboard was exercised through a number of steady-state and transient test conditions in a thermal vacuum chamber with the following controlled variables: 1) number (1 or 2) and locations of control sensors (side panels, conductively tied heater plate, radiatively tied heater plate), 2) control temperature level, 3) type of controller (proportional, on/off) and number (1 or 2), 4) simulated external environment (cold, nominal, hot), 5) simulated internal dissipations (low, nominal, high), and 6) passive control. After an initial shakedown period, the tests ran in a very predictable fashion, with control being maintained at all test conditions.[4]

The thermal breadboard was designed as development hardware to verify that the heat-pipe concept for the thermal control canister is capable of providing very accurate temperature control under a variety of imposed boundary conditions. Some specific test objectives were as follows:

1) To demonstrate that the canister can be isothermalized to less than $\pm1^\circ$C for nominal canister heat fluxes over the entire range of mission external environments. The flight canister can reject 300 W of heat from two radiators. Because of the mismatch in panel area between the canister (12 m^2), breadboard (2.5 m^2), and heater plates (1.8 m^2), the heat flux incident on the simulated side walls of the breadboard had to be increased to allow this high total heat rejection from the radiators. The incident heat flux had to be increased further to account for nonuniform heat dissipation in the heater plates, and, during the thermal vacuum test, flux densities of up to 195 W/m^2 (18.1 W/ft^2) were imposed on the side panels. The 1°C allowable gradient in flight hardware therefore would equate to approximately 8°C in the test hardware.

2) To demonstrate that a 15°C canister temperature level can be maintained within $\pm1^\circ$C for extreme variations in orbital environments and experiment power levels. For these proof-of-concept tests, the minimum environment was chosen so that the ammonia in the heat pipes would not freeze.

3) To demonstrate that the canister can be maintained at any temperature between 0° and 20°C by adjusting the control

logic of the reservoir heater. The variation in set point of
a VCHP is dependent upon the initial noncondensable gas mass
charge. Having selected a gas charge for a condition of 16°C
active vapor temperature, -10°C reservoir temperature, and a
3-W reservoir control heater, then the ability to run at lower
controller set points is dictated by the ability to obtain a
lower reservoir temperature, whereas running at higher temperatures is limited by the ability of the feedback-controlled
heater to support the $\sigma\,T^4$ radiation capability of the reservoir fin.

4) To investigate the effect on temperature control of
using the temperature of an experiment node to control the
reservoir heaters.

5) To investigate proportional and on-off feedback controller logic. A 28-V proportional controller with variable
set point within the range -50° to +50°C and a control band
of 1°C was selected as the prime controller. A 28-V on-off
controller, also with variable set point and a control band of
±1/2°C, was used at two points in the test sequence.

6) To investigate degradation in control band resulting
from the loss of feedback controller. Passive VCHP operation
then is entirely dependent upon the environment temperature
variation and the experiment dissipation.

7) To exercise the thermal breadboard model through
a typical shuttle mission. NASA GSFC Thermal Systems Branch
has proposed flying a thermal control canister as part of
the Orbital Flight Test Vehicle 4 (OFT-4) payload. A con-

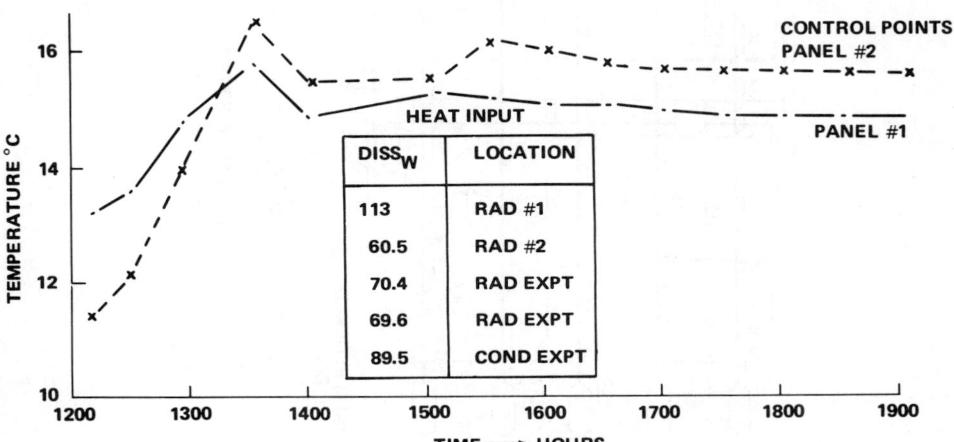

DISS$_W$	LOCATION
113	RAD #1
60.5	RAD #2
70.4	RAD EXPT
69.6	RAD EXPT
89.5	COND EXPT

Fig. 3 Approach to steady-state point.

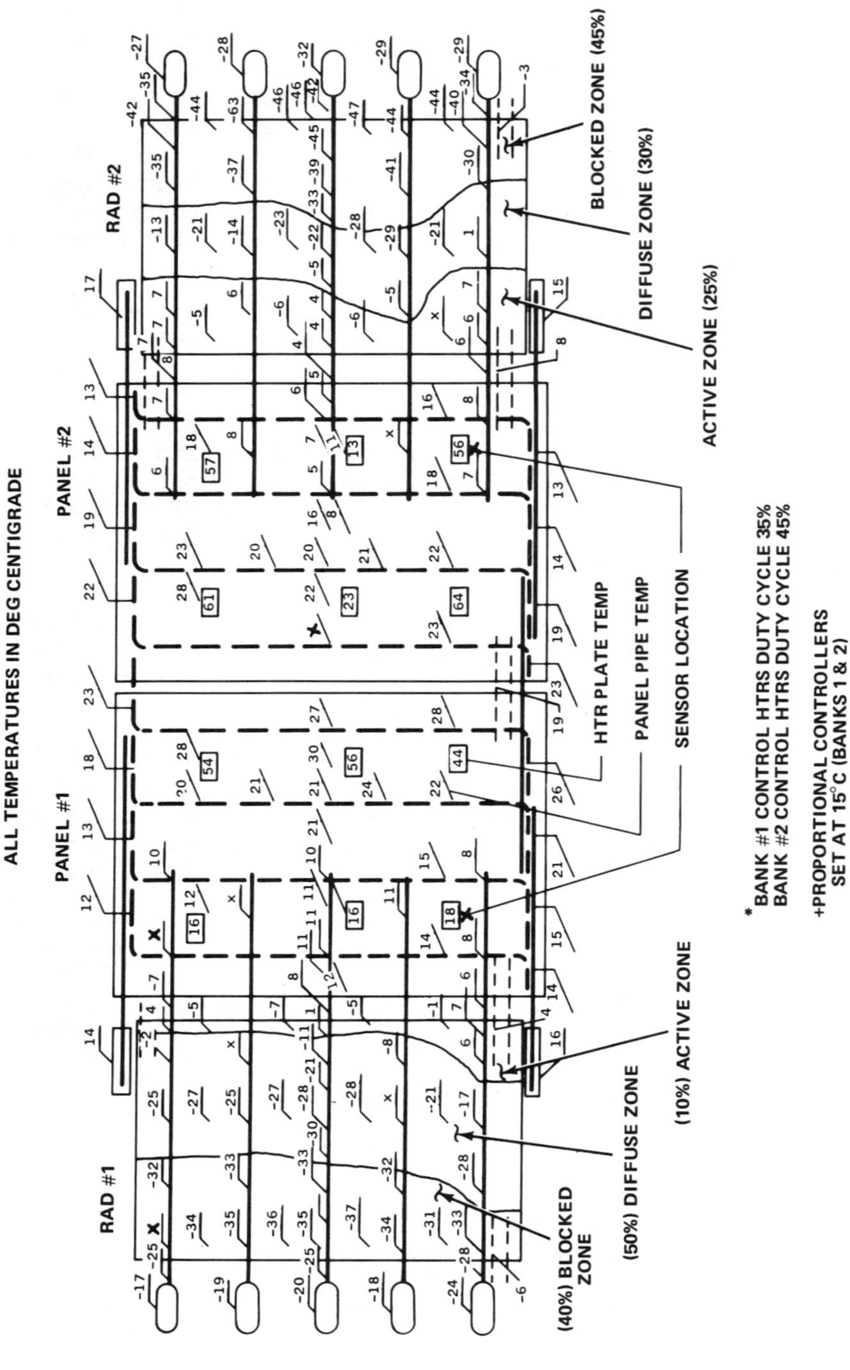

Fig. 4 Instantaneous temperature distribution.

densed version of the OFT-4 environment profile was selected
for simulation.

IV. Discussion

Steady-State Control

For this steady-state test point, the environment fluxes
absorbed by the two radiators and two banks of reservoir fins
were set at nominal hot case, orbital average conditions. A
total instrument dissipation of 230 W was simulated, divided
among three heater plates on the conductive experiment (30 W
per plate) and four heater plates on the radiative experiment
(35 W per plate). Two proportional controllers were used with
the sensors on the simulated side walls (TS 1 and 2) and each
controller set to control one bank of reservoir heaters. Both
controllers were set at 15°C, and, during this steady-state
control point, controller 1 duty cycled at approximately 35%
and controller 2 at approximately 45%. The temperature
history of the two controlled points up to the steady-state
point is shown in Fig. 3.

The separation in temperature of the two control points
is due to the relative locations of the thermocouples and
control sensors, the tolerance in setting the two controllers,
and the difference in power densities of the heater plates.
The temperature distribution recorded at the steady-state
point is given as Fig. 4. The individual isothermalizer heat

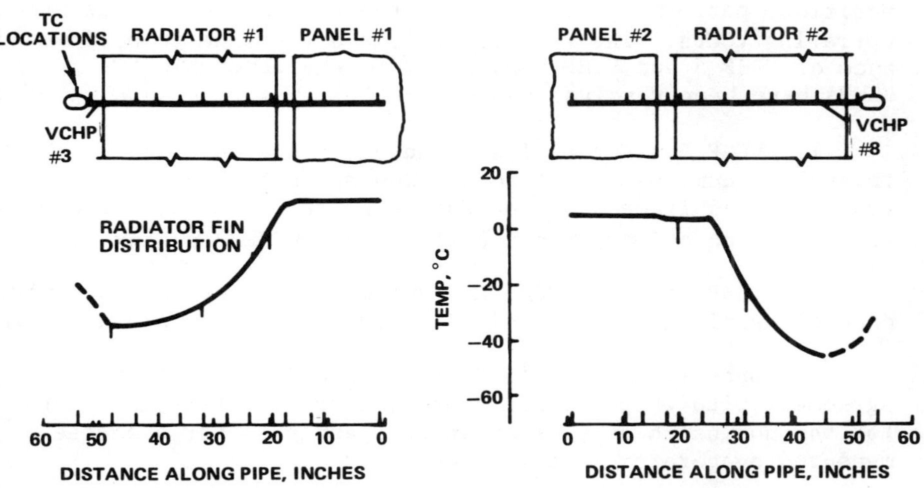

Fig. 5 Instantaneous temperature distribution along VCHPs
(steady-state point).

pipes were effective in isothermalizing along their length, but
there was a significant gradient within the panel between
pipes. These high gradients are considered to be due to
clamping the pipes to the panels (rather than bonding as with
flight hardware) and to the high heat flux simulated in the
test (approximately 200 W/m^2). The computed temperature
difference between the midpoint between two heat pipes and the
heat pipes themselves is 8°C at 200 W/m^2, compared to less
than 1°C at 25 W/m^2.

Temperature distributions along two VCHPs are plotted in
Fig. 5. The upstream edges of the gas fronts in the VCHPs of
Radiator 1 are located in the adiabatic or transport sections
of the heat pipes, but the corresponding points in Radiator
2 are well within the radiator section. This figure dramatically illustrates the long gas-vapor interface commonly
found in aluminum, axial-grooved heat pipes. Within this
region there is some diffusion of the noncondensable gas
(nitrogen) into the working fluid (ammonia) and very low
vapor velocity. The dominant mode of heat transport within
this region is by conduction along the wall of the pipe and
through the radiator fin. Temperature distributions such as
those shown in Fig. 5 were used to locate the upstream and
downstream edges of the gas-vapor diffusion region for each
VCHP, and these regions are superimposed on Fig. 4 to define
the "diffuse zones" of the radiators. The zones are relatively
flat except for two pipes: VCHP 5, which opens up more than
the other pipes on the radiator, and VCHP 9, which is shut
down more than the other pipes on this radiator. It is in-
teresting to note that VCHP 1, which was only clamped in
position, performed very close to the performance of the other
operating VCHPs. The reasons for the difference in performance of VCHP 5 and VCHP 9 compared to the other eight
VCHPs have been resolved as follows:

1) VCHP 5: During instrumentation checkout in the
thermal vacuum chamber, a conduction strap between the
reservoir and frame was removed and never replaced, allowing
VCHP 5 to run colder than the other four VCHPs in bank 1.

2) VCHP 9: This VCHP inadvertently was overcharged with
noncondensable gas by 15 psi (difference between psia and psig)

The operation point for each VCHP was in excellent
agreement with what one would predict from flat-front theory
for the nominal heat-pipe dimensions and gas charge at the
measured evaporator temperatures.

A number of steady-state test points were run to verify
1) operation over a temperature range of 6° to 20°C (panel

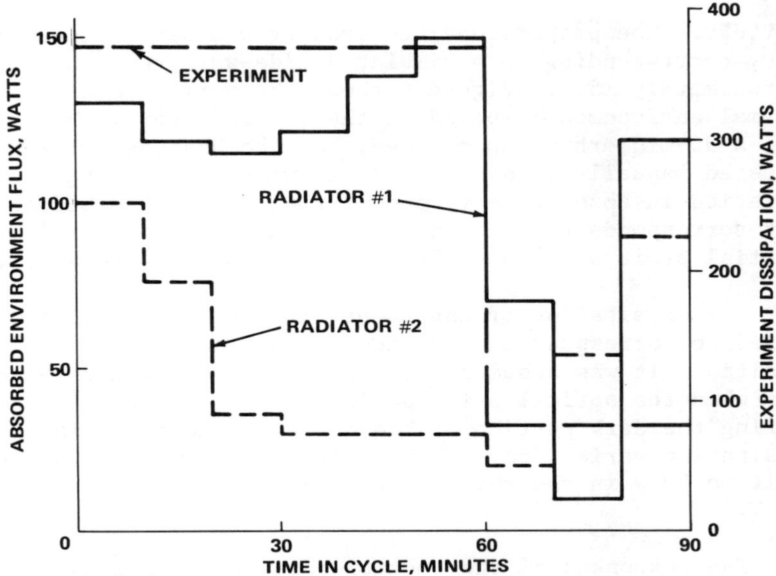

Fig. 6 Transient input power profile.

temperature), 2) operation over a range of equivalent environment conditions (60 to 240 W/m^2), 3) control of the temperature
of an instrument node rather than a canister node, and 4) control using one controller and sensor to control two banks of
VCHPs. After an initial familiarization period, the testing
proceeded very smoothly, and control was achieved in all cases.

Passive control can be documented from one test where the
absorbed flux simulated by the reservoir fin environment heaters was set too high to allow the reservoir to cool down and
control to the attempted 0°C set point. The pressures in the
reservoirs were too high to allow the VCHPs to open, the simulated side walls could not cool down, and the VCHPs operated in
the passive mode, maintaining a temperature of 11°C. The fact
that an 11°C canister temperature was maintained was fortui-
tious, as, depending upon environment, it could have been any-
where between 0°C (set-point temperature) and 15°C (design
point at hottest environment with zero control heater
dissipation).

Transient Operation Control

A number of test runs were made to simulate varying
orbital environment fluxes and varying instrument dissipation.
During one of these test runs, one sensor on a simulated
instrument node (a heater plate) was used to control both banks

of VCHPs. The proportional controller was set to control to
29°C, corresponding to a simulated side-wall temperature of
approximately 15°C. Figure 6 shows the variation of absorbed
thermal environment imposed on the two radiators. For simplicity, a 90-min orbit was assumed, and the absorbed flux was
adjusted manually in steps every 10 min to track the orbital
variation in absorbed environment flux. The absorbed environment corresponds to a nominal hot case, beta 45°, stellar
inertial orbit with the solar vector normal to radiator 1.

The dissipation of the reservoir fin heaters also was
cycled to correspond to this nominal hot case. For this test
condition, it was assumed that the instrument would be shut
down when the optical axis was looking directly at the Earth
(during the dark portion of the orbit). The total instrument
dissipation varied from 371 W in the sunlit portion of the
orbit to 80 W in the shadow, an orbital average dissipation of
270 W.

The responses of the control point and the local panel are
shown in Fig. 7. This figure shows that the variation of dissipation of the heater plate (control point) was from 20 to 5 W
(total instrument dissipation 380 to 80 W). The temperature of
the control point varied by approximately $\pm 2\frac{1}{2}°$C, whereas the
panel varied by $\pm 1°$C. This control was very encouraging considering that the dissipation of the control point varied from
approximately 120 to 39 W/m^2. The observations noted from the
steady-state runs also apply to the transient performance: the
isothermalizer heat pipes are very effective in isothermalizing
along their length; significant gradients are built up within
the panels between isothermalizers; VCHP 5 opens more than the
other four VHCPs on Radiator 1, whereas VCHP 9 opens less than
the other VCHPs on Radiator 2; and each VCHP has a long diffusion zone. Transient heat input cases were run at other control
temperatures and with the control point moved to the simulated
sidewalls. Control was achieved in all cases.

OFT-4 Mission Simulation

It is proposed to fly a thermal control canister on
Orbital Flight Test Vehicle #4 (OFT-4). This would be a
rigidly mounted canister pointing out of the space shuttle bay.
During the 120 hr of experimentation, a total of 20 kW-hr of
electrical energy is available to support the canister thermal
checkout.[5] The total available electrical energy was matched
to the instantaneous dump capability of the radiators and a
power profile established for the OFT-4 mission. This mission
was simulated on the thermal breadboard model.

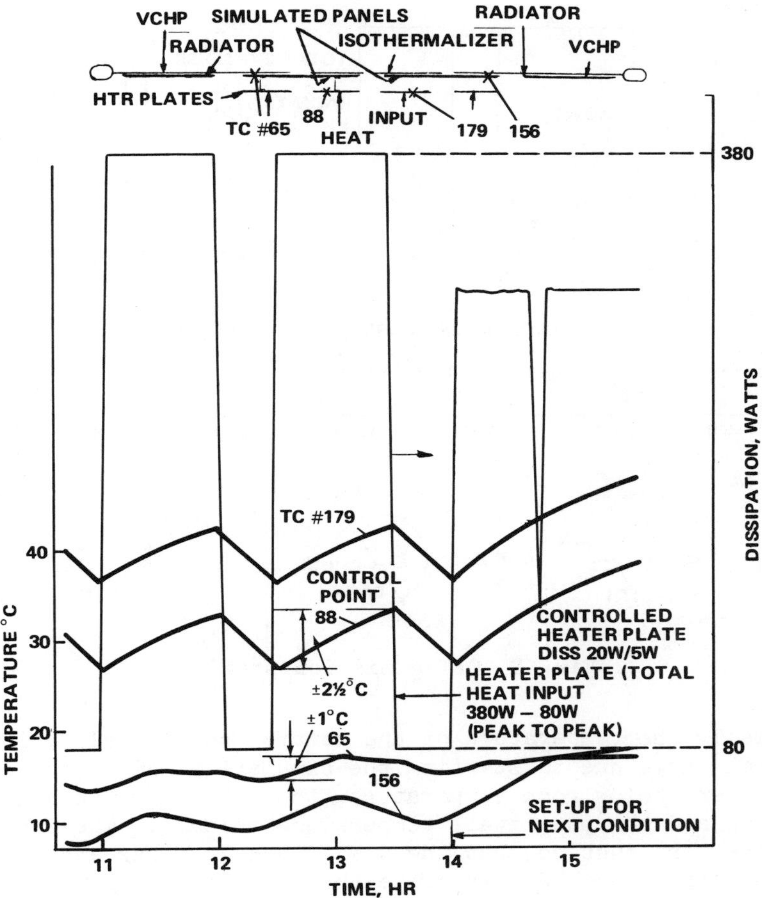

Fig. 7   Temperature time history.

For the OFT-4 mission simulation, the entire breadboard
was brought up to room temperature (equivalent to the canister
in the shuttle bay prior to opening doors). Two proportional
controllers were set to 15°C, the sensors being on the panels.
The radiators then were allowed to cool down with only $1W/ft^2$
of absorbed environment flux. The dissipation in the heater
plates was increased progressively to 360 W. This mode was
held for 2½ hr, and then the barbecue mode was simulated for
1½ hr. A 10-min cycle was assumed for the barbecue mode.
Phase 2 then was simulated for 3 hr (two orbits), followed by
phase 3 for 4½ hr (three orbits). The equivalent environment
flux absorbed by Radiator 1 is shown in Fig. 8, together with
the assumed instrument power profile. The temperature histories of the two control points also are shown in Fig. 8.

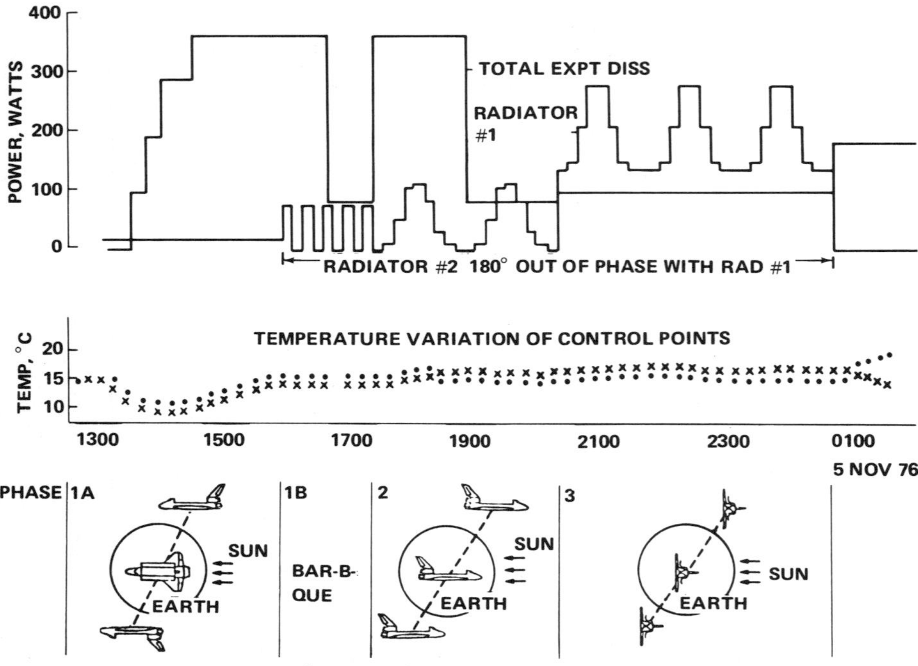

Fig. 8   OFT-4 mission profile.

Initially the temperature of the control points fell below the control range due to starting the simulation while the reservoirs were below room temperature, thus allowing the VCHPs to open up and the panel wall temperature to fall. The control heaters were enabled, but the 3 W of heater power available to each reservoir could only drive the reservoir temperature up slowly, and the VCHPs stayed open for some time. Eventually, however, the reservoir temperatures were driven high enough to close down the VCHPs, and control was re-established. This loss of control would not have occurred if the reservoir temperature had been at room temperature at the start of the simulation. At approximately 4½ hr into the simulation, the control points fell slightly below the control band due to the step change in instrument dissipation from 360 to 80 W which occurred at this time. This step change was considerably higher than the design conditions. Control was re-established and maintained for the rest of the simulation.

V.   Conclusions

The thermal breadboard was subjected to a rigorous test series in a thermal vacuum chamber. Both steady-state and transient operation were investigated, the test conditions covering a wide range of simulated external environments and internal instrument dissipations. The canister control temperature was varied within the range 6° to 20°C. Control of an experiment node also was demonstrated. The conclusions to be drawn from this test series are as follows:

1) The temperature of the side panels was controlled to within ±1°C at set points of 6°, 15°, and 20°C over a wide range of simulated radiator environments and experiment dissipations.

2) The temperature of a simulated electronic box was held to ±2½°C over a wide range of orbital environments and experiment dissipations.

3) An abbreviated OFT-4 mission was simulated, and control was maintained at all times except for a short period 1 hr into the test when loss of control was the result of incomplete setup for the test.

4) Operation of the VCHPs in the passive mode was demonstrated in a hot-case environment.

5) Control with both proportional and on-off controller logic was demonstrated.

From additional tests performed but not reported in this paper, the following comments can be drawn:

1) During a shuttle re-entry mission simulation, the VCHPs were shut down by rapidly driving up the reservoir temperatures. No reversals of the VCHPs were detected, and the temperature of the simulated sidewalls continued to decrease.

2) A number of design improvements have been identified for inclusion in flight hardware. These include a) low-conductance sections between the evaporator and condenser sections of each VCHP; b) redesign of the reservoir to eliminate the welded fin; c) improved method of attachment of the isothermalizers to the canister walls; and d) additional circumferential isothermalizers.

References

[1] Bohlin, R.C. (ed.), "Small Astronomy Payloads For Spacelab," NASA Publ. X-692-75-91, 1975.

[2] "Feasibility Study To Use Heat Pipes To Control Temperatures of SIPS Canister," Grumman Aerospace Corp., Final Rept., NASA Contract NAS5-22335, 1975.

[3]"Transient Thermal Response of a Thermal Control Canister," Grumman Aerospace Corp., Final Rept., NASA Contract NAS5-22570, 1976.

[4]"Thermal Vacuum Tests on a Thermal Control Canister Breadboard," Grumman Aerospace Corp., Final Rept., NASA Contract NAS5-22980, 1977.

[5]"Definition of a Shuttle Thermal Canister Experiment," Rept. SD76-SA-0150, prepared by Rockwell International for NASA under Contract NAS5-23203, 1977.

A PRECISE SATELLITE THERMAL CONTROL SYSTEM
USING CASCADED HEAT PIPES

W. H. Steele*
McDonnell Douglas Astronautics Company, St. Louis, Mo.

and

H. B. McKee[†]
Frito-Lay, Inc., Irving Texas

Abstract

A cascaded, dry reservoir, variable-conductance heat-pipe
system was tested. Results show passive temperature control
within $\pm 0.5°F$ of the desired set point for a wide range of heat
input and effective space environment temperatures. The use
of long capillary tubes to isolate the reservoir and prevent
set-point temperature change due to cyclic heat loads and/or
cyclic environment temperature was demonstrated. Orbit set-
point temperature control feasibility was investigated using
variable-volume control-gas reservoirs. Set-point temperature
adjustment over a range from $50°$ to $90°F$ was achieved successfully with high control accuracy.

Summary and Introduction

This precision thermal control system development program
using cascaded variable-conductance heat pipes has demonstrated
successfully two innovations in heat-pipe system operation:
1) the use of a long capillary tube to achieve set-point temperature drift control, and 2) application of an adjustable
volume reservoir to provide on-orbit set-point temperature
adjustment.

Presented as Paper 77-777 at the AIAA 12th Thermophysics ©
Conference, Albuquerque, N. Mex., June 27-29, 1977. Copyright
American Institute of Aeronautics and Astronautics, Inc., 1977.
All rights reserved.

*Senior Technical Specialist, Technology.
[†]Technical Manager, Process Design.

Long capillary feasibility was demonstrated by set-point temperature control for both precise control (level II) and radiator control (level I) heat pipes which agreed with predictions within 0.1°F after eight days of testing. The testing included simulation of payload operation, station pass type operations, and quiescent periods without equipment power dissipation. Orbital environment variations were simulated by varying the effective radiator environment. Control gas reservoirs were charged with pure helium, rather than a helium-ammonia mixture, to intensify testing of the capability of long capillaries to provide isolation between the control-gas reservoir and the heat-pipe condenser.

Tests conducted with adjustable volume reservoirs demonstrated good temperature-bellows position repeatability, fine adjustment capability, and capability to compensate, if a need should arise, for a change in set-point temperature.

The test results demonstrate that the cascade heat-pipe precision thermal control system is a viable option for spacecraft thermal control. With the adjustable volume reservoir option, the ability of the system to provide temperature control adjustment in orbit as required makes it extremely versatile.

Set-Point Temperature Drift is Principal Problem

The principal problem was drift in the set-point temperature when the system was subjected to many heat load and/or environment temperature cycles. The drift in the set-point temperature occurs as a result of a change in the working fluid partial pressure in the reservoir which is caused by movement of the working fluid vapor into or out of the noncondensable gas reservoir. Techniques were conceived to eliminate set-point temperature drift and provide a capability for on-orbit adjustment. Testing and evaluation of a long capillary tube between the gas reservoir and the heat-pipe condenser and an adjustable volume reservoir is the subject of this paper.

Passive Variable-Conductance Heat Pipes Provide Simple Control Option

Various designs have been proposed for variable-conductance heat pipes.[1-4] Several "feedback control" designs have been proposed to overcome or utilize the influence of gas reservoir temperature changes to improve system temperature control accuracy.[1,2] The feedback systems require electrical power or additional control. The dry reservoir without electrical control devices offers the greatest opportunity for

totally passive control. It is simple in concept and requires
no electrical power for control.

Spacecraft Application Requires Integrated System Design

The conceptual layout of a spacecraft application for
the cascaded heat-pipe system is shown in Fig. 1. The basic
thermal control elements include external surface coatings,
insulation, and a space radiator thermally connected to the
equipment compartment with a cascade heat-pipe system. In
order to provide +2°F equipment compartment temperature control, a more precise control requirement of +0.5°F was established for the heat-pipe control system. This finer control
allows a temperature gradient to occur between the heat pipes
and remote parts of the equipment compartment.

The thermal coatings are selected to bias the outer surface of the spacecraft warmer than the desired control temperature. This generates a small heat leak into the spacecraft

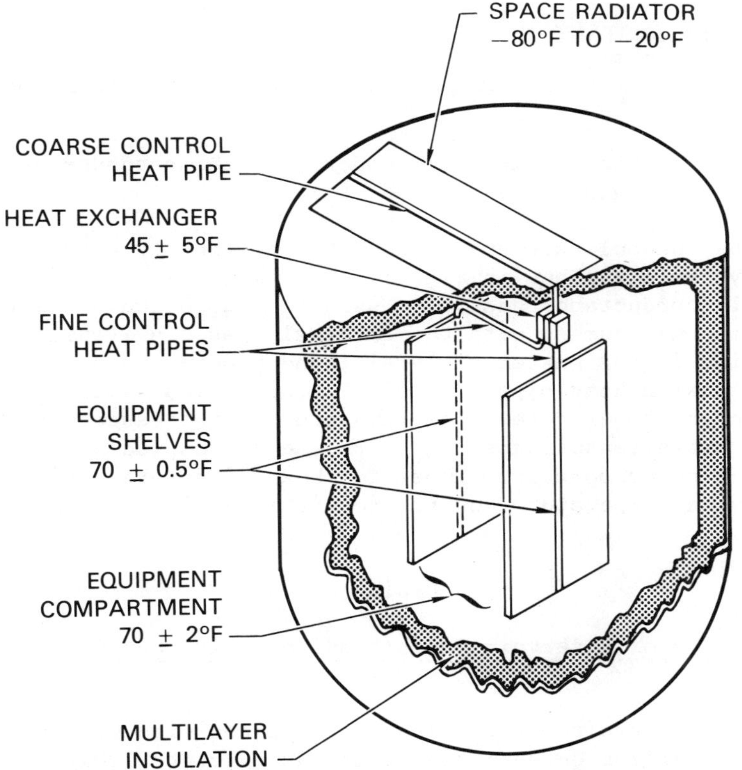

Fig. 1 Spacecraft precision thermal control.

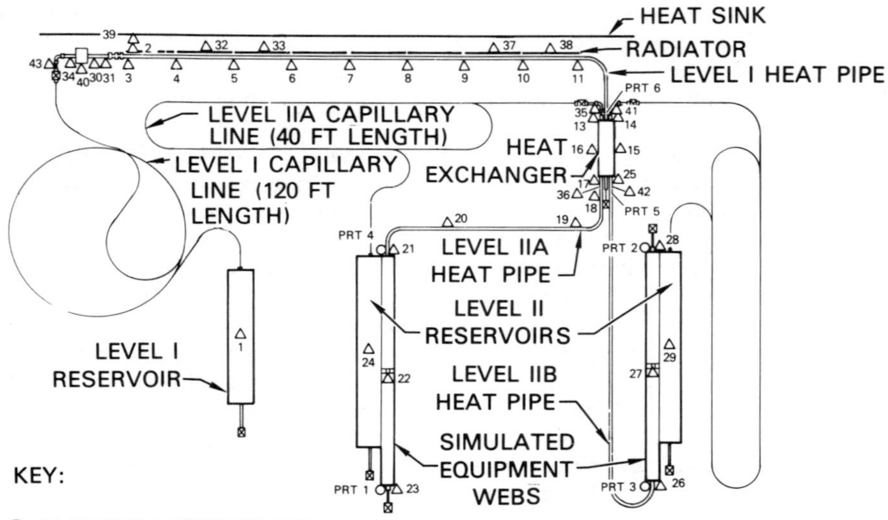

Fig. 2   Cascade heat-pipe schematic and nomenclature.

which is utilized to maintain control of the cascaded heat-pipe
system when equipment heat load is zero.

The nomenclature and a schematic of the cascaded heat-
pipe system are shown in Fig. 2. The system includes a single
variable-conductance heat pipe used to couple the spacecraft
internal environment with the external space environment. This
heat pipe is designated as level I and provides a heat sink for
the precision heat pipes (level II) controlled within +5°F of
a selected temperature. The remainder of the cascade system is
two variable-conductance heat pipes, each coupled to the level
I heat-pipe evaporator. These two heat pipes are designed
to provide temperature control to within +0.5°F of a selected
value.

Design

Elimination of Reservoir Concentration Variations Enhances Control

Gas reservoir composition interactions with heat pipes
and their effect on set-point temperature control have been
researched extensively.[3,5] Hinderman[3] was concerned only with

static isolation between the reservoir and the heat pipe and
proposed a capillary tubing sized to prevent diffusion between
them. McKee[5,6] showed that tubing sized to such criteria was
inadequate to prevent interaction occurring as a result of
environmental or duty cycle variations. He proposed instead
that reservoir charging composition be determined to match the
predicted average heat-pipe inactive condenser section composition during operation. Although feasible for known environ-

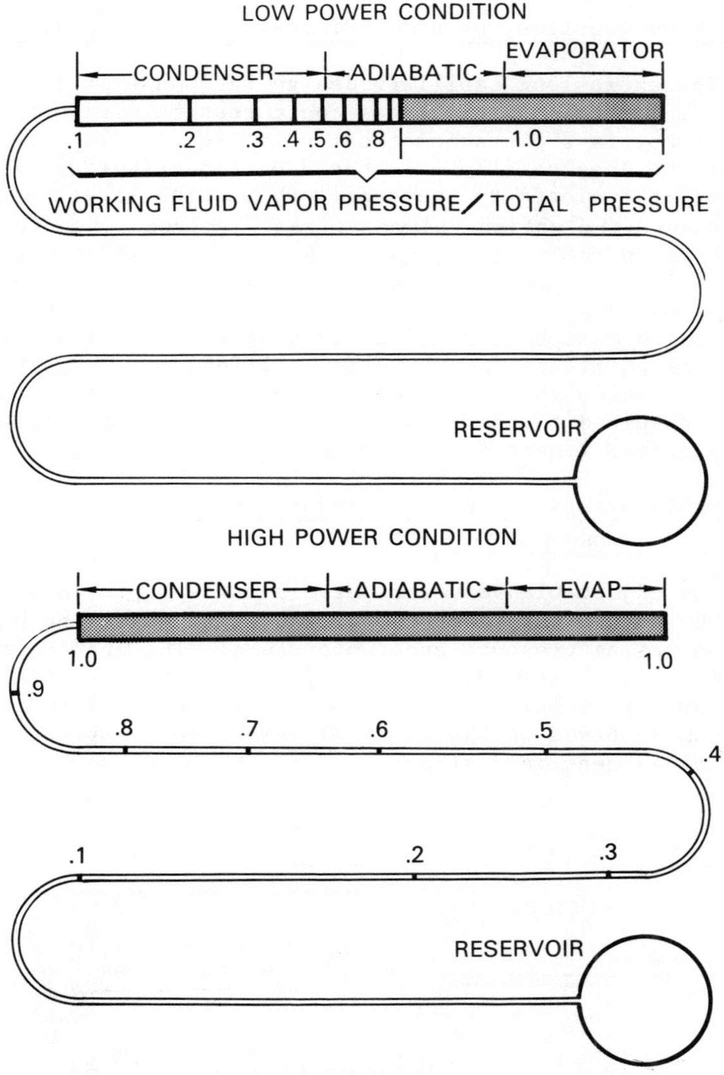

Fig. 3 Typical reservoir isolation using long capillary.

ment and operating conditions, the success of this technique is
compromised when random duty cycles or environmental variations
are expected. The use of long capillary tubes sized to contain
the vapor/noncondensable gas mixture displaced from the heat
pipe toward the reservoir and the use of variable-volume reservoirs were conceived as ways to prevent set-point temperature
drift or to control it when operating conditions might prove
either too variable or too unpredictable for reservoir charging
procedures alone to be adequate.

Extra-Long Capillary Provides First-In Last-Out Control

The extra long capillary design rationale is selected to
insure a first-in last-out gas concentration within the tube.
The concept is shown schematically in Fig. 3. The design siz-
ing of the long capillary is based on two criteria. First,
the volume enclosed by the capillary must be large enough to
accommodate the entire volume of gas displaced when the interface between the noncondensable gas and the active portion of
the heat pipe moves from the minimum heat-rejection position to
the maximum heat-rejection position. In addition, the capillary length must be sufficient to control diffusion axially
along the capillary for the mission lifetime. These two conditions provide the basis to compute the length of capillary
tubing required to maintain the set-point temperature within
the specified limits for the mission duration.

Adjustable Reservoir Provides Added Control of
Set-Point Temperature

The adjustable volume reservoir concept provides a technique to adjust the reservoir pressure and hence the heat-
pipe operating pressure and temperature. The evaporator saturation pressure is approximately inversely proportional to
the reservoir volume. The range in reservoir volumes desired
thus can be based on the range of evaporator temperatures over
which adjustment is desired.

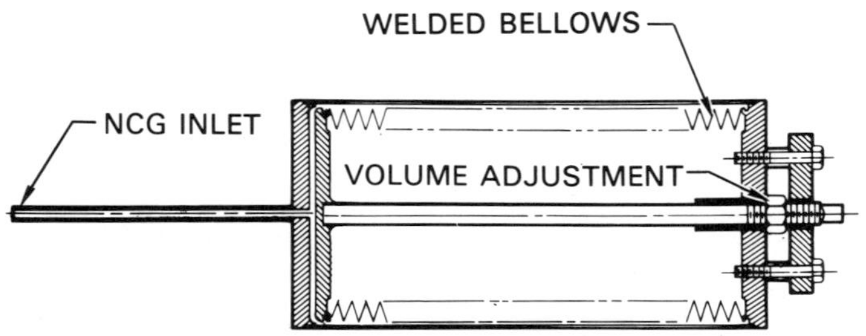

Fig. 4   Bellows reservoir schematic.

For the adjustable volume reservoir test, welded stainless-steel bellows reservoirs of the design shown schematically in Fig. 4 were installed. The volume range was selected to provide an adjustment capability of $\pm 20°F$ for each heat pipe without recharging.

Test Facilities and Installation

All heat-pipe tests were conducted in a thermal vacuum chamber. The vacuum chamber internal wall temperature was not controlled but remained within a few degrees of nominal room ambient temperature throughout the test. The vacuum chamber pressure was maintained below 10^{-5} Torr throughout the testing.

A simulated deep-space heat sink and the leveled cascade heat-pipe test hardware were suspended from a cantilever framework attached to the chamber door. The space radiator was aligned parallel to the liquid-nitrogen-cooled heat sink, separated by a gap of about 0.25 in.

In previous testing, the heat pipes had been operated with the evaporator 0.5 in. above the condensers to assure that liquid return was by wicking. No performance degradation was observed during these tests which verified satisfactory wick operation.

Separate strip heaters attached to the level IIA and IIB evaporator webs provided heat input over a range from less than 1 to 100 W. Two Nichrome ribbon strip heaters attached to the back side of the radiator on either side of the heat pipe were used to control radiator temperature as required for environmental simulation. A heater was installed on the level I reservoir to provide temperature control if desired. Automatic switching and monitoring equipment was assembled to control the cyclic environment and power input and to keep track of the time and cycles accumulated.

The heat exchanger, reservoirs, and the level II heat pipes were covered with multilayer insulation to minimize heat loss to surroundings. The radiator and heat sink were wrapped with MLI to minimize the impact of their lower temperature on adjacent components and to prevent a long-term effect on chamber temperature.

Charging Procedures

Heat-pipe and reservoir charging procedures were defined to establish the desired set-point temperature accurately. The selected procedure employed the pressurization of the gas-

controlled reservoir with helium to a predetermined level,
based on the control-gas mass balance as corrected for the
difference between charging and operational temperatures. The
charging pressure thus determined is slightly above the in-
tended operational pressure. This pressure increment provides
the control gas required for the inactive section of the condenser. The heat pipes were charged by evaporating double
distilled ammonia from a small container into the heat-pipe.

Test Instrumentation

The test hardware was instrumented with six platinum-
resistance thermometers where a high degree of accuracy was
required and with 43 copper/constantan thermocouples where
large temperature changes were expected and less accuracy was
necessary. The system temperature control performance was monitored using the high-accuracy data provided by the platinum-

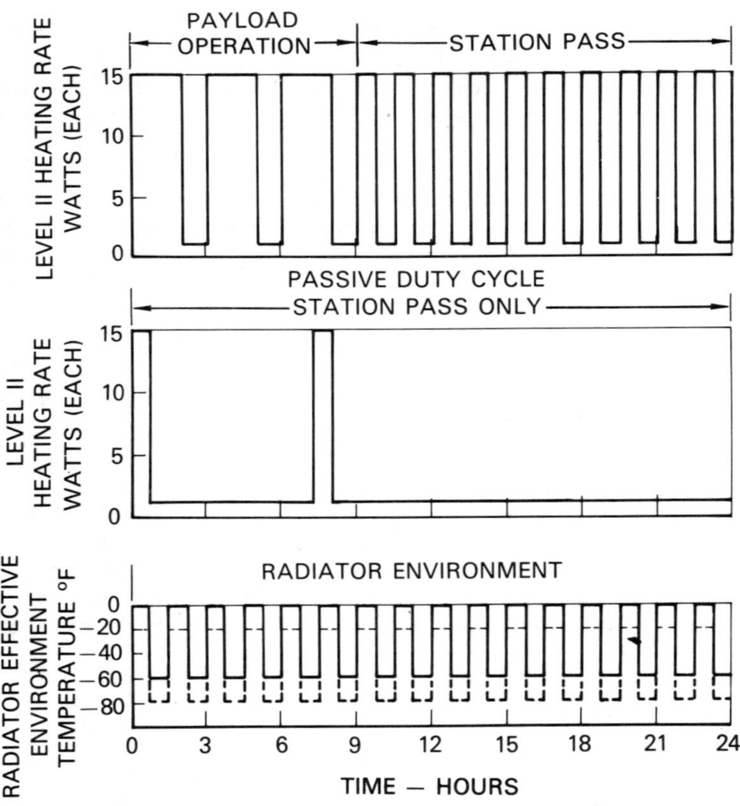

Fig. 5 Duty and environmental cycles.

resistance thermometers. The thermometers were calibrated before and after the heat-pipe tests. Sensor locations are shown in the system schematic (Fig. 2). Thermocouple and thermometer output, heater power parameters, and time were recorded continuously at 1-, 2-, 5-, or 10-min intervals on magnetic tape for ease of data processing and on paper tape for real-time monitoring.

Long Capillary Tests

Environment and Duty Cycle Simulation

Test conditions for the long capillary evaluation were selected to provide a test representative of operational conditions and a rigorous test of the heat-pipe system. A 60°F amplitude for the cyclic effective radiator environment temperature was selected to simulate the most extreme transient anticipated during orbit operation. Such cycling in the absence of the long capillary tubes was shown previously to provide a severe test of proper reservoir charging and to lead to

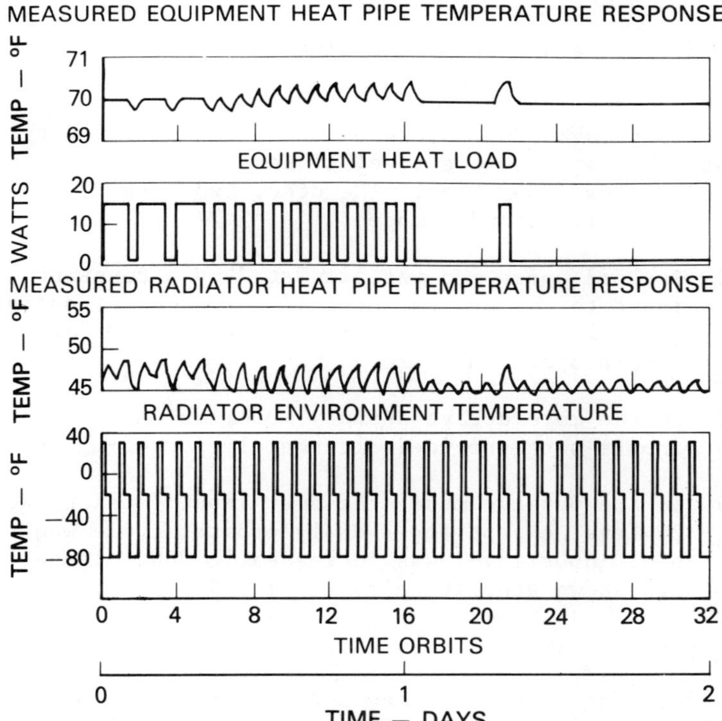

Fig. 6 Response of heat pipes to heat load and environment.

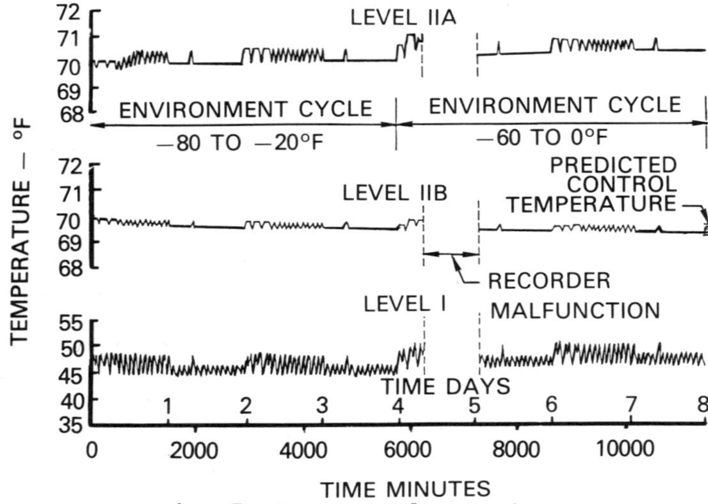

Fig. 7   Summary of test data.

set-point temperature drift for improper reservoir gas composition.

Three types of duty cycle events were simulated by varying
the heat load input to the level II heat pipes. The first
simulated payload operation. The other two types simulated
active and passive standby modes:

1)   Payload operation was simulated by a constant 30W
     input to the system for 2-1/4 hr, corresponding to a
     gradual release of thermal energy stored during the
     20-min operating time.

2)   In the more active standby mode, 30- and 2-W heat
     inputs were provided alternately for 45-min periods.

3)   In the passive mode, the heat input remained at the
     2W level except for a daily orbital cycle when the
     heat input level was increased to 30W. These cycles
     are shown schematically in Fig. 5.

Test Results

Typical data for a two-day period are shown in Fig. 6.
The radiator environment was increased for the first 21 min as
shown to warm the inactive radiator sections to steady-state

temperatures in a short time. This facilitated separate examination of the slower radiator transients associated with the
increased heat input to the level II heat pipes.

Summary data for the task 1 testing are shown in Fig. 7.
Details of temperature responses are similar to those shown in
Fig. 6. The test was conducted in two sections of four days
each. In the first, the radiator environment was varied between nominal limits of -80° and -20°F. For the last four
days, these limits were increased 20°F.

Discussion

Temperature control without the long capillaries would
have been totally different from the observed data. Temperature shifts would have occurred both initially if the reservoir composition was incorrect and at the time of environment
change. Fig. 8 illustrates the kind of drift observed previously when the initial reservoir composition included too
much ammonia. To intensify the severity of the test, the reservoirs of all three of the heat pipes were charged with pure
helium. Using analytical techniques previously derived[6] it can
be shown that this would have led to a temperature increase
of about 13°F for level I in the first 10 cycles with a 7-ft
capillary. That such set-point drifts did not take place
provides a clear verification of the effectiveness of the long
capillary tubes.

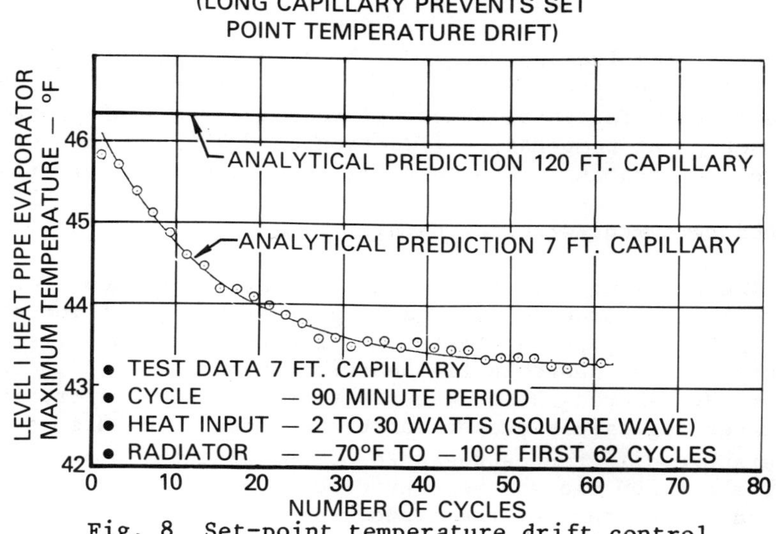

Fig. 8 Set-point temperature drift control.

There are several features of Fig. 7 which merit additional discussion. These include the apparent downward drift of the level IIB temperature, the amplitude of the level IIA oscillations, and the change in control temperature when the average environment temperature was changed.

The level IIB data show an apparent downward trend. Since the reservoir initially was charged only with helium, a decrease in control temperature could occur only if some control gas leaked out or the heat-transfer process to the level I heat pipe through the heat exchanger became much more efficient. The latter effect was rejected, since no similar effect ever was observed with the level IIA heat pipe. To test the leak hypothesis, a Veeco helium leak detector was connected to the 8-ft vacuum chamber and all evacuation performed through the Veeco vacuum system. A leak of 9.2×10^{-5} std cm^3/sec was detected. (The nominal accuracy of such measurement is $\pm 50\%$.) When the task 1 testing was completed, the Veeco was attached directly to the IIB reservoir filling port, and, leakage rate of 8.7×10^{-5} std cm^3/sec was measured. Measurement made at the filling ports of the levels I and IIA reservoirs indicated leakage rates two orders of magnitude lower. It thus was concluded that the entire leak was from the IIB reservoir. The IIB reservoir leak was determined to be caused by a defective valve. (This valve facilitates test operations but would not be included as a part of spacecraft hardware.) The temperature change associated with such a leakage rate may be determined from

$$\frac{dT}{dt} = \phi \, \frac{(14.7 \text{ psia})}{V_R} \, \frac{dT}{dP(T)}$$

where ϕ is the leakage rate, V_R the reservoir volume, and $dP(T)/dT$ is the gradient of saturation pressure with respect to temperature. The total effect of such leakage would correspond to a set-point temperature reduction of $0.38°F$ at the conclusion of the testing. The level IIB temperature after eight days of testing using this prediction is shown in Fig. 7 to be in good agreement with the data.

The amplitude of the temperature variation shown for the level IIA heat pipe is noticeably greater than that for the level IIB heat pipe for test times greater than the first 500 min. This change in IIA behavior at 500-700 min is shown in Fig. 9. Whereas both heat pipes at the start of testing possessed negligible evaporator temperature gradients, such temperature differences began to occur for the IIA heat pipe after about 280 min and were evidenced primarily by an increased temperature of the evaporator at the warm end. The reason for this behavior is believed to be localized behavior

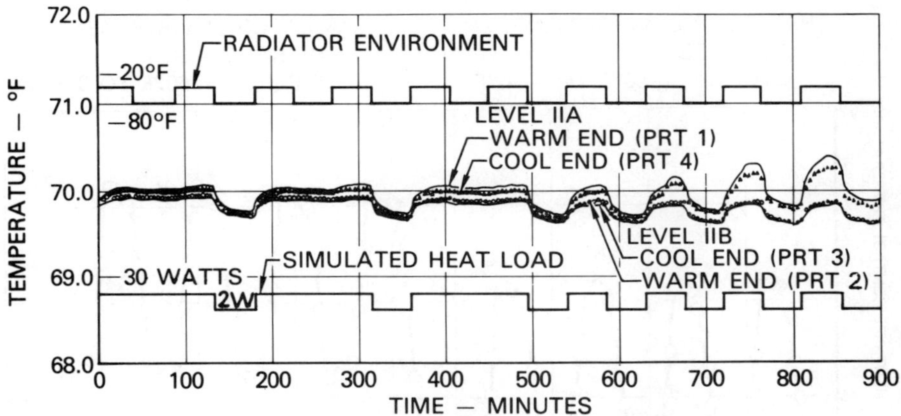

Fig. 9   Level IIA evaporator ΔT development.

over the heat-pipe evaporator surface which probably is associated with the fit of the wall screen to the wall. The temperature differences do not get worse with time and clearly
are not a dry-out phenomenon. In previous testing, the appearance of such a temperature difference was attributed tenta-
tively to contamination of some unexplained type. In the current tests, nonvolatile residue tests of the working fluid
before and after testing have demonstrated that 0.9999 purity
has been maintained. Therefore, it no longer is felt that any
of the temperature variation is associated with contamination.
The slight increase in control temperature of the level IIA is
believed to be associated with an imperfectly fabricated heat
pipe. Techniques since have been developed to improve fabrication standards so that this behavior can be avoided on future
hardware.

A slight change in control temperature took place when
the radiator environemnt was changed at the test midpoint.
Those changes have been evaluated for both the level I and
level II heat pipes and found to be consistent with that expected from the environment change. The level I heat-pipe
average temperature increased when the environment was warmed.
The transient characteristic is shown in Fig. 7. Stabilized
conditions for cycling actively before and after the change in
environment are shown in Fig. 10. The observed increase in
control temperature of about 1.9°F compares exceptionally well
with a calculated value of 1.89°F obtained using the charging
equations for set-point pressure. Such a change occurs for
two reasons. First, a large active radiator area is required
to reject the heat load because of the reduced temperature
difference between the heat-pipe evaporator and the effective
environment temperature. This causes a front movement and an

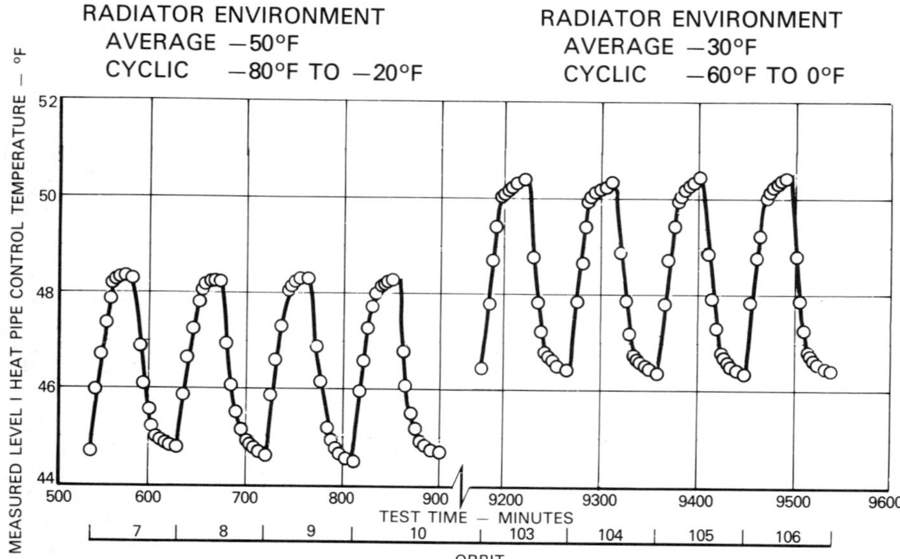

Fig. 10   Impact of radiator environment temperature upon set-point control temperature.

accompanying temperature increase. The second cause arises from the change in gas composition in the inactive section of the condenser. At the higher temperature, the partial pressure of the ammonia is higher, so that less helium is required to fill the inactive section. The helium that previously occupied the section now is forced into the reservoir; this leads to increased reservoir pressure. This temperature change would be present for a radiator heat-pipe system with or without the long capillaries. Without the capillaries, there would be an additional change associated with the transfer of working fluid vapor into the reservoir.

An operational aspect of variable-conductance heat-pipe systems is the prevention of vapor condensation in the capillaries. This is necessary to keep liquid from being injected into the gas reservoir. In a spacecraft installation, this would be accomplished by allowing a slight heat leak from a structural connection to the capillary. For the current tests, this heat input has been applied directly using heaters placed a few inches from the end of the heat pipe. For the task 1 testing, these heaters were set at nominal power levels of about 0.2 W each. These levels were set somewhat arbitrarily but kept there to prevent incurring a change associated with adjustment until the eight days of testing had been completed. At that time, the heat input levels, were reduced to 0.1 W per

elbow heater. In addition, the main level II heat inputs were
reduced so that the total heat input, including the elbow heat
levels, was 2.0 W. The principal effect of this change was
a reduction in the level I trap temperature of about 13°F.
This led to a level I temperature reduction of 0.32°F compared
to a predicted change of 0.285°F based on the reduced trap
temperature. No significant changes were observed in either
of the level II heat pipes.

The implications of this special test pertain principally
to the conclusion that a total minimum heat load as low as
2.0 W is sufficient to operate the cascade heat-pipe system.
The 0.1-W elbow heat inputs were used for the subsequent bellows reservoir tests.

Adjustable Volume Reservoir Test Setup and Results

The objective of these tests was to demonstrate the capability to adjust the level II heat-pipe set-point temperature over a range of 70° +20°F upon command. The results show
that the objective was achieved and that versatility is improved with such a system. Adjustable volume reservoirs were
tested on all three heat pipes.

The test setup used for the adjustable volume reservoir
differed only slightly from that used for the long capillary

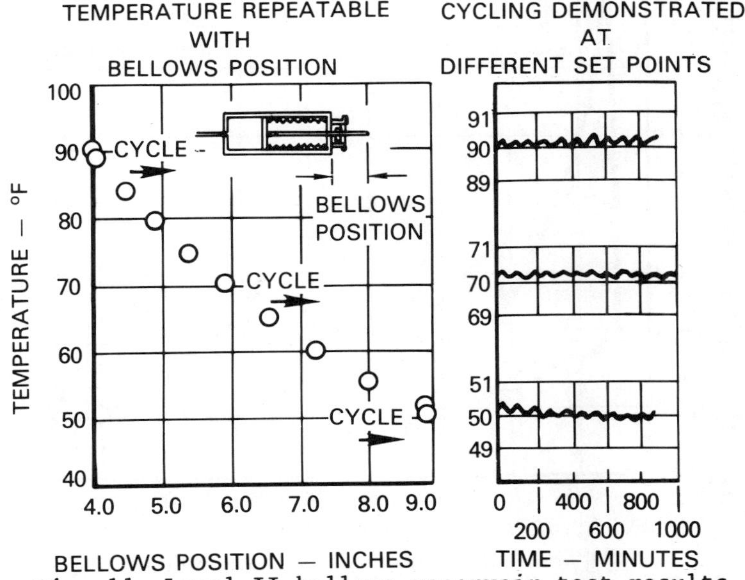

Fig. 11 Level II bellows reservoir test results.

tube testing. The long capillary tubes were disconnected from
the original fixed-volume reservoirs and attached to a pass-
through in the chamber wall. The bellows reservoirs were installed outside the chamber. The reservoirs were installed
on the chamber door platform. A thermocouple was mounted on
the side of each reservoir at the midpoint. Each reservoir was
wrapped with a foam insulation to minimize convective heat
transfer.

Test Procedure and Results

Test conditions for bellows evaluation were selected to
permit examination of the adjustment features separately from
cyclic operation. The principal selections included radiator
operation with constant effective temperature of -80°F and
constant level II heat inputs. The bellows initially were set
at positions corresponding to reservoir volumes of 60 in.^{3}
(level I) and 90 in.^{3} (level II) used in prior tests. This,
coupled with charging pressures identical to those used for
the task I testing, provided a verification that no erroneous
changes had occurred. The test consisted of varying the control temperature in increments of about 5°F from 70° to 90° to

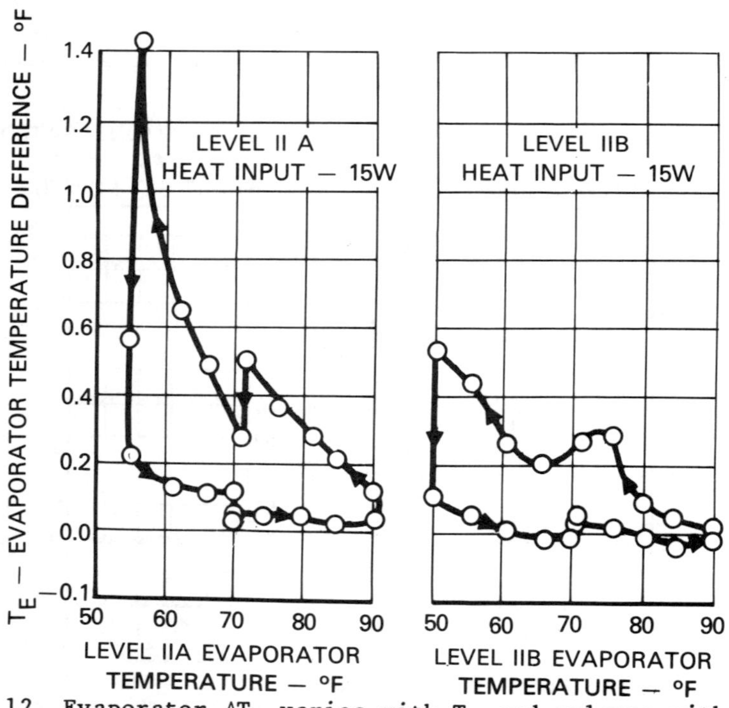

Fig. 12   Evaporator ΔT_E varies with T_E and relaxes with time.

50°F and back to 70°F. Overnight cycling was performed at 50°,
70°, and 90°F using the active cycle with 2-and 30-W heat inputs alternated at 45-min intervals in conjunction with a -80°
to -20°F radiator cycle.

The bellows reservoir results for the level IIB heat pipe,
which are summarized in Fig. 11, verify bellows feasibility and
demonstrate excellent repeatability. Testing was initiated at
70°F, and the cycling data at 70°F were collected prior to bellows adjustment. The radiator environment then was stabilized
at -80°F, the heat input to each level II heat pipe was set
at 15 W, and the temperature-position points up to 90°F were
obtained. The two data circles at 89.6° and 90.1°F demonstrate a fine tuning performed to achieve the desired temperature prior to conducting the 90°F cycling. Following cycling,
each previous point was repeated with such agreement that repeated points overlay the original data. The test procedure

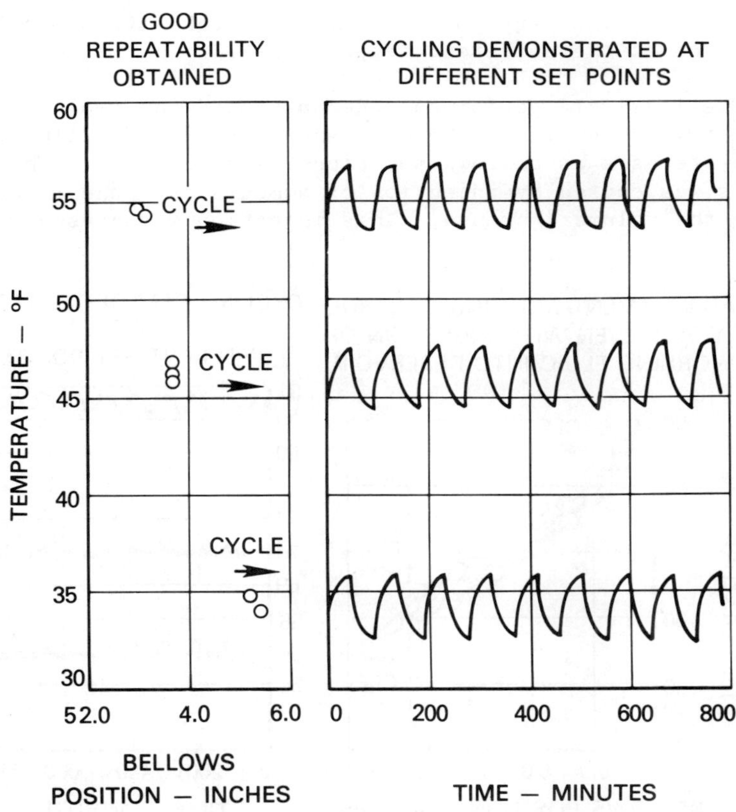

Fig. 13 Level I bellows reservoir test results.

was continued to yield the temperatures down to 50°F, cycling
at 50°F, and the return to 70°F.

The repeatability and preciseness of the bellows adjust-
ments and correlation with temperature are excellent, as shown
in the cycling performance at both 70° and 90°F. There is no
drift, and cycles are uniform. At 50°F, the cycling shows a
slight downward drift during the first seven cycles. This
effect probably is associated with an observed evaporator
end-to-end temperature difference ΔT_E, which varied with
evaporator temperature. This ΔT_E behavior is shown in Fig. 12.
This phenomenon apparently is related to a working fluid
noncondensable gas interaction, which is intensified when the
control temperature is reduced substantially from a previous
control level.

The level I heat pipe was not adjusted as extensively
during this phase of testing as the level II heat pipes. The
+10°F temperature position adjustment is shown in Fig. 13.
Repeatability was very good, and cycling behavior shows little
difference at the different temperatures.

The level IIA performance demonstrated the real capability
of the bellows design. The data shown in Fig. 14 illustrate
some hysteresis as bellows positions were repeated. The hysteresis was caused by injection of ammonia into the reservoir
during the volume increase. This injection was caused by two

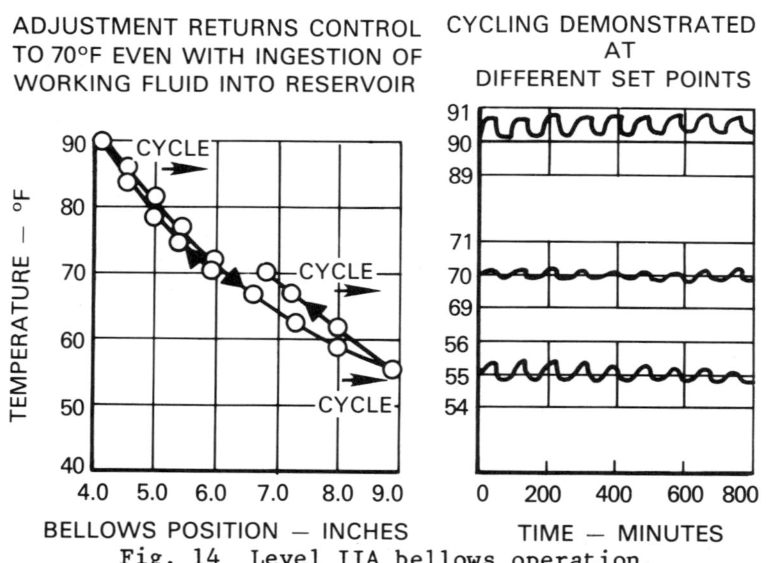

Fig. 14   Level IIA bellows operation.

phenomena: too rapid an expansion, and the marginal accept-
ability of the fabrication of the level IIA heat pipe. With
the rapid expansion, ammonia vapor was forced into the reservoir before it could condense in the heat pipe. In spite of
these flaws, the bellows adjustment provided a capability to
restore control to the 70°F set-point temperature, as shown by
the cycling performance at 70°F which was obtained after the
other variations had been conducted.

These tests demonstrate that the use of a bellows reservoir provides a tremendous increase in gas-controlled variable-
conductance heat-pipe versatility. On-orbit temperature control is possible either as a control option or as a redundancy
to provide backup recovery for heat-pipe malfunction or compen-
sation for overall thermal control system variance from anticipated operating conditions.

References

[1] Ekern, W. F. and Hollister, M. P., "Performance of a Precision
Thermal Control System Using Variable Conductance Heat Pipes,"
AIAA Progress in Astronautics: Thermal Control and Radiation,
Vol. 31, edited by C. L. Tien, MIT Press, Cambridge, Mass.,
1973, pp. 67-82.

[2] Bienert, W., Brennan, P. J., and Kirkpatrick, J. P., "Feedback
Controlled Variable Conductance Heat Pipes," AIAA Progress
in Astronautics and Aeronautics: Fundamentals of Spacecraft
Thermal Design, Vol 29., edited by J. W. Lucas, MIT Press,
Cambridge, Mass., 1972, pp. 463-485.

[3] Hinderman, J. D., Waters, E. D., and Kaser, R. V., "Design
and Performance of Noncondensable Gas Controlled Heat Pipes,"
AIAA Progress in Astronautics and Aeronautics: Fundamentals of
Spacecraft Thermal Design, Vol. 29, edited by J. W. Lucas, MIT
Press, Cambridge, Mass., 1972, pp. 445-462.

[4] Marcus, B. D., "Theory and Design of Variable Conductance
Heat Pipes," NASA CR-2018, April 1972.

[5] McKee, H. B., "Thermal Control Range Associated with Heat
Pipe Cycling," Proceedings of the 1974 Heat Transfer and Fluid
Mechanics Institute, edited by L. R. Davis and R. E. Wilson,
Standford University Press, Standford, Calif., 1974, pp. 57-72.

[6] McKee, H. B., "Heat Pipe Thermal Control Set Point Shift,"
Journal of Spacecraft and Rockets, March 1975, pp. 191-192.

CONTROLLABILITY ANALYSIS FOR PASSIVELY AND
ACTIVELY CONTROLLED HEAT PIPES

A. M. Lehtinen*
Rockwell International, Downey, California

Abstract

An analytical technique was developed for steady-state
and pseudotransient control analysis of variable-conductance
heat pipes (VCHP) and feedback-controlled heat pipes (FCHP).
The approach uses a modified vapor temperature profile and a
simple five-node thermal network. This approach differs from
past techniques in that it accounts for gas blockage of the
adiabatic section, and the set-point temperature is referenced
to the control-point node rather than the vapor node. In FCHP
systems, the gas inventory is determined at a design set-point
temperature and held constant for analysis of varying controller set-point temperatures. The pseudotransient analysis
integrates the reservoir response time equations with the
steady-state control equations. The most significant findings
were that reservoir volume increases due to controller set
point, response time, and reservoir temperature limitations;
and minimum and maximum controller set-point temperatures
exist when reservoir temperature limitations exist.

Nomenclature

A	=	area
A'	=	film coefficient area per unit length
A_{cs}	=	cross-sectional area of adiabatic section
A_{rad}	=	radiator area
C	=	capacitance
K	=	conductance
L	=	section length

Presented as Paper 77-776 at the AIAA 12th Thermophysics
Conference, Albuquerque, N. Mex., June 27-29, 1977. Copyright
© American Institute of Aeronautics and Astronautics,, Inc.,
1977. All rights reserved.
*Member of the Technical Staff.

$P(T_a)$	=	vapor pressure at temperature T_a
$P\{T_a\}$	=	$[P(T) - P(T_a)]$ at specified conditions
$\dot{Q}$	=	heat-transfer rate
$\dot{Q}_{div}$	=	heat leak with $\alpha = 0$ and $\beta = 1$
R	=	universal gas constant
T	=	temperature
V	=	volume
h	=	film coefficient of heat transfer
k	=	thermal conductivity
n	=	moles of noncondensable gas
Δ	=	difference operator
α	=	$(L_b - L_c)/L_a$ at cold-case conditions
α'	=	$(L_b - L_c)/L_a$ at hot-case conditions
β	=	L_b/L_c at cold-case conditions
β'	=	L_b/L_c at hot-case conditions
δ	=	temperature tolerance
ε	=	emissivity
η_f	=	radiator fin efficiency
η_{rad}	=	radiator off efficiency
Θ_1	=	slowest response time
Θ_2	=	fastest response time
σ	=	Stefan-Boltzmann constant
ϕ	=	V_a/V_c

Subscripts

a	=	adiabatic; dummy subscript
c	=	condenser; cooling
ccx	=	cooling at $T_{cs\ max}$
cd	=	cooling at T_{ds}
ce	=	control point to evaporator
cs	=	controller set point
ds	=	design set point
e	=	evaporator
ev	=	evaporator to vapor
h	=	reservoir heater; heating
hcx	=	heating at $T_{cs\ max}$
hd	=	heating at T_{ds}
$hvmcx$	=	heating to $T_{v\ min}$ at $T_{cs\ max}$
max	=	maximum conditions

```
min     =   minimum conditions
R       =   reservoir
Rmcx    =   minimum reservoir temperature at T
```
$T_{cs\ max}$
```
Rmd     =   minimum reservoir temperature at T
```
T_{ds}
```
Rxcx    =   maximum reservoir temperature at T
```
$T_{cs\ max}$
```
Rxd     =   maximum reservoir temperature at T
```
T_{ds}
```
rad     =   radiator
rs      =   radiator to space
s       =   effective sink conditions
v       =   vapor
vr      =   vapor to radiator
```

Introduction

An analytical technique was developed for steady-state
and pseudotransient control analysis of variable-conductance
heat pipes (VCHP) and feedback-controlled heat pipes (FCHP).
VCHP/FCHP controllability is dependent upon reservoir temperature range and volume. The analytical approach uses a modified vapor temperature profile and a five-node thermal network.
This approach differs from past techniques in that it accounts
for gas blockage of the adiabatic section, and the set-point
temperature is referenced to the control-point node rather than
the vapor node. The reason for considering blockage of the
adiabatic section was to predict the minimum heat leak under
cold-case conditions as a function of total blockage length.
The set-point temperature is referenced to the control point
rather than to the vapor node to account properly for the
temperature drops between the sensor and the vapor temperature
for the maximum and minimum conditions.

Another significant variation in the model was in the
analysis of feedback-controlled systems with variable controller set points. In this model, the number of moles of gas
is computed at the design set-point temperature and held constant for varying controller set-point temperatures. This
differs from previous analyses in which the required number of
moles in the reservoir is computed for each set-point temperature by virtue of substitution for the gas molar density.

The pseudotransient analysis integrates the steady-state
control equations with reservoir response equations for a
specified response time. The specified response time is determined for best-case cooling response of the control point. For
low-capacitance control points, very large increases in the
required reservoir volume can result.

Noncondensible Gas-Controlled Heat Pipes

There are two general classifications of variable-conductance heat pipes using a noncondensible gas; these are wicked reservoir VCHP's and nonwicked reservoir VCHP's. Within each of these classifications are four subclassifications. These subclassifications are passive VCHP, actively assisted FCHP, isothermal reservoir VCHP, and active FCHP.

The passive VCHP is designed for a single set-point temperature and does not require an active control system. All of the other VCHP/FCHP systems can be designed for either a single set-point temperature or for a variable set-point temperature and generally require an active control system (isothermal reservoir being the possible exception). Also, one particular VCHP/FCHP may fall into different classifications for different applications.

Originally the analysis was developed for an active wicked-reservoir FCHP. However, the general analytical technique used for this system is applicable for the remaining subclassifications. Also, this technique can be used for

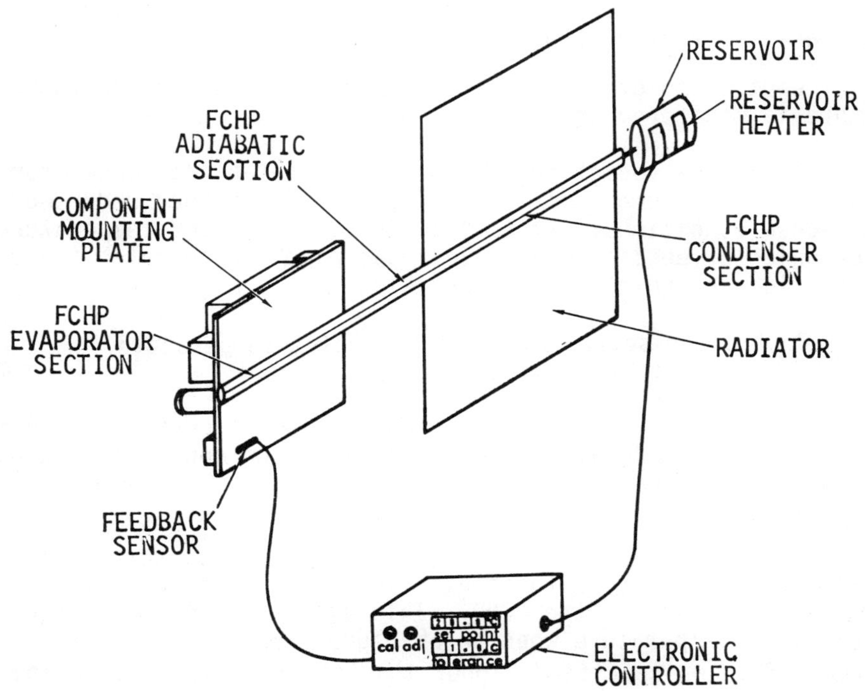

Fig. 1 Basic FCHP system design.

analysis of non-wicked reservoirs, except for transient
analyses where transient effects due to diffusion must be
accounted for. This paper uses an active FCHP with a wicked
reservoir to demonstrate the analysis.

FCHP Baseline Design

An active ammonia-argon FCHP with a stainless-steel
wicked reservoir is used to demonstrate the analytical technique (Fig. 1). It consists of an ATS-F axial groove heat
pipe[1] with 38.1-cm evaporator section, 12.7-cm adiabatic
section, and an 76.2-cm condenser section. The adiabatic
section has its heat-pipe cross-sectional area reduced to
0.205-cm^2. Total radiator heat-rejection area is 0.186-m^2.
A conductance of 10 W/K was assumed between the control point
and the evaporator, and a conductance of 29 W/K was assumed
between the evaporator and the vapor. Finally, a condenser
film coefficient of 5676 W/m^2-K was assumed.

Definitions

During the development of the analysis, new terms were
defined and other currently used terms were redefined to provide a concise analytical terminology. These terms are
summarized briefly below:

Control point: The point in the system where the temperature is monitored and controlled.

Design set-point temperature: The set-point temperature
for which the number of moles of gas is determined under worst-
hot-case conditions. It is the minimum set-point temperature
for which a feedback-controlled heat pipe can operate with the
specified set-point design conditions.

Controller set-point: The set-point temperature for
which the feedback-controller is set.

Set-point temperature tolerance: The allowable bilaterial
variation in temperature above and below the controller set-
point temperature.

Minimum set-point temperature: The minimum set-point
temperature for which the controller can be set and still
maintain control of the FCHP. This differs from the design
set-point temperature when an absolute minimum reservoir
temperature is specified higher than the reservoir temperature
under the design set-point temperature condition.

Maximum set-point temperature: The maximum set-point temperature for which the controller can be set to maintain control. This condition occurs when the maximum reservoir temperature is specified at a specific value.

Radiator-off efficiency: The efficiency of the radiator for the condition when the condenser is blocked completely by gas. The reference temperature is taken at the point where the heat enters the radiator.

Fin efficiency: The normal fin efficiency for the active portion of the condenser. The reference temperature is the root temperature in the active portion of the condenser.

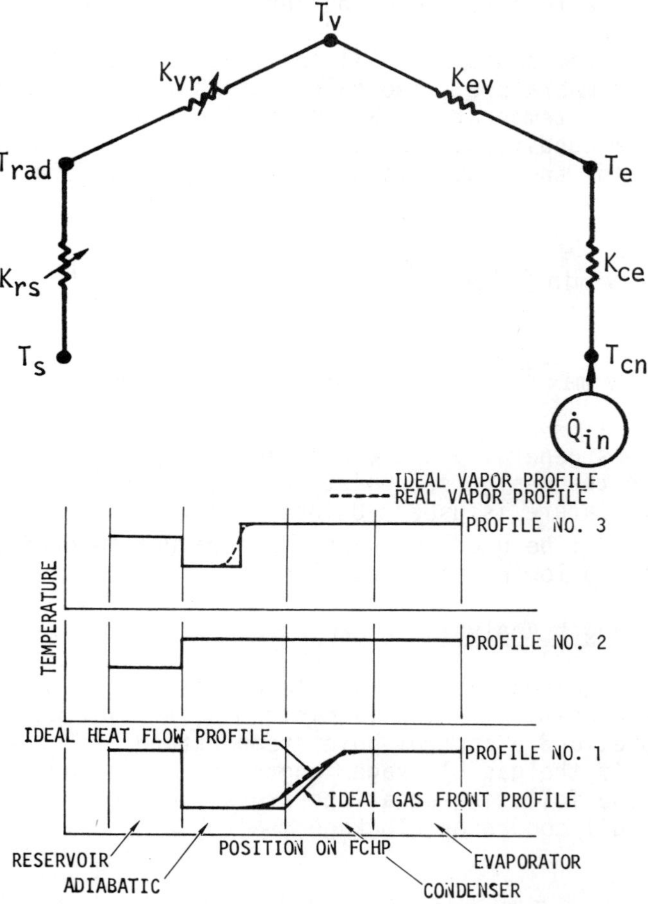

Fig. 2 Thermal network and vapor temperature profiles.

Analytical Approach for Steady-State Model

The steady-state model is a modified flat-front analysis
that allows analysis of gas front entry into the adiabatic
section. The assumed temperature profiles used in this technique are shown in Fig. 2. These profiles are used in the
calculation of the heat flow and gas inventory condition of
the FCHP. The dashed lines indicate what the real temperature
profiles might look like. In a VCHP system, the real case will
show a slight increase of the design set-point temperature[3].
In a FCHP system, the predicted reservoir temperatures will be
slightly higher than in the real case for the same controller
set-point temperature. The second part of the model is the
simple five-node thermal network shown in Fig. 2. This network is integrated with the assumed temperature profiles to
calculate the temperature drops throughout the FCHP system.

A central concern to all analyses is the calculation of
the vapor temperature. The calculations for the minimum and
maximum vapor temperatures are made with respect to the minimum
and maximum conditions. This results in the following general
equations for the minimum and maximum vapor temperature:

$$T_{v\ min} = T_{cs} - \delta - \dot{Q}_{min} \left[1/K_{ce} + 1/K_{ev} \right] \tag{1}$$

$$T_{v\ max} = T_{cs} + \delta - \dot{Q}_{max} \left[1/K_{ce} + 1/K_{ev} \right] \tag{2}$$

where T_{cs} is generally the controller set-point temperature;
but, under the design set-point condition, the design set-
point temperature is used. Unless otherwise stated, these
equations will be used for vapor temperature determination in
all of the following analyses.

Blockage Length Analysis

The required gas blockage length is determined for the
cold-case condition, which is minimum heat load, minimum sink
temperature, and a control-point temperature of $T_{cs} - \delta$. To
determine if the gas blockage length enters the adiabatic
section, the following equations are solved to find the heat
leak for full condenser blockage:

$$\dot{Q}_{div} = \eta_{rad}\ \varepsilon\ \sigma\ A_{rad} \left[T_{v\ div}{}^4 - T_{s\ min}{}^4 \right] \tag{3}$$

where

$$T_{v\ div} = T_{cs} - \delta - \dot{Q}_{div} [1/K_{ce} + 1/K_{ev}] \tag{4}$$

For a minimum heat load less than $\dot{Q}_{div}$, adiabatic blockage occurs, and the following equation calculates the blockage length:

$$L_b = (kA_{cs}/\dot{Q}_{min}) [T_{v\ min} - T_{rad}] + L_c \tag{5}$$

For a minimum heat load greater than $\dot{Q}_{div}$, condenser blockage occurs, and the following equations are used to find the blockage length:

$$f(\beta) = \dot{Q}_{min} - n_f\ \sigma\ \varepsilon\ A_{rad}\ (1 - \beta)[(T_{v\ min} -$$
$$\dot{Q}_{min}/(hA\ L_c\ (1 - \beta))^4 - T_{s\ min}^4] = 0 \tag{6}$$

where

$$\beta = L_b/L_c \text{ for } L_b < L_c \tag{7}$$

Many of the following analyses will show β to be a convenient parameter for condenser blockage. Another convenient parameter for adiabatic blockage is α, which is defined as

$$\alpha = (L_b - L_c)/L_a \text{ for } L_b \geq L_c \tag{8}$$

Two other relationships exist between α and β; these are

$$\text{if } \alpha \geq 0; \text{ then } \beta = 1.0$$

and

$$\text{if } \beta < 1; \text{ then } \alpha = 0.0$$

Design Set-Point Temperature Condition

The design set-point temperature condition is important to all analyses. The design set-point temperature is defined as the set-point temperature for which the FCHP system is exposed to the maximum heat load, maximum sink temperature, and minimum reservoir temperature at $T_{cs} + \delta$. Under these conditions, the molar gas density of the reservoir is

$$n/V_R = (P\ (T_{v\ max}) - P\ (T_{R\ min}))/(R\ T_{R\ min}) \tag{9}$$

The reservoir gas density remains constant until the design set-point temperature conditions are changed.

Reservoir Temperature Analysis

For a given reservoir volume, the maximum and minimum reservoir temperature requirements are determined for a specified controller set-point temperature. Parametric values of maximum reservoir temperature as a function of reservoir volume can be generated by the following equations:

$$P\,(T_{R\ max}) = P\,(T_{v\ min}) - \psi\,T_{R\ max}/(V_R/V_c) \qquad (10)$$

$$T_{v\ min} = T_{cs} - \delta - \dot{Q}_{min}\,[1/K_{ce} + 1/K_{ev}] \qquad (11)$$

where

$$\psi = R(n/V_R)(V_R/V_c) - \beta\,P\{T_{s\ min}\}/T_{s\ min}$$

$$-\,2\,\alpha\,\phi'\,P\,\{(T_{v\ min} + T_{s\ min})/2\}/(T_{v\ min} + T_{s\ min}) \qquad (12)$$

Parametric values of minimum reservoir temperature as a function of reservoir volume are generated by the following equations:

$$P\,(T_{R\ min}) = P\,(T_{v\ max}) - (\psi'\,T_{R\ min})/(V_R/V_c) \qquad (13)$$

$$T_{v\ max} = T_{cs} + \delta - \dot{Q}_{max}\,[1/K_{ce} + 1/K_{ev}] \qquad (14)$$

where

$$\psi' = R\,(n/V_R)\,(V_R/V_c) - \beta'\,P\,\{T_{s\ max}\}/T_{s\ max} -$$

$$(2.0\,\alpha\,\phi\,P\,\{(T_{v\ max} + T_{s\ max})/2\})/(T_{v\ max} + T_{s\ max}) \qquad (15)$$

Reservoir Volume Analysis

Generally, the reservoir volume required to maintain control is determined at a controller set-point temperature with the molar gas density determined at the design set-point temperature. Analysis of the reservoir volume also is dependent on the restraints placed on the reservoir temperature.

Analysis without reservoir temperature restraints

The required reservoir volume for control of FCHP systems

that allow the maximum reservoir temperature to equal the minimum vapor temperature for a particular controller set-point temperature is defined as follows:

$$(V_R/V_C) = [\beta \, P \, \{T_{s\ min}\}/T_{s\ min} + (2.0 \, \alpha \, \phi \, P \, \{(T_{v\ min} + T_{s\ min})/2\})/(T_{v\ min} + T_{s\ min})]/R \, (n/V_R) \tag{16}$$

Analysis with reservoir temperature restraints

The general control equation for the required reservoir volume is

$$(V_R/V_C) = [\beta \, P \, \{T_{s\ min}\}/T_{s\ min} + (2.0 \, \alpha \, \phi \, P \, \{(T_{v\ min} + T_{s\ min})/2\})/(T_{s\ min} + T_{v\ min})]/[R \, (n/V_R) - P \, \{T_{R\ max}\}/T_{R\ max}] \tag{17}$$

Notice that this equation is dependent on the maximum reservoir temperature only, but both maximum and minimum reservoir temperatures have an effect on the controller set-point temperature regime for which the solution of the Eq. (17) is valid.

Definition of Valid Controller Set-Point Regimes

Restraints placed on the reservoir temperature will cause the controller set-point temperature to be limited to certain regimes for which controllability of the FCHP exist. It has been found that a restraint on the minimum reservoir temperature defines the lower limit of the regime, whereas the restraint on the maximum reservoir temperature defines the upper limit of the regime. The lower and upper limits to the controller set-point temperature are referred to as the minimum and maximum controller set-point temperatures.

The minimum controller set-point temperature is generally the design set-point temperature, except when the restrained reservoir temperature is higher than the reservoir temperature at the design set-point. The equations for defining the minimum controller set-point temperature are Eqs. (13-15), with the following substitution for V_R/V_C:

$$(V_R/V_C) = (V_R/V_C)_{min} = [\beta \, P \, \{T_{s\ min}\}/T_{s\ min} + (2.0 \, \alpha \, \phi \, P \, \{(T_{v\ min} + T_{s\ min})/2\})/(T_{v\ min} + T_{s\ min})]/R \, (n/V_R) \tag{18}$$

This is the equation for the minimum reservoir volume, which means that all of the gas is located in the condenser and adiabatic sections. As the reservoir volume increases above the minimum, the minimum controller set-point will increase, as defined in Eqs. (13-15), until it reaches the value defined for an infinite reservoir. This is defined by the solution of the following equations:

$$P \, (T_{v \, max}) = P \, (T_{R \, min}) + R \, T_{R \, min} \, (n/V_R) \tag{19}$$

$$T_{v \, max} = T_{cs \, min} + \delta - \dot{Q}_{max} \, [1/K_{ce} + 1/K_{ev}] \tag{20}$$

A similar development can be used to define the maximum controller set-point temperature as defined by Eqs. (10-12). For an infinite reservoir volume, the maximum controller set-point temperature is found by the following equations:

$$P \, (T_{v \, min}) = P \, (T_{R \, max}) + R \, T_{R \, max} \, (n/V_R) \tag{21}$$

$$T_{v \, min} = T_{cs \, max} - \delta - \dot{Q}_{min} \, [1/K_{ce} + 1/K_{ev}] \tag{22}$$

This condition means that, for increasing set-point temperatures, compression of the noncondensable gas must occur, since the pressure at the minimum vapor temperature continues to increase, whereas the partial vapor pressure and the partial gas pressure in the reservoir remain constant. In other words, the rate of change of reservoir gas pressure with respect to controller set-point temperature becomes negative.

The maximum controller set-point temperature decreases with decreases in reservoir volume as defined by Eqs. (10-12). This continues until the reservoir volume reaches the minimum reservoir volume for the maximum reservoir temperature, which corresponds to a controller set-point temperature of

$$T_{cs \, max} = T_{R \, max} + \delta + \dot{Q}_{min} \, [1/K_{ce} + 1/K_{ev}] \tag{23}$$

This corresponds to a condition of the maximum reservoir temperature equal to the minimum vapor temperature at the maximum controller set-point temperature.

Analytical Approach for Pseudotransient Model

The pseudotransient model integrates the steady-state controllability model with a reservoir transient response

model. This model calculates the minimum and maximum reservoir temperatures based on controllability and response
requirements. For low-capacitance control points, this
analytical approach becomes an important consideration in
preventing control-point temperature excursions outside the
temperature tolerance band. This approach shows that tremendous increases in required reservoir volume occur for rapidly
responding systems.

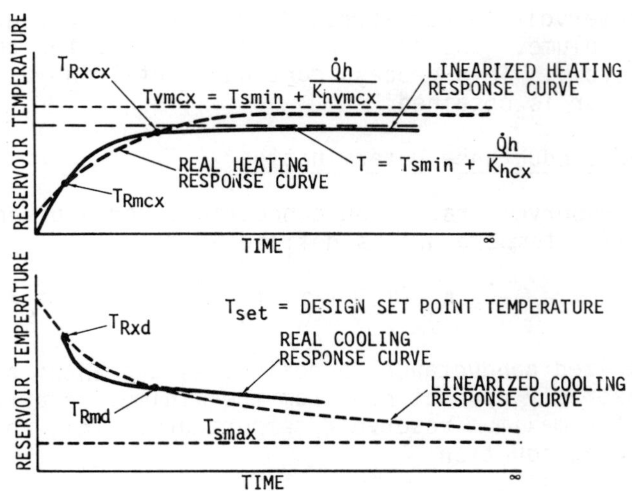

Fig. 3   Real and linearized thermal response curves.

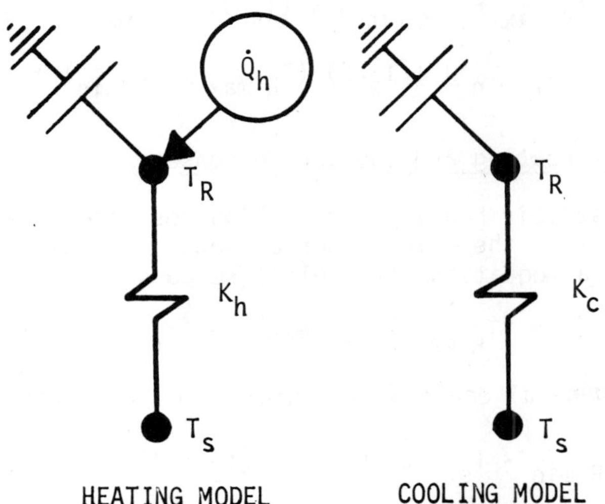

Fig. 4 Simple reservoir thermal networks for heating and cooling.

The reservoir transient response model uses a linearized conductance technique to simulate the transient radiation response. Fig. 3 illustrates the response of the real reservoir with the response using the linearized technique. This shows the linearized response to be dependent on the endpoint temperatures.

Simultaneously, the steady-state controllability analysis is performed, based on the calculated minimum and maximum reservoir temperatures, to determine the required reservoir volume. The new volume then is used for the next solution cycle. This process continues until the desired maximum error is obtained.

Linearized Conductance Determination

The reservoir radiation conductance for a given reservoir and sink temperature is defined as

$$K_T = \eta_R \, \sigma \, \varepsilon \, A_R \, (T_R^3 + T_s \, T_R^2 + T_s^2 \, T_R + T_s^3) \qquad (24)$$

The linearized conductance is defined as the integrated average of the reservoir conductance from the minimum reservoir temperature to the maximum reservoir temperature. This results in the following equation:

$$K_L = \eta_R \, \varepsilon \, \sigma \, A_R \, [(1/4) \, (T_{R\,max}^3 + T_{R\,min} \, T_{R\,max}^2 +$$
$$T_{R\,min}^2 \, T_{R\,max} + T_{R\,min}^3) + (T_s/3) \, (T_{R\,max}^2 + T_{R\,min} \, T_{R\,max} +$$
$$T_{R\,min}^2) + (T_s^2/2) \, (T_{R\,max} + T_{R\,min}) + T_s^3] \qquad (25)$$

Reservoir Heating and Cooling Response

Reservoir heating and cooling response equations were derived using the simple thermal networks shown in Fig. 4. The general equation for cooling response is

$$T_R = (T_{R\,max} - T_s) \, \exp \, [(- \, (K_c/C) \, \Theta)] + T_s \qquad (26)$$

and the general equation for heating response is

$$T_R = (T_{R\,min} - T_s - \dot{Q}_h/K_h) \, \exp \, [(- \, K_h/C) \, \Theta] + T_s + Q_h/K_h \qquad (27)$$

Pseudotransient Controllability

Pseudotransient controllability of the FCHP requires that

the reservoir be able to respond under worst-case cooling and worst-case heating conditions, such that the control-point temperature remains within the temperature tolerance band. Simultaneous solution of the heating and cooling response equations and the controllability equations for a given design and specified maximum controller set-point temperature was done iteratively.

The specified time response should be based on the best-case cooling response of the control point. Considerations that should be accounted for are environmental exposure time, time at minimum and maximum heat loads, thermal control system design, and the control-point capacitance.

The worst-case cooling and worst-case heating condition corresponds to natural radiation cooling at the design set point temperature. Worst-case heating response is governed by the applied reservoir heater power. It generally is desirable that the worst-case heating response be matched to the worst-case cooling response. This corresponds to a minimum reservoir heater power condition. Matching requires that the reservoir heater provide, for worst-case heating, the same potential that the maximum sink temperature provides for worst-case cooling. Since the worst-case heating response occurs at the specified maximum controller set-point temperature, then the corresponding minimum vapor temperature should be the upper limit for maximum reservoir temperature, which would correspond to the maximum heater power being applied constantly. Therefore, the maximum reservoir heater power is defined by the following equations:

$$\dot{Q}_h = K_{hvmcx} \left(T_{Rmcx} - T_{smin} \right) \tag{28}$$

and

$$K_{hrmcx} = f \left\{ T_{Rmcx}, T_{vmcx}, T_{smin} \right\} \tag{29}$$

where $f\{a_1, a_2, \ldots, a_i\}$ is a generalized function of the variables $a_1, a_2, \ldots, a_i$.

Notice the end-point temperatures in the calculation of the linearized conductance. These are the minimum reservoir and minimum vapor temperatures. This is because the response calculations based on linearized conductances are valid only for the end points. Figure 3 shows that, if K_{hcx}, were used in place of K_{hvmcx}, then both the upper limit of the reservoir temperature and the heater power would be low.

Equation (28) is dependent indirectly upon T_{Rmcx}, which means that it must be solved simultaneously with the following basic equations for solution of the pseudotransient controllability:

$$T_{Rmd} = (T_{Rxd} - T_{smax}) \exp [(- (K_{cd}/C) \, \theta_1)] + T_{smax} \qquad (30)$$

$$T_{Rmcx} = (T_{Rxcx} - T_{smax}) \exp [(- (K_{ccx}/C) \, \theta_2)] + T_{smax} \qquad (31)$$

$$T_{Rxcx} = (T_{Rmcx} - T_{smin} - \dot{Q}_h/K_{hcx}) \exp [- (K_{hcx}/C) \, \theta_1] +$$
$$T_{smin} + (\dot{Q}_h/K_{hcx}) \qquad (32)$$

$$T_{Rxd} = (T_{Rmd} - T_{smin} - \dot{Q}_h/K_{hd}) \exp [- (K_{hd}/C) \, \theta_2] +$$
$$T_{s \, min} + (\dot{Q}_h/K_{hd}) \qquad (33)$$

$$K_{cd} = f \, \{T_{Rmd}, \, T_{Rxd}, \, T_{smax}, \, A_{Res}\} \qquad (34)$$

$$K_{ccx} = f \, \{T_{Rmcx}, \, T_{Rxcx}, \, T_{smin}, \, A_{Res}\} \qquad (35)$$

$$K_{hd} = f \, \{T_{rmd}, \, T_{Rxd}, \, T_{smin}, \, A_{Res}\} \qquad (36)$$

$$K_{hcx} = f \, \{T_{Rmcx}, \, T_{Rxcx}, \, T_{smin}, \, A_{Res}\} \qquad (37)$$

$$K_{hvmcx} = f \, \{T_{Rmcx}, \, T_{vmcx}, \, T_{smin}, \, A_{Res}\} \qquad (38)$$

$$\dot{Q}_h = K_{hvmcx} \, (T_{vmcx} - T_{smin}) \qquad (39)$$

$$T_{vxd} = f \, \{\dot{Q}_{hd}\} \qquad (40)$$

$$(n/V_R)_{ds} = f \, \{T_{vxd}, \, T_{Rmd}\} \qquad (41)$$

$$T_{vmcx} = f \, \{\dot{Q}_{mcx}\} \qquad (42)$$

$$L_{bcx} = f \, \{\dot{Q}_{mcx}, \, T_{smin}, \, T_{vmcx}\} \qquad (43)$$

$$(n/V_R)_{cx} = f \, \{(n/V_R)_{ds}, \, \dot{Q}_{xcx}, \, \alpha', \, \beta'\} \qquad (44)$$

$$T_{Rmcx} = f\ \{(n/V_R)_{cx}\} \tag{45}$$

$$(V_R/V_c) = f\ \{T_{Rmcx},\ T_{vmcx},\ T_{smin}\} \tag{46}$$

Solution of these equations was accomplished with the aid of a computer program using an iterative technique.

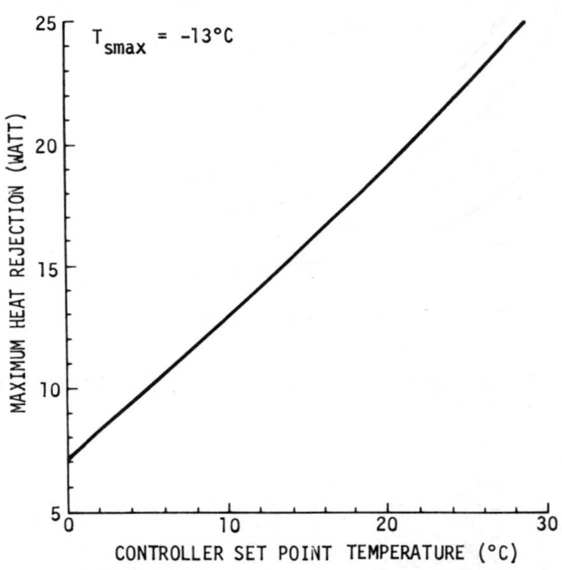

Fig. 5 FCHP system heat rejection capacity.

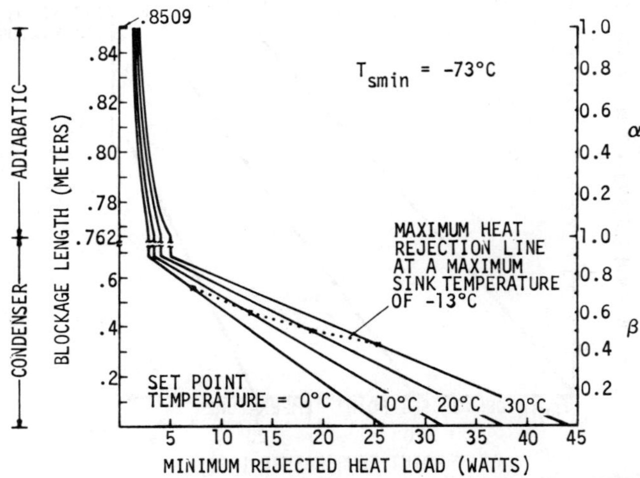

Fig. 6 Required blockage length for worst case cold condition.

A. M. LEHTINEN

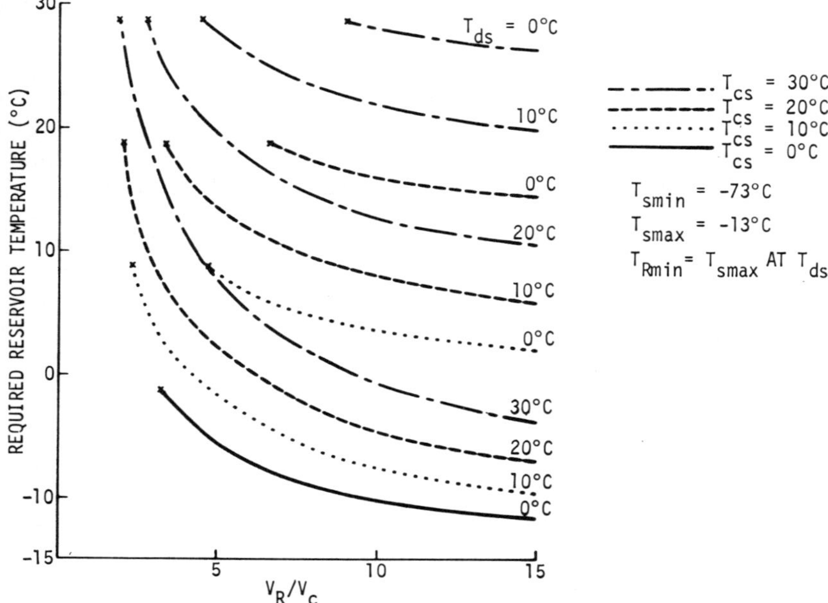

Fig. 7 Maximum required reservoir temperature.

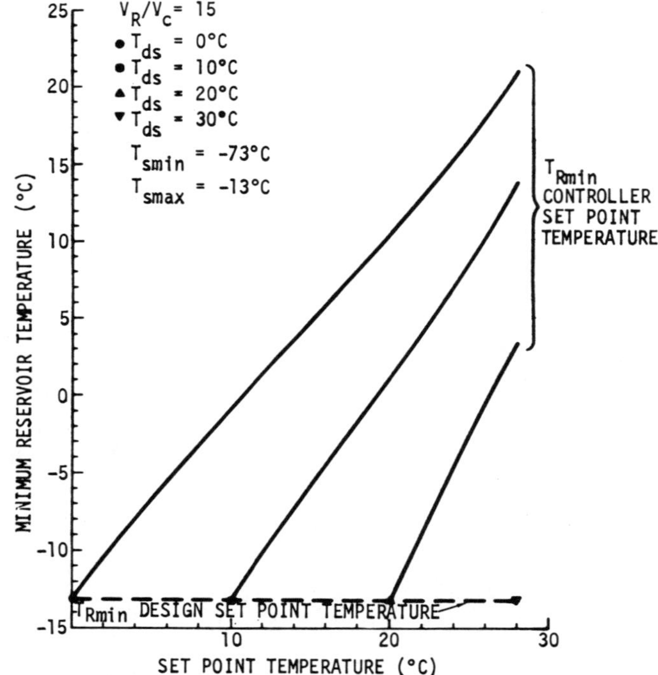

Fig. 8 Minimum reservoir temperature map for $V_R/V_C = 15$.

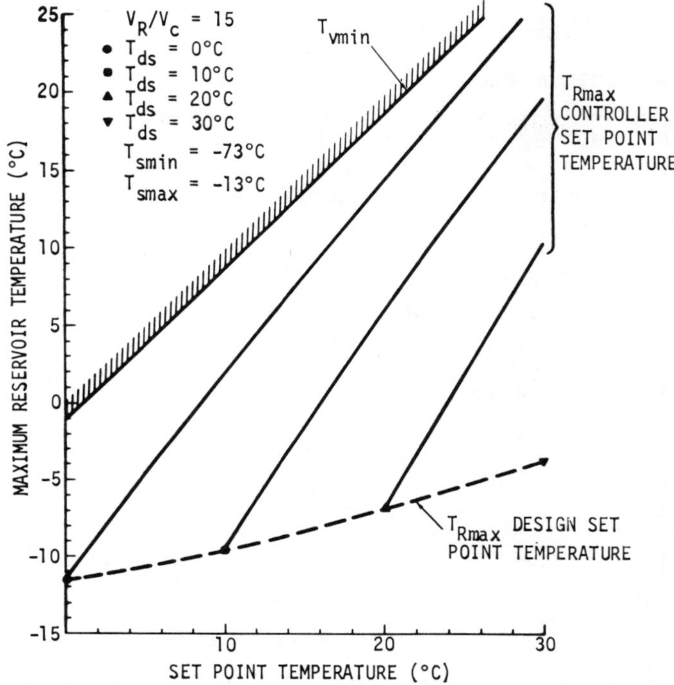

Fig. 9   Maximum reservoir temperature map for V_R/V_C = 15.

Analytical Results

Both the steady-state and pseudotransient controllability
analyses were based on maintaining control with the blockage
length varying from zero to the full condenser-adiabatic
blockage. For a blockage length of zero, the FCHP system heat-
rejection capacity is found as a function of controller set-
point temperature (Fig. 5). Similarly, for full condenser-
adiabatic blockage, the minimum heat load is found as a
function of controller set-point temperature. Figure 6 shows
the required blockage length as a function of minimum rejected
heat load for worst-case cold condition. The discontinuity in
the curve is due to the change in the analytical techniques
used to determine the condenser and adiabatic blockage lengths.
The minimum heat load variations applicable to the following
studies are found on Fig. 6 for the condition α = 1. Temperature control of the FCHP system was specified to be within
+ 1.0°C of the set-point temperature. The radiator and the
reservoir were exposed to a minimum environmental sink temperature of -73°C and a maximum environmental sink temperature
of -13°C.

Reservoir Temperature Results

The maximum excursion of the reservoir temperature is from T_{Rmin} to T_{vmin} at any controller set-point temperature. A minimum required reservoir volume exists for a specified design set-point temperature and maximum controller set-point temperature, which will require the reservoir temperature to

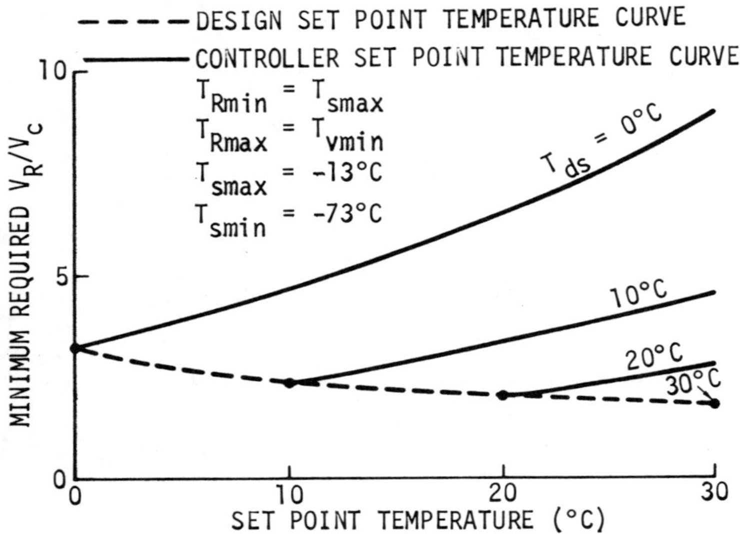

Fig. 10   Reservoir volume requirements without T_{Rmax} restrained.

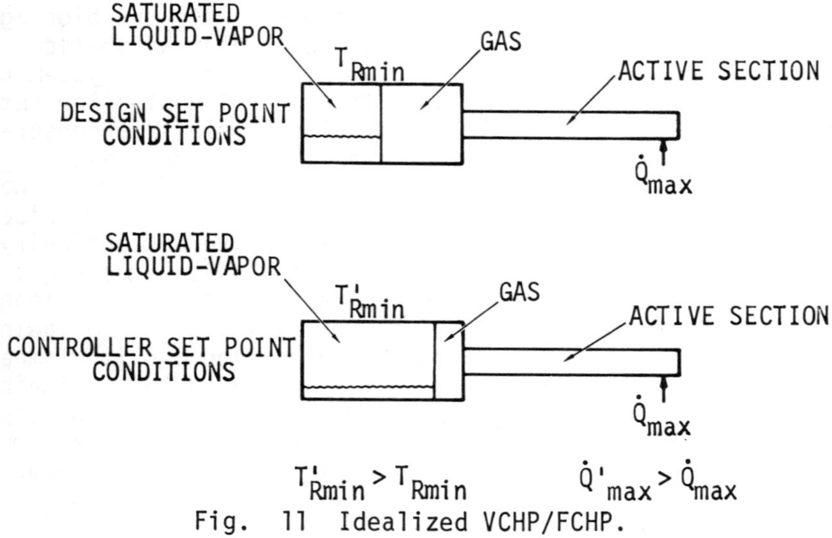

Fig. 11   Idealized VCHP/FCHP.

vary from T_{Rmin} to T_{vmin}. In Fig. 7, the minimum required
reservoir volumes are indicated by the X's. Figure 7 shows
that, for reservoir volumes greater than the minimum, a
decrease in the maximum required reservoir temperature for
control occurs. Notice that the reservoir temperature decrease
is most pronounced when the specified design and controller
set-point temperatures are equal, and least pronounced for the
maximum separation of the specified design and maximum controller set-point temperatures.

Parametric reservoir temperature maps of the design and
controller set-point temperatures can be generated for a
specific reservoir volume (Figs. 8 and 9). Figures 8 and 9
are used by following along the design set-point line until
the design set-point temperature is reached, then following
parallel to the controller set-point lines until the controller set-point temperature is reached, and then reading the
reservoir temperature. In reality, both the minimum and
maximum reservoir temperatures would be expected to be
slightly lower, since less gas would be transferred into the
condenser and adiabatic sections of the FCHP. This is

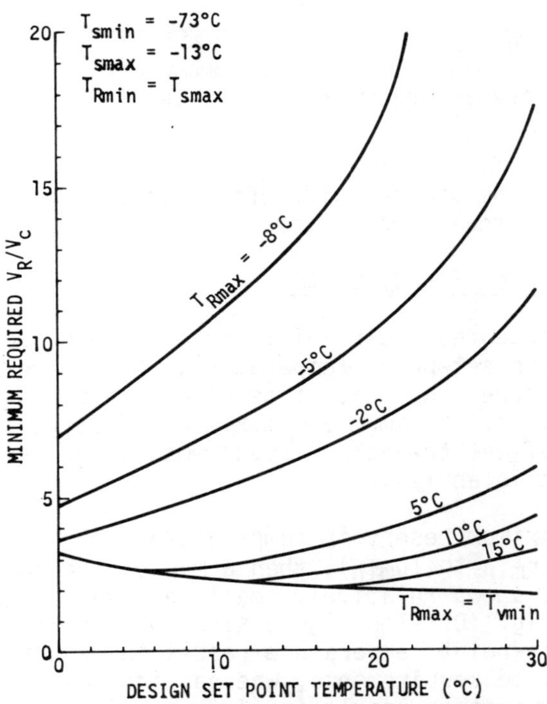

Fig. 12 Reservoir requirements at T_{ds} with T_{Rmax} restrained.

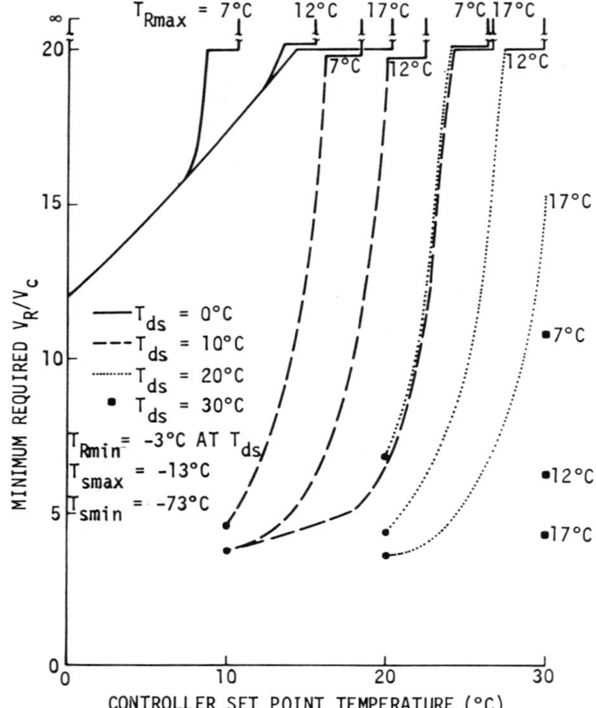

Fig. 13 Reservoir volume requirements at T_{cs} with T_{Rmax} restrained.

equivalent to the upward shift effect in the design set-point temperature of real VCHP systems.

Reservoir Volume: Steady State

Steady-state reservoir volume requirements are dependent upon the design set-point temperature, the controller set-point temperature, and restraints placed on the maximum reservoir temperature. Minimum and maximum effective sink temperatures also affect the volume requirements but have been held constant in this analysis.

If the maximum reservoir temperature excursion is allowed (i.e., from T_{Rmin} to T_{vmin}), then a map of the reservoir volume requirements, shows relatively small reservoir requirements, as shown in Fig. 10. The figure also shows that increases in the design set-point temperature result in decreases in the reservoir volume requirements, whereas increases in controller set-point-temperature result in increases in reservoir volume requirements.

Decreases in reservoir volume for increasing design set-point temperatures are due to the higher molar gas density in the reservoir. This means that more gas is available for blockage per unit volume. The gas density in the gas-blocked portion of the condenser and adiabatic sections also increases, but generally this is to a lesser extent when the sink temperatures are below the minimum reservoir temperature.

Increases in reservoir volume for increasing controller set-point temperature are due to a lower apparent molar gas density is in no way real, since the actual molar gas density is still the same as at the design set temperature.

Consider the following: idealize the gas and liquid-vapor phases of the reservoir as two separate pure systems with the gas zone as a buffer zone (Fig. 11). Now, under worst-case hot conditions, as the reservoir temperature is raised, the vapor pressure in the reservoir rises, which will cause the system total pressure to rise. This results in compression of the pure-gas zone to a higher density. Therefore, the original molar gas density takes on the character of having a lower apparent density as the system total pressure increases.

The lower apparent molar gas density results in a lesser ability for gas blockage per unit volume. Also, as the controller set-point increases, the condenser and adiabatic sections require more gas for blockage. These two conditions compound the problem, and the net result is an increased reservoir volume.

If the maximum reservoir temperature is restrained, very large increases in reservoir volume result for both design and controller set-point temperatures. Figure 12 shows the reservoir requirements for the design set-point temperature. These are not as severe as those for controller set-point temperatures (Fig. 13). The explanation for this is basically the same reasoning as explained previously. The only difference is that, once the minimum vapor temperature reaches the specified maximum reservoir temperature, then control is maintained by increases in reservoir volume alone. This is seen clearly by the separation of the curves in both Fig. 12 and 13.

Figure 13 also can be read to determine the maximum controller set-point temperature as a function of reservoir volume. When these temperatures are reached all of the gas in the FCHP is located in the condenser and adiabatic sections.

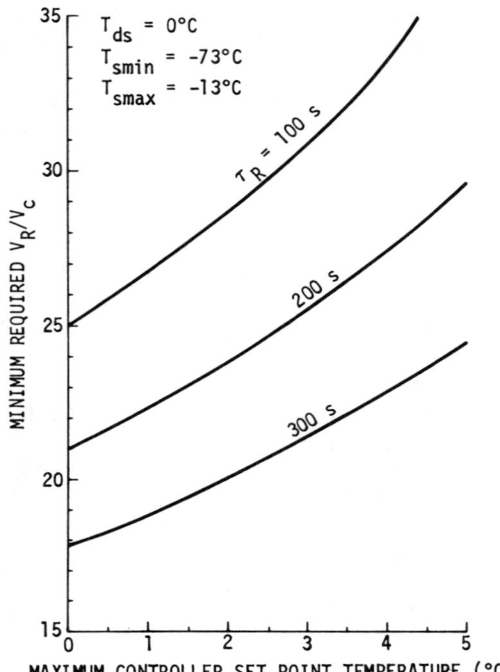

Fig. 14 Reservoir volume requirements for specific response times at T_{ds} = 0°C.

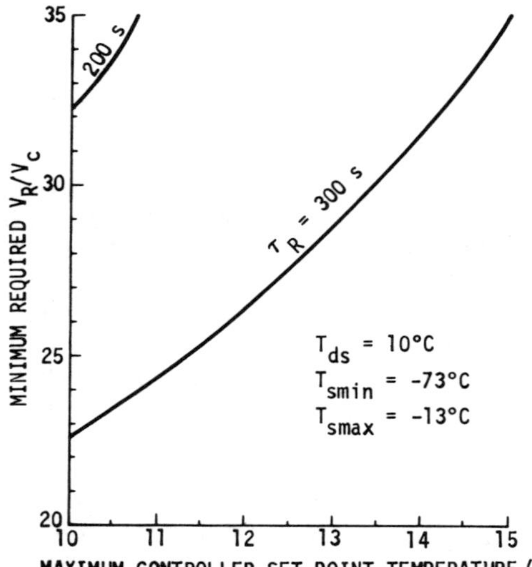

Fig. 15 Reservoir volume requirements for specified response times at T_{ds} = 10°C.

Additional increases in reservoir temperature theoretically would result in compression of the gas front, with no effect on set-point temperature. In reality, though, an unstable condition would exist, and vapor temperature oscillation would be expected. Notice also that the design set-point temperature affects the location of the maximum controller set-point temperature for the same maximum reservoir temperature.

At the top of Fig. 13, the maximum controller set-point temperature is found for an infinite reservoir. This is the point at which the sum of the partial gas pressure and partial vapor pressure in the reservoir equals the vapor pressure at the minimum vapor temperature. Control is not possible beyond this point because compression of the gas would occur.

Reservoir Volume: Pseudotransient

Reservoir volume requirements, for FCHP systems requiring rapid response, may increase several times that of steady-state

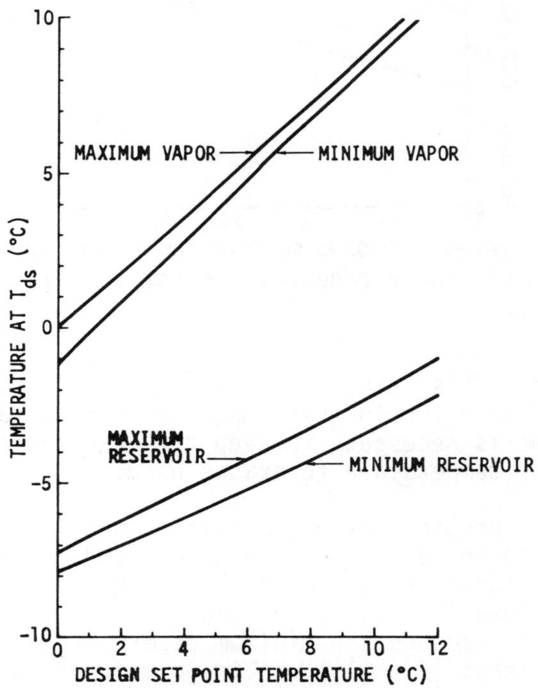

Fig. 16 Reservoir and vapor temperatures at T_{ds} for τ_R = 200 sec.

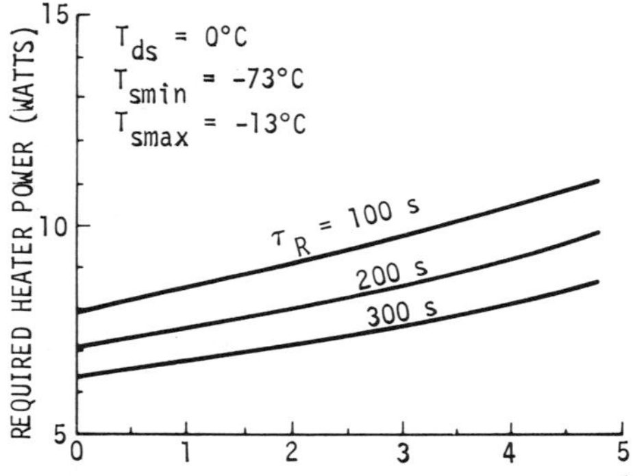

Fig. 17 Heater power requirements for specific response times at T_{ds} = 0°C.

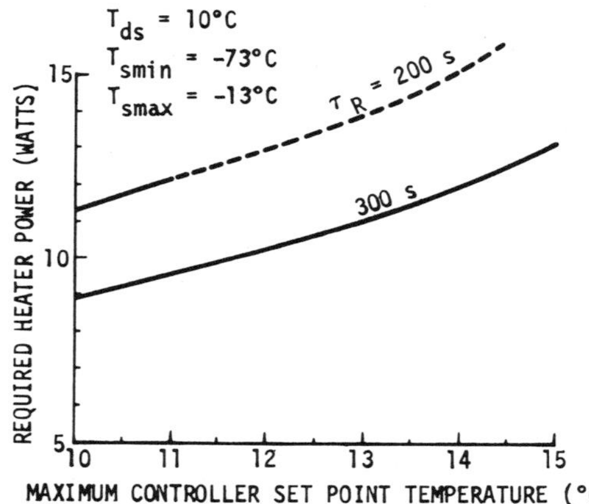

Fig. 18 Heater power requirements for specific response times at T_{ds} = 10°C.

FCHP systems. This is particularly applicable to low-capacitance control-point systems, for which a rapid specified response time is necessary to keep the FCHP system within the control-point temperature tolerance band.

For the pseudotransient analysis, the reservoir volume increases with increases both in design and controller set-point temperatures (Figs. 14 and 15). The reason for the reservoir volume increase as a function of design set-point temperature is because the minimum reservoir temperature is not held constant (Fig. 16), as is done in the steady-state analysis. The increase in the minimum reservoir temperature results in larger reservoir volumes being required for control,

as is the case for maximum reservoir temperature restraints.
Also notice that, for higher design set-point temperatures, in
Fig. 16, the effects of higher cooling rates on the reservoir
temperature range are minimal. As the specified response time
increases (Figs. 15 and 16), the reservoir requirements
decrease. The limiting case would ultimately be the steady-
state results for control.

The reservoir heater power requirements (minimum power
requirements) is shown in Figs. 17 and 18. As would be
expected from the reservoir volume requirements, the minimum
reservoir power requirements increase with both design and
controller set-point temperatures. Also, the minimum power
requirements decrease with increases in the reservoir response
time.

Conclusions

An analytical technique for steady-state and pseudotransient controllability has been developed. The steady-
state analytical technique uses a modified flat gas front
analysis, which consists of a modified vapor temperature
profile and a simple five-node thermal network. For the
pseudotransient analysis, the response time of the reservoir
is solved simultaneously with the steady-state control
analysis. The results of the analytical technique have shown
a number of significant conclusions. It was found that
reservoir volume requirements generally decrease with increasing design set-point temperature, whereas reservoir volume
requirements generally increase with controller set-point
temperature. If limitations are placed on the maximum reservoir temperatures, then large increases in reservoir volume
will result in some areas of design and controller set-point
temperatures. In addition, limitations placed on minimum and
maximum reservoir temperatures will result in a controller
set-point temperature regime for which control can be
maintained. This exists for both finite and infinite reservoir volumes. The results of the pseudotransient analysis
have shown that for rapidly responding systems the reservoir
volume may need to be increased several times that calculated
for steady-state conditions. In general, the analytical
technique presented has proven to be a viable tool in parametric analyses of VCHP/FCHP systems.

Acknowledgements

The author would like to express his appreciation to
J. P. Wright, D. E. Wilson and T. T. Cafferty, who have, in
discussion of this work, provided a greater understanding of
the analytical results.

References

[1]Schlitt, K. R., Kirkpatrick, J. P., and Brennan, P. J., "Parametric Performance of Extruded Axial Grooved Heat Pipes from 100° to 300°K," _AIAA Progress in Astronautics and Aeronautics: Heat Transfer with Thermal Control Applications_, Vol. 39, edited by M. Yovanovich, New York, 1975, pp. 215-234.

[2]"Transient Thermal Response of a Thermal Control Canister," Grumman Aerospace Corp., Contract NAS5-2270, 1976.

[3]Marcus, B. D., "Theory Design of Variable Conductance Heat Pipe," TRW Systems Group, NAS CR-2018, April 1972.

LOW-TEMPERATURE PHASE-CHANGE MATERIAL PACKAGE

P. J. Brennan [*] and H. J. Suelau [+]
B & K Engineering, Inc., Towson, Md.

and

R. McIntosh [#]
NASA Goddard Space Flight Center, Greenbelt, Md.

Abstract

Test data are presented which were obtained for a low-temperature phase-change material (PCM) canister. The canister was designed to provide up to 30 W-hr of storage capacity at approximately -90°C with an overall thermal conductance that is greater than $8^O W/^O C$. N-heptane, which is an n-paraffin and has a -90.6°C freezing point, was used as the working fluid. The canister was fabricated from aluminum and has an aluminum honeycomb core. Its void volume permits service temperatures up to 70°C. Results obtained from component and system's tests indicate well-defined melting and freezing points, which are repeatable and within 1°C of each other. Subcooling effects are less than 0.5°C and are essentially negligible. Measured storage capacities are within 94 to 88% of the theoretical.

Introduction

Passive thermal control can be accomplished by utilizing the solid/liquid-phase transition as a mechanism for the storage and release of heat. Phase-change materials (PCM) provide temperature stability by absorbing or rejecting heat as they melt or freeze without appreciable temperature change. This control technique has been adapted successfully to space-

---

Presented as Paper 77-762 at the AIAA 12th Thermophysics Conference, Albuquerque, N. Mex., June 27-29, 1977. Copyright © American Institute of Aeronautics and Astronautics, Inc., 1977. All rights reserved.

*Manager.
+Project Engineer.
#Member Thermal Systems Branch, Systems Division.

craft thermal control systems subjected to diurnal temperature cycling.

A considerable amount of laboratory and flight exper-ience has been obtained with ambient temperature systems.[1-3] To date, however, the only flight data for a low-temperature system were derived from a unit with dual phase-change materials which was used to attenuate the warming trend that resulted from degradation of the optical coating of a passive-ly cooled satellite.[4] In this system, normal heptane and methyl ethyl ketone (with respective melting points of -90.6° and 85.9°C) were contained in separate compartments, which were integral with a radiator. The increased heat input that resulted from the continuous degradation of the radiator's second surface mirrors was absorbed during the phase transi-tion of these materials.

The purpose of this paper is to describe the development of a low-temperature, high-thermal-conductance PCM canister that is used as an energy storage device in the heat-pipe experiment package (HEPP).[5] The HEPP was developed for flight aboard the Tiros-N spacecraft; however, it now is being considered for flight on the long-duration exposure facility (LDEF) to be placed in orbit by the Space Shuttle. The canister consists of an aluminum box containing partially expanded aluminum honeycomb core and n-heptane (PCM). This

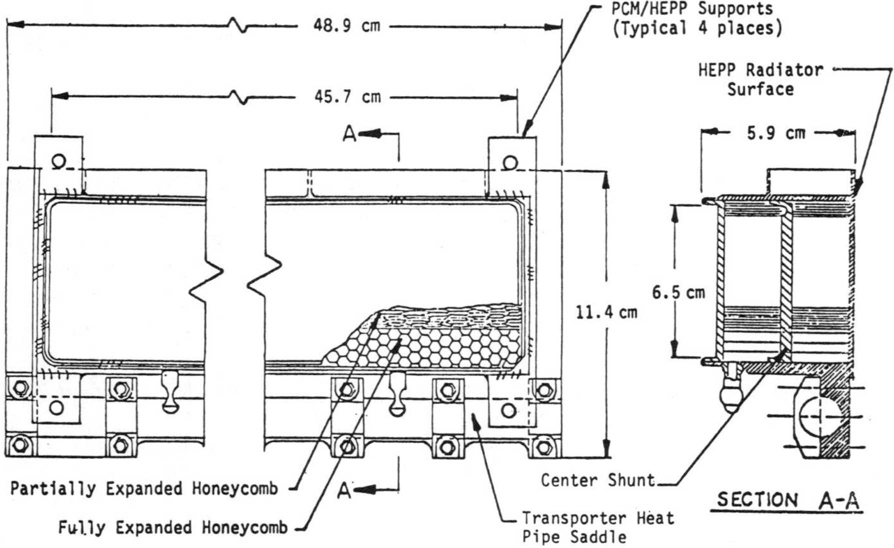

Fig. 1 PCM canister assembly.

canister is designed to provide 30 W-hr of energy storage and to limit the temperature drop across the box to 5°C with a 40-W heat input.

Canister Design

The design objectives identified in the development of the PCM canister were as follows:

1) Select a fluid that provides stability at the desired operating temperature of the HEPP radiator system (approximately -90°C).
2) Obtain the maximum thermal capacity within the allowable envelope of the HEPP.
3) Provide the maximum thermal conductance while minimizing the weight and material volume requirements.
4) Provide a container that is leak-tight and capable of maintaining pressure retention integrity over the operating temperature range (-150° to +50°C).

PCM Selection

The fluid chosen for the canister was n-heptane, a readily available n-paraffin material with a well-defined melting point (-90.6°C), consistent with experiment objectives. In addition, it has a high latent heat, low toxicity, a low vapor pressure at ambient temperatures, and is compatible with aluminum. As an n-parrafin, it is generically similar to the octadencane used in the ATFE PCM package,[2] which has been in operation onboard the ATS-6 spacecraft since June 1974. To date, the ATFE PCM has operated as predicted with no apparent performance degradation after undergoing more than 1100 freeze/thaw cycles.

Container Design

Design and fabrication procedures developed for the ATFE canister were used extensively in this application. The canister consists of a TIG welded aluminum rectangular box divided into two identical compartments by an aluminum center shunt. This assembly is illustrated in Fig. 1. The walls and center shunt are sized for a minimum conductance of 8 W/°C across the box. A saddle to be used for attachment of the transporter heat pipe (TPHP) in the HEPP is integral with the PCM canister. This enhances the thermal conduction between these components. Each compartment is filled with 0.19-cm (3/16-in.) cell aluminum honeycomb core, with partially expanded core accounting for 80% of the volume and fully expanded core for the remainder. The void density of the honeycomb occupied compartment is approximately 90%. Adhesive sheets (Hysol EA 934) are used on the center shunt and the top and bottom surfaces to hold the honeycomb in place. The assembly is pressed together so that the honeycomb cuts through the adhesive layer to establish metal-to-metal contact.

The honeycomb core provides high-conductance paths to the low-conductivity PCM. The amount of honeycomb is sized to yield a maximum 1°C temperature drop from the side walls into the center of either PCM compartment with a 40-W heat input. The multiplicity of cells provides a high surface contact area to PCM fluid volume ratio, which, in combination with the uniform straight path conductance in the cell direction (heat flow direction), results in a high thermal diffusivity. This characteristic minimizes transient temperature drops between the heat input/output zone and the freeze/thaw front.

Each cell contains two 0.16-cm (1/16-in.) holes to permit charging of the PCM and also liquid expansion during melting. A series of 25 0.238-cm (3/32-in.) holes is located along the length of the central shunt to allow for charging and communication between the individual compartments. The void space, which permits expansion of the heptane up to 70°C, contains fully expanded honeycomb to reduce sloshing effects during launch or orbital maneuvers. The void is located near the heat input side to avoid any local compressibility effects during melting. In 0 g, the liquid will tend to fill the partially expanded honeycomb as it cools because of their higher capillary forces vs those associated with the fully expanded core. Thus, when the liquid melts, it will expand into a void instead of being restricted and causing a local overpressure. The canister also was designed to permit testing in 1 g. The void volume is located at the top when

Table 1 PCM canister design summary

Envelope	TIG welded 6061-T6 aluminum assembly per Fig. 1
Core	0.19-cm (3/16-in) cell by 0.005-cm-thick 5052 aluminum honeycomb
PCM	753 g of n-heptane
Adhesive	Hysol adhesive, EA934
Total weight	2733 g

integrated with the HEPP in the system test configuration.

Pressure retention does not present a problem with the present canister design. The maximum internal pressure attained in a space environment will be less than 2 psia (vapor pressure of n-heptane at 50°C), and on the ground a maximum external pressure of 1 atm will be experienced. The container walls are relatively heavy because of thermal conduction considerations, and the honeycomb and center shunt act as strength members in compression. The single PCM cavity was employed to minimize the length of welds required. All seals were TIG-welded and helium leak checked prior to PCM charging. A summary of the PCM canister design is presented in Table 1.

Test Program

The PCM Canister has been subjected to a series of thermal vacuum tests at the component and at the system's level.

Component Tests

Test setup

The PCM canister was installed in a thermal vacuum test chamber with a liquid nitrogen cold wall, as shown in Fig. 2. The canister was oriented at a 28-deg angle with respect to the HEPP. The fully expanded honeycomb core is located at the top to assure that this volume is the effective void

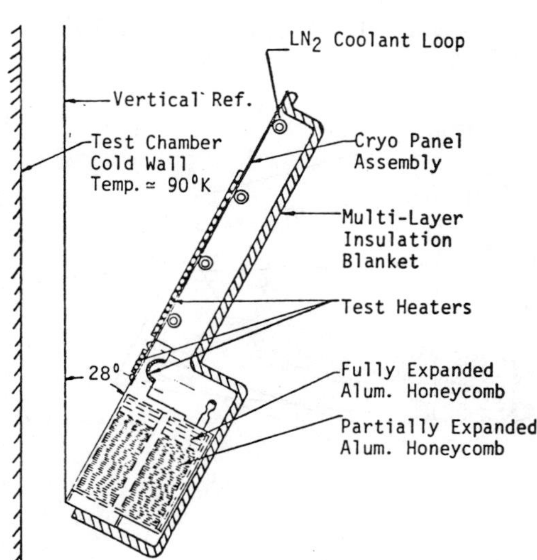

Fig. 2 Thermal vacuum test setup.

volume in the gravity field. An aluminum fin (cryopanel
assembly) was attached to the PCM box to simulate the HEPP
radiator. A liquid nitrogen coolant loop attached to the fin
permitted rapid cooldown during thermal cycle tests. Similarly rapid heatup was accomplished with a heater attached to the
fin. Heat-pipe inputs also were simulated with an electrical
heater attached to the aluminum heat-pipe saddle, which is
integral with the canister. The underside of the test
assembly was insulated with 22 thicknesses of multilayer
insulation (MLI).

The test assembly was instrumented with a total of 33
copper constantan thermocouples. Five strain gages were
attached to the canister to monitor deflections during thermal
cycling. A quadropole mass spectrometer was interfaced with
the thermal vacuum chamber to perform periodic leak checks.

Test procedure

The PCM canister was cycled over a temperature range of
-150° to 40°C every 5 hr for a total of 24 cycles. Before, during,
and after these cycling tests, the PCM canister was exposed to
different heat loads between 15 and 45 W to determine freeze/
thaw characteristics, thermal conductance, transient behavior,
and the effect of temperature cycling. A cold soak test was
conducted to simulate startup mode behavior, and a calibration
test was run to determine the heat-rejection capacity of the
test system.

The sequence of thermal tests performed on the PCM
canister was as follows: 1) cold soak (LN_2 cold wall only),
2) simulated heat-pipe performance (15, 25, 35, 45 W),

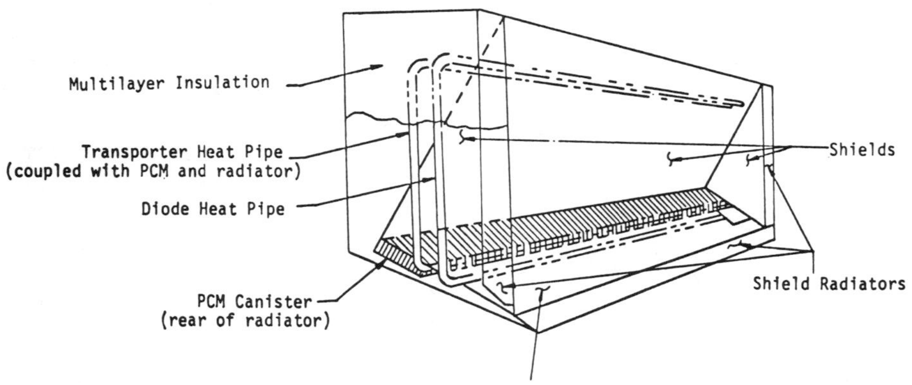

Fig. 3 PCM canister integrated with HEPP for systems test.

3) thermal cycling (12 cycles), 4) steady-state radiator calibration test, 5) simulated heat-pipe performance (15, 45 W), 6) thermal cycling (12 cycles), and 7) simulated heat-pipe performance (15, 25, 35, 45 W).

System Tests (HEPP)

After the PCM was qualified as a component, it was integrated with the HEPP, as shown in Fig. 3, and system testing was performed. The function of the PCM canister in this configuration is to provide temperature stability during heat-pipe transport tests. During the experiment testing, the PCM was subjected to various heat loads and underwent several melt/freeze cycles.

Test Results

The PCM canister was subjected to a cold soak during both component and system testing. The temperature vs time cooldown profiles are shown in Fig. 4. A -90.8°C freezing point was exhibited during freezing in component tests and 93.1°C during the HEPP system tests. This compares with -90.6°C for the n-heptane freezing point which is noted in published data.[4] Since different test setups and instrumentation were used for these tests, it is possible that the location or calibration of the thermocouples is the cause of this discrepancy. Further investigation is required to determine whether the difference is due to the test setup or to an actual shift in physical properties. The n-heptane showed negligible subcooling behavior, with 0.5°C measured during component tests and 0.2°C during system tests. The difference in the freezing intervals (4½ hr for component tests vs 5 hr for system test) is due to the different heat-rejection capacity of the two different test systems. The PCM/cryopanel capacity at the melting point was determined to be 6.1 W during steady-state calibration tests. The HEPP

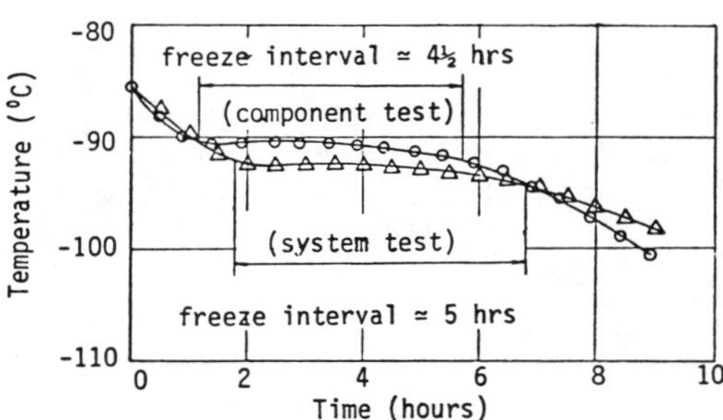

Fig. 4 PCM temperature profile during passive cooldown.

radiator has a net capacity under the simulated Tiros-N environment of 5.2 W. This latter value was calculated using a nodal network analysis for the HEPP and subsequently was verified during system tests. The influence of the larger radiator capacity can be observed in Fig. 4, where the component temperature profile shows a more rapid rate of decrease. The storage capacity of the PCM canister corresponding to the respective freeze intervals and radiator capacities is 27.5 W-hr for the component test and 26.0 W-hr for the HEPP system test. The theoretical storage capacity of the PCM canister, based on the 753-g n-heptane charge with a latent heat of fusion of 33.5 cal/g, is 29.4 W-hr. The values determined experimentally are within 6 and 12% of the predicted storage capacity derived from the measured charge and published physical properties. This difference may be due to the test setups and/or the analytical techniques.

The PCM canister was subjected to heat loads of 15, 25, 35, and 45 W during the component thermal performance tests to determine melting characteristics. A fixed power level was applied to the PCM which was frozen completely until melting was complete. Typical melting profiles are shown in Fig. 5 for the 25- and 45-W tests. Melting is initiated when the temperature of the PCM reached -90.4°C in both cases. The average canister temperature rises to -89° and -87.5°C before melting is completed with the 25- and 45-W inputs, respectively. This rise is due to an increase in the temperature gradient across the box as the melt front advances from the heat input side to the bottom of the canister. The average melt temperature for the 10 melt tests was -90.4°C, compared with an average melt temperature of -92.4°C exhibited during the HEPP tests and a published melting point of -90.6°C. Again, this discrepancy may be due to different test setups and calibrations or to a shift in physical properties.

A thermal analysis of the experimental results of the thermal performance component tests indicates that the average

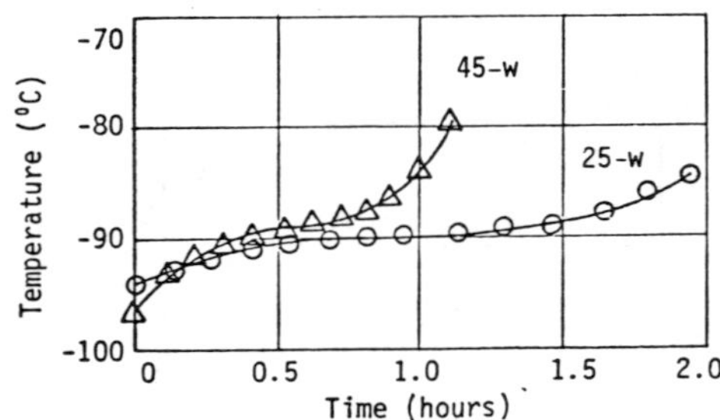

Fig. 5 PCM temperature profile during melt tests.

storage capacity of the PCM is 26.5 W-hr. This analysis is
essentially an energy balance accounting for heater input,
radiation to the chamber cold wall, the sensible heat
absorbed by the PCM/radiator assembly, and the storage
capacity of the PCM. Data obtained from 10 melt tests at
various heat inputs were utilized in this analysis. Thermal
gradients across the PCM canister were measured in each of
the tests, and the thermal conductances were determined at
the corresponding heat loads. The average thermal conductance
across the canister was 10.5 W/°C. The minimum conductance
was 8.8 W°C, which satisfies the design goal of 8.0 W/°C.

The PCM canister was subjected to 24 thermal cycles
between -150° and 40°C during the component tests. Each
cycle lasted approximately 5 hr and included accelerated
melting and freezing of the n-heptane. The purpose of these
tests was to simulate the effect of extended spaceflight
operation. The thermal performance tests were conducted prior
to, at the midpoint, and after the cycling tests in order to
observe any degradation effects. The average melting point
for each series of melt tests deviated less than 0.2°C from
the -90.4°C average, with no apparent degradation trend.
Strain-gage measurements obtained throughout the course of
the component testing indicated that the maximum deflection of
the welded aluminum assembly was less than 0.05 mm. This is
consistent with the low-pressure variations to which the box
is subjected (less than 1 atm) and the structural design of
the box. Mass spectrometer tests showed a leak rate of
2×10^{-7} atm-cm^3/sec at -150°C, increasing to 6×10^{-5} atm-cm^3/
sec at 40°C. This profile remained constant throughout the
thermal cycling tests, indicating that no degradation of the

Table 2 PCM canister performance summary

	Component test	System test	Predicted
$\overline{T}_{melt}$, °C	-90.4	-92.4	-90.6[4]
$\overline{T}_{freeze}$, °C	-90.8	-93.1	-90.6[4]
$\overline{Q}_{fusion}$, W-hr	27.5[a] 26.5[b]	26.0[a]	29.4
$\overline{C}$, W/°C	10.5	....	8.0

[a]Based on analysis of cold soak test data.
[b]Based on analysis of melt test data.

weld zones occurred. The higher leak rates at high temperatures may be due to increased outgassing of contaminants.

Summary

The objective of this development program was to design, fabricate, and test a low-temperature PCM canister for integration with the HEPP flight experiment. Qualification criteria for the canister included the ability to provide approximately 30 W-hr of temperature stability at -90°C, with a minimum thermal conductance of 8 W/°C, and to assure component reliability in terms of well-defined repeatable thermal behavior and structural integrity. Extensive thermal testing of the PCM canister, both on the component and the system's level, was conducted to verify conformance to the preceding criteria.

A summary of the PCM thermal performance characteristics derived from experimental results is presented in Table 2 and compared with corresponding design goals and published information. The measured melt and freeze points of the PCM were within 0.7°C of each other. In each case, the melt temperature was higher. Since the temperatures are measured by externally located thermocouples, this difference may be due to the temperature drop across the wall. This gradient is reversed for melt and freeze conditions and higher during melt tests due to larger heat fluxes. The measured and published phase transition temperatures are within 2.5°C. The experimentally determined PCM storage capacities are 6 to 12% lower than the predicted 29.4 W-hr. Further investigation is required to determine conclusively whether these variations are due to the test setup and/or analytical techniques or to an actual shift in physical properties. The thermal conductance across the PCM canister was 31% higher than the 8.0-W/°C design goal. The maximum amount of subcooling was 0.5°C. This compares with a 5°C subcooling that was noted previously for a n-heptane system.[4] The improvement is due to the numerous nucleation sites provided by the honeycomb core. Strain-gage and mass spectrometer measurements indicate that the canister is leak-tight and structurally sound.

The PCM canister has demonstrated repeatable thermal behavior in two different test configurations after being subjected to accelerated thermal cycling. The empirically determined thermal characteristics of the canister correspond closely to design goals and predicted values. The PCM currently is undergoing vibration testing within the HEPP for flight qualification. Postvibration testing is scheduled to evaluate further the performance of the HEPP/PCM radiator system.

References

[1]Humphries, W. F., "Performance of Finned Thermal Capacitors,"
NASA TND-7690, July 1974.

[2]Kirkpatrick, J. P. and Brennan, P. J., "Advanced Thermal
Control Flight Experiment," AIAA Progress in Astronatics and
Aeronautics: Thermophysics and Spacecraft Thermal Control,
Vol 35, edited by R. G. Hering, New York, 1974, pp. 409-430.

[3]Abhat, A. and Groll, M., "Investigation of Phase Change
Material (PCM) Devices for Thermal Control Purposes in
Satellites," AIAA Paper 74-728, July 1974.

[4]Keville, J. F., "Development of Phase Change Systems and
Flight Experiment on an Operational Satellite," AIAA Paper
76-436 July 1976; also AIAA Progress in Astronautics and
Aeronautics: Thermophysics of Spacecraft and Outer Planet
Entry Probes, Vo. 56, edited by A. M. Smith, New York, 1977,
pp. 19-36.

[5]Brennan, P. J. and Suelau, H. J., "Thermal Design of TIROS-N
Heat Pipe Experiment Package - HEPP," B & K Engineering, Inc.,
Final Rept., Feb. 1977.

Index to
Contributors to Volume 60